Springer Theses

Recognizing Outstanding Ph.D. Research

Aims and Scope

The series "Springer Theses" brings together a selection of the very best Ph.D. theses from around the world and across the physical sciences. Nominated and endorsed by two recognized specialists, each published volume has been selected for its scientific excellence and the high impact of its contents for the pertinent field of research. For greater accessibility to non-specialists, the published versions include an extended introduction, as well as a foreword by the student's supervisor explaining the special relevance of the work for the field. As a whole, the series will provide a valuable resource both for newcomers to the research fields described, and for other scientists seeking detailed background information on special questions. Finally, it provides an accredited documentation of the valuable contributions made by today's younger generation of scientists.

Theses are accepted into the series by invited nomination only and must fulfill all of the following criteria

- They must be written in good English.
- The topic should fall within the confines of Chemistry, Physics, Earth Sciences, Engineering and related interdisciplinary fields such as Materials, Nanoscience, Chemical Engineering, Complex Systems and Biophysics.
- The work reported in the thesis must represent a significant scientific advance.
- If the thesis includes previously published material, permission to reproduce this must be gained from the respective copyright holder.
- They must have been examined and passed during the 12 months prior to nomination.
- Each thesis should include a foreword by the supervisor outlining the significance of its content.
- The theses should have a clearly defined structure including an introduction accessible to scientists not expert in that particular field.

More information about this series at http://www.springer.com/series/8790

Matthew John Kirk

Charming New Physics in Beautiful Processes?

Doctoral Thesis accepted by
the Durham University, Durham, UK

Author
Dr. Matthew John Kirk
Dipartimento di Fisica
La Sapienza, University of Rome
Rome, Italy

Supervisor
Prof. Alexander Lenz
Department of Physics
Durham University
Durham, UK

ISSN 2190-5053   ISSN 2190-5061   (electronic)
Springer Theses
ISBN 978-3-030-19199-3   ISBN 978-3-030-19197-9   (eBook)
https://doi.org/10.1007/978-3-030-19197-9

This Springer imprint is published by the registered company Springer Nature Switzerland AG
The registered company address is: Gewerbestrasse 11, 6330 Cham, Switzerland

Supervisor's Foreword

The Standard Model (SM) of Particle Physics comprises all our knowledge about the subatomic world. It has been confirmed by thousands of experimental measurements with a high precision. The latest big success of the SM was the discovery of the Higgs Particle at the Large Hadron Collider (LHC) at CERN in 2012.

Despite these achievements the SM leaves many questions unanswered, like what is the nature of dark matter or what is the origin of the matter–antimatter asymmetry in the Universe. Identifying a breakdown of the SM, as well as finding the correct extension of the SM that will answer the abovementioned questions, is a major motivation for current research in particle physics.

There are two principal strategies for finding physics beyond the Standard Model (BSM): direct BSM searches try to directly create new unknown elementary particles by building more powerful accelerators—a strategy that was very successful in the past (e.g. the discovery of the Higgs particle in 2012 at LHC, CERN or the discovery of the top quark in 1995 at Tevatron, Fermilab). Indirect searches for BSM effects use comparisons of high-precision SM calculations with extremely precise measurements. Any discrepancy here can give first glimpses of new effects, that cannot be accommodated by the SM. This strategy was also very successful in the past: electroweak precision measurements before 2012 have shown that there should be a Higgs particle with a mass of the order of 100. The existence and the value of mass of the top quark could be deduced from the measurement of B mixing in 1986 (UA1, CERN) and 1987 (ARGUS, DESY).

The thesis of Dr. Matthew Kirk studies very different aspects of indirect new physics searches within the subfield of Flavour Physics—this field studies the decays of hadrons containing heavy charm or heavy bottom quarks. These systems are very promising for understanding CP violation, which seems to be the key to answer the origin of the matter–antimatter asymmetry. The main difficulty in doing indirect BSM searches within Flavour Physics is a control over hadronic effects, that dominate the decays of the heavy mesons. Any misjudgement of these QCD effects might else be interpreted as unambiguous BSM signals. Dr. Kirk will present a phenomenological study of the possible maximum size of unknown hadronic effects (denoted by duality violation) in B mixing and D mixing, as well as in the lifetimes of the B and D mesons. Moreover, he will present a determination of non-perturbative objects that govern neutral B and D meson mixing as well as B and D meson lifetime ratios. The corresponding Heavy Quark Effective Theory (HQET) sum rules require a perturbative 3-loop calculation. For meson mixing, the results are competitive with the most recent lattice evaluations and for the lifetimes they are the only available results based on QCD.

Besides these high-precision SM calculations, Dr. Kirk investigated also how hypothetical dark matter particles might couple to heavy charm quarks. He will present in his thesis a comprehensive investigation of bounds on this possibility stemming from relic densities of particles in the Universe, flavour bounds measured at particle physics colliders, direct dark matter detection bounds from underground experiments, indirect dark matter detection bounds from satellite experiments and dark matter bounds from the LHC.

Finally, Dr. Matthew Kirk worked also on BSM models that might describe the anomalies that are currently observed in the decays of b quarks by the LHCb collaboration. A confirmation and understanding of these flavour anomalies would revolutionise our understanding of the subatomic world. It also would give direct hints for where exactly to look for the new particles in accelerators. He performed a model-independent study using effective theory methods and he could show within this approach that an observation of a momentum dependence of the anomalies would not be an unambiguous signal for a hadronic origin of the discrepancy, as typically claimed in the literature, but could also be due to new BSM effects. In another study, he has shown several of the discussed BSM possibilities to explain the flavour anomalies are ruled out by a precise study of B mixing.

In summary, this thesis contains an extraordinarily broad range of topics related to indirect flavour searches for BSM effects and it is an essential read for newcomers in this field.

Durham, UK   Prof. Alexander Lenz
March 2019

Abstract

In this thesis, we study quark flavour physics and in particular observables relating to B meson mixing and lifetimes. Meson mixing arises due to the nature of the weak interaction, and leads to several related observables that are highly suppressed in the Standard Model (SM). Alongside meson mixing, lifetimes provide an insight into rare B processes which can shed light on possible new physics.

Both the calculations are based on an Effective Field Theory (EFT) framework, in particular the Weak Effective Theory. This framework allows us to separate the high-scale effects which are calculable in perturbation theory from the low-energy matrix element, which are determined through other means. Within this framework, the observables are expanded using the Heavy Quark Expansion (HQE) technique, which utilises the relatively large masses of b and c quarks to reveal a further hierarchy of corrections. The basics of EFTs and the HQE are explored in detail as an entry point to the majority of the work in this thesis.

In the rest of the thesis, we take aim at pushing the accuracy of our SM predictions further: by testing the underlying assumption of Quark-Hadron duality in the HQE; by studying possible new physics models that can explain the long-standing problem of dark matter as well as recently seen anomalies; and by using alternative approaches to determining the low-energy constants associated with mixing and lifetimes in order to provide independent and state-of-the-art results.

Publications Related to This Thesis

- T. Jubb, M. Kirk, A. Lenz and G. Tetlalmatzi-Xolocotzi, *On the ultimate precision of meson mixing observables, Nucl. Phys.* **B915** (2017) 431–453, [1603.07770].

- T. Jubb, M. Kirk and A. Lenz, *Charming Dark Matter, JHEP* **12** (2017) 010, [1709.01930].

- S. Jäger, K. Leslie, M. Kirk and A. Lenz, *Charming new physics in rare B-decays and mixing?, Phys. Rev.* **D97** (2018) 015021, [1701.09183].

- M. Kirk, A. Lenz and T. Rauh, *Dimension-six matrix elements for meson mixing and lifetimes from sum rules, JHEP* **12** (2017) 068, [1711.02100].

- L. Di Luzio, M. Kirk and A. Lenz, *Updated B_s-mixing constraints on new physics models for $b \to s \ell^+ \ell^-$ anomalies, Phys. Rev.* **D97** (2018) 095035, [1712.06572].

Acknowledgements

I would like to thank my supervisor Alex Lenz, for his expert guidance, advice and support throughout my Ph.D., without which I would not have reached this stage. He has been an invaluable guide in the world of flavour physics, and he has been an outstanding supervisor over the past 4 years.

I must also thank the IPPP for offering me the chance to continue studying the subject I love, and STFC for the funding that enabled me to do so.

Throughout my Ph.D., my work has been done in collaboration with others and so I must extend my thanks to Gilberto Tetlalmatzi Xolocotzi, Thomas Jubb, Sebastian Jäger, Kirsty Leslie, Thomas Rauh and Luca Di Luzio for being enjoyable to work with. Along with Alex, they have helped shape this thesis. I must also single out and thank Alex, Rachael, Danny, Andrew, Jonny and Duncan for agreeing to proofread parts of this thesis, large or small.

Over my time in Durham, my fellow students at the IPPP (and in particular my office mates past and present) have made it a joy to come to work every day. While the atmosphere of the office has changed over the past 3 years, it has always been great. I make my apologies to those in other offices (particularly OC118), who have been subjected to my bored wanderings.

Beyond work, I cannot begin to describe in this short space how much I have enjoyed my time as part of Grey MCR. It has shaped me, and made these past few years some of the best of my life. It has been a pleasure and a privilege to be a part of the amazing community at Grey, both on the MCR Exec and otherwise—I hope I have, in one way or another, conveyed this to everybody I have met here. I cannot name you all, but I feel I must pick out and thank a few individually. Bear: for everything from your passing suggestion of dinner to our many hours of gaming. Darren: for the sheer joy physics inspires in you 24 hours a day, your mad banter and the many helpful conversations. Sarah: for being such an amazing friend. And Rachael: for these past 10 months.

Finally of course, my parents, for supporting me my entire life.

Contents

Chapter 1
Introduction

Particle physics can be described as the area of physics which concerns itself with describing the fundamental building blocks of the universe. Its aim is no less lofty than the construction of a model that can, with minimal input, generate correct predictions for the interactions on the smallest scales, and allow us to build up physical laws we can use to describe our world. Our current best working model of this type is known as the Standard Model (SM)—the nature of the SM will be described in the rest of this chapter, alongside a brief historical overview of its construction. Finding ways to clearly test the SM and probe possible extensions to it is the work which the remainder of this thesis consists of.

1.1 The Standard Model

The SM, as a complete theory, has been developed over an extended period of time. The start of the journey could well be considered to be the development of Quantum Electrodynamics (QED) over several decades, from Dirac [1] to Tomonaga [2], Schwinger [3, 4], Feynman [5–7], and many others. The other major constituent parts of the SM were developed in the 1960s and 1970s—the Brout-Englert-Higgs (BEH) mechanism [8–10]; the unified theory of electroweak interactions by Glashow [11], Weinberg [12], and Salam [13]; the Quantum Chromodynamics (QCD) Lagrangian by Fritzsch, Gell-Mann, and Leutwyler [14], and the nature of asymptotic freedom in QCD by Gross and Wilczek [15], and Politzer [16].

The SM is formulated as a quantum field theory (QFT)—the dynamics of the theory are characterised by the Lagrangian density $\mathcal{L}_{\mathrm{SM}}$.[1] The SM Lagrangian can be written rather succinctly in the form[2]

---

[1]Typically referred to as just the Lagrangian, which we will do from here on out.

[2]Available on t-shirts, mugs, etc. from all good gift shops.

© Springer Nature Switzerland AG 2019
M. J. Kirk, *Charming New Physics in Beautiful Processes?*,
Springer Theses, https://doi.org/10.1007/978-3-030-19197-9_1

Table 1.1 Field content of the SM

Field	$SU(3)_c$	$SU(2)_L$	$U(1)_Y$
Q_L	**3**	**2**	$1/6$
u_R	**3**	**1**	$2/3$
d_R	**3**	**1**	$-1/3$
L_L	**1**	**2**	$-1/2$
e_R	**1**	**1**	-1
H	**1**	**2**	$1/2$

$$
\begin{aligned}
\mathcal{L}_{\mathrm{SM}} = & -\frac{1}{4} F_{\mu\nu} F^{\mu\nu} \\
& + i\bar{\psi}\slashed{D}\psi \\
& + \psi_i y_{ij} \psi_j H + \text{h.c.} \\
& + |D_\mu H|^2 - V(H),
\end{aligned}
\tag{1.1.1}
$$

where successive lines describe, respectively, the kinetic and self-interactions of gauge fields; the kinetic terms of fermions and their interaction with gauge fields; the interactions of the fermions with the Higgs field; and the kinetic and self-interactions of the Higgs field. The SM is a gauge theory, with the Lagrangian having a $SU(3)_c \times SU(2)_L \times U(1)_Y$ gauge symmetry. It is also Lorentz invariant, as is required to be consistent with special relativity. The power of symmetries can be appreciated here—given we want a renormalisable theory which has $SU(3)_c \times SU(2)_L \times U(1)_Y$ gauge symmetries, $SO(1, 3)$ Lorentz symmetry, and the field content in Table 1.1, Eq. 1.1.1 is the only possible Lagrangian we can write.[3]

In the next two sections, we will break the SM Lagrangian down differently to show the different gauge groups and their properties.

1.1.1   QCD

QCD is a non-abelian gauge theory, meaning the generators of the symmetry don't commute. The symmetry is $SU(3)_c$, where the subscript stands for *colour*—the label we use for the gauge charges of QCD. Quarks live in the fundamental representation of $SU(3)$, while gluons live in the adjoint (from which we need only take that there are three colours of quarks, but eight colours of gluon).

Singling out the pure QCD parts of Eq. 1.1.1, we have two relatively simple parts: the QCD gauge field tensor and the covariant derivative that couples quarks to the gluon field. The gauge field tensor in QCD can be written as

---

[3]The issue of a possible right handed neutrino ν_R, which could in principle be added, is briefly discussed later in Sect. 1.3.3.

$$G^a_{\mu\nu} = \partial_\mu G^a_\nu - \partial_\nu G^a_\mu + g_s f^{abc} G^b_\mu G^c_\nu \,,$$

where G^a is a gluon field, g_s is the coupling constant of QCD (also called the strong coupling constant), and f^{abc} is the antisymmetric structure constant of $SU(3)$. The covariant derivative for QCD, acting on quarks which exist in the fundamental representation of $SU(3)$, is

$$(D_\mu)_{ij} = \partial_\mu \delta_{ij} - i g_s G^a_\mu t^a_{ij} \,, \tag{1.1.2}$$

where t^a are the generators of $SU(3)$ which obey the following useful relations:

$$t^a_{ij} t^a_{jk} = C_F \delta_{ij} \equiv \frac{N^2_c - 1}{2 N_c} \delta_{ij} \,,$$

$$t^a_{ij} t^a_{kl} = \frac{1}{2} \left(\delta_{il} \delta_{kj} - \frac{1}{N_c} \delta_{ij} \delta_{kl} \right) \,.$$

(The second result can be thought of as a Fierz relation in colour space, see Appendix A for details.)

1.1.2 Electroweak Theory and the Higgs Mechanism

The other half of the SM is the electroweak sector—the unification of the weak and electromagnetic interactions into a $SU(2)_L \times U(1)_Y$ gauge group, where the subscript L stands for *left* and the subscript Y for *weak hypercharge*. The respective gauge field strengths are

$$W^a_{\mu\nu} = \partial_\mu W^a_\nu - \partial_\nu W^a_\mu + g \varepsilon^{abc} W^b_\mu W^c_\nu \,,$$

where ε^{abc} is the standard three dimensional Levi-Civita tensor, and

$$B_{\mu\nu} = \partial_\mu B_\nu - \partial_\nu B_\mu \,.$$

The $SU(2)$ group is labelled left as the electroweak sector explicitly distinguishes between left and right chiralities—left handed fermions sit in the doublet representation of $SU(2)$, while right handed fermions are $SU(2)$ singlets; and as seen in Table 1.1 the left and right handed fields have different charges under the $U(1)$ group. This makes the SM a parity violating theory—we will discuss this more in Sects. 1.2.2 and 1.3.2.

Since the different chiralities sit in different representations of the group, the covariant derivative acts differently on them—for left handed particles, it takes the form

$$D_\mu = \partial_\mu - igW_\mu^a \frac{\sigma^a}{2} - ig'Y_L B_\mu , \qquad (1.1.3)$$

where σ^a are the Pauli matrices, while for right handed particles only the weak hypercharge field acts and it takes the form

$$D_\mu = \partial_\mu - ig'Y_R B_\mu , \qquad (1.1.4)$$

and we have explicitly distinguished the weak hypercharge $Y_{L,R}$ for left and right handed fields to remind the reader that they are not equal. The chiral nature of the electroweak theory means that a standard Dirac mass term like $m\psi_L \psi_R$ cannot be simply included, as the left and right handed fields transform differently under the gauge symmetry. Mass terms for the vector bosons W_μ^a and B_μ, like $M_V^2 V_\mu V^\mu$, are also not gauge invariant; and yet we know the W and Z bosons are definitely not massless. The resolution of both these problems is the BEH mechanism [8–10]. By adding a complex scalar field with a particularly shaped potential, we can arrange for the field to acquire a vacuum expectation value (VEV), spontaneously breaking the symmetry obeyed by the SM Lagrangian down to $SU(3)_c \times U(1)_{\text{EM}}$.

A form of potential that achieves our aims is

$$V(H) = -\mu^2 (H^\dagger H) + \lambda (H^\dagger H)^2 ,$$

with $\mu^2, \lambda > 0$—often known as the "Mexican hat" potential. It is easily seen that with those choices for the sign of the Higgs potential parameters, the field has a minimum at

$$|H| = v, \text{ where } v = \sqrt{\frac{\mu^2}{2\lambda}} \neq 0 \text{ and has dimensions of mass} .$$

The non-zero VEV breaks the electroweak symmetry $SU(2)_L \times U(1)_Y$ down to $U(1)_{\text{EM}}$, i.e. just the symmetry associated with the conservation of electric charge. After symmetry breaking we can, in the unitary gauge, write the Higgs doublet in the form

$$H = \frac{1}{\sqrt{2}} \begin{pmatrix} 0 \\ v+h \end{pmatrix} , \qquad (1.1.5)$$

where h is the field associated with the Higgs boson. In this form, it is most straightforward to see the origin of the fermion and gauge boson mass terms. We leave the discussion of the origin of fermion masses to Sect. 1.2.1 as this is a crucial part of the broad spectrum of flavour phenomenology. For the gauge bosons, if we expand the covariant derivative terms in the broken Higgs phase, we find terms with two gauge fields and factors of v, g, g' as coefficients—these look exactly like gauge boson mass terms. With this procedure we find mixed terms however, like $v^2 W_\mu^3 B^\mu$. If we diagonalise the mass matrix for the B^μ and $W^{3,\mu}$ fields, we find one massless eigenstate and one massive eigenstate, which we will suggestively call A^μ and Z^μ

respectively. Defining also the combination $W_\mu^\pm \equiv \frac{1}{\sqrt{2}}(W_\mu^1 \mp i W_\mu^2)$, we end up with the following mass terms:

$$
\begin{aligned}
\mathcal{L}_{\text{SM}} &\supset -\frac{v^2 g^2}{4} W^{+\mu} W_\mu^- - \frac{v^2(g^2 + g'^2)}{4} Z^\mu Z_\mu \\
&= -\frac{v^2 g^2}{4} W^{+\mu} W_\mu^- - \frac{v^2 g^2}{8 \cos^2 \theta_W} Z^\mu Z_\mu \\
&\equiv -M_W^2 W^{+\mu} W_\mu^- - \frac{M_Z^2}{2} Z^\mu Z_\mu .
\end{aligned}
$$

As seen in experiment, we have massive W and Z bosons while the photon stays massless.

As a final point to round out this brief discussion, we mention the different gauge choices for the Higgs field. While unitary gauge (which leaves us the form shown in Eq. 1.1.5) is convenient for demonstrating the mass generation mechanism, for calculational purposes Feynman gauge is generally better, as it improves the convergence of diagrams with virtual massive electroweak bosons. In this gauge, along with the $W^\pm$ and Z bosons we also have charged Goldstone scalars $\phi^\pm$ and a neutral Goldstone scalar ϕ^Z. These couple to fermions in a similar way as the corresponding $W^\pm$ and Z bosons, and so are important for loop corrections; in the calculation of meson mixing for example (see Sect. 2.4.2 for more detail) it is important to consider the Goldstone diagrams.

1.2 Flavour

While the field content of the SM as detailed in Table 1.1 might at first glance seem relatively modest, there is an complication. We have found an extra two copies of the up, down, electron, and electron neutrino particles, whose fundamental properties are exactly the same except for their masses. We say that there are three generations of quarks and leptons, and refer[4] to the six different types of quarks and leptons as *flavours*. The different generations give rise to a huge variety of phenomenology, and the study of quark flavour will be the primary focus of this thesis (although see Sect. 1.3.4 and Chap. 7 for some interesting signs involving different lepton flavours). In flavour physics, processes that change the flavour of particles can occur through tree-level interactions of the charged $W^\pm$ bosons, or through rarer, so called Flavour Changing Neutral Current (FCNC), interactions. Why FCNCs are not seen at tree level in the SM, and are instead much rarer due to being loop-suppressed is a result of the specific nature of how the BEH mechanism arises in the SM, and we will discuss this in the next section.

---

[4]After a fortuitous trip to a Baskin-Robbins shop [17, 18] by Harald Fritzsch and Murray Gell-Mann.

1.2.1 CKM

The method by which fermions gain their mass via the BEH mechanism is slightly more complex than that for the Z and W bosons, due to the existence of multiple generations. The generic form of the Yukawa interaction [19] between the Higgs field and quarks is

$$
\begin{aligned}
\mathcal{L} &\supset Y^u_{ij}\overline{Q^i_L}\widetilde{H}u^j_R + Y^d_{ij}\overline{Q^i_L}Hd^j_R + \text{h.c.} \\
&\supset \frac{v}{\sqrt{2}}Y^u_{ij}\overline{u^i_L}u^j_R + \frac{v}{\sqrt{2}}Y^d_{ij}\overline{d^i_L}d^j_R + \text{h.c.}
\end{aligned}
\tag{1.2.1}
$$

where $\widetilde{H} = i\sigma^2 H^*$, i, j are indices in generation space, and we have replaced the Higgs field with its form in the unitary gauge (see Eq. 1.1.5) and dropped terms without the VEV. As $Y^{u,d}$ need not be diagonal (and certainly not both simultaneously), we use singular value decomposition to rotate to the basis of quark mass eigenstates

$$
M^u = \frac{v}{\sqrt{2}}U^u_L Y^u (U^u_R)^\dagger, \qquad M^d = \frac{v}{\sqrt{2}}U^d_L Y^d (U^d_R)^\dagger,
\tag{1.2.2}
$$

where $U^{u,d}_{L,R}$ are four unitary matrices,[5] and the mass matrices $M^{u,d}$ are diagonal in the quark masses: $M^u = \text{diag}(m_u, m_c, m_t)$, $M^d = \text{diag}(m_d, m_s, m_b)$. Now that our mass Lagrangian is diagonal, what effect does this have on the other terms in the SM Lagrangian? The gluon, photon, and Z boson fields only couple fermions to their conjugate states, and so our change of basis has no effect. For photons and gluon, this result is a consequence of gauge invariance—up and down type quarks exist in different gauge representations and so they cannot be coupled together in the kinetic term, where interactions with the gauge bosons arise. For Z bosons, the story is slightly trickier since the Z is a gauge boson of a broken symmetry. But in the SM the Z coupling is a combination of electric charge and weak isospin; electric charge is an unbroken symmetry, and all particles with the same weak isospin happen to have the same electric charge; hence no FCNC arise for the Z boson. This is why there are no FCNCs at tree level—the form of the SM prevents them. Extending this very specific flavour structure to BSM models is the principal of Minimal Flavour Violation (MFV) [20, 21]—all new flavour changing effects follow the pattern shown in the SM, and are governed by the known Yukawa and CKM structures.

On the contrary, the charged W bosons couple up and down quarks together, and so the change of basis looks like

$$
\overline{u_L}\gamma^\mu d_L W^+_\mu \rightarrow \overline{u_L}(U^u_L)\gamma^\mu (U^d_L)^\dagger d_L W^+_\mu \equiv \overline{u^i_L}\gamma^\mu V_{ij}d^j_L W^+_\mu
$$

---

[5]We get four, rather than two as might be expected from standard matrix diagonalisation, as we are doing singular value decomposition since we need our mass eigenvalues to be ≥ 0.

where we have defined the matrix $V \equiv U_L^u (U_L^d)^\dagger$. This matrix is known as the Cabibbo–Kobayashi–Maskawa or CKM matrix [22, 23]. Since the indices of the matrix are related to the quark generations, we often write the elements of V as

$$
V = \begin{pmatrix} V_{ud} & V_{us} & V_{ub} \\ V_{cd} & V_{cs} & V_{cb} \\ V_{td} & V_{ts} & V_{tb} \end{pmatrix} \approx \begin{pmatrix} 0.97 & 0.23 & 0.0037e^{-1.1i} \\ -0.22 & 0.97 & 0.042 \\ 0.0086e^{-0.39i} & -0.041 & 1 \end{pmatrix}, \quad (1.2.3)
$$

where we show the approximate size of the elements of the CKM matrix using a recent [24] set of inputs from the CKMfitter group [25, 26].[6] There are two common ways of parameterising the CKM matrix—the "standard" parameterisation [29] and the "Wolfenstein" parameterisation [30]. A general 3×3 unitary matrix has nine "degrees of freedom" in total, which can be broken down into six phases and three real parameters. However, we can absorb all but one of the phases into the quark fields, leaving us with just four independent parameters. The standard parameterisation is given in terms of three mixing angles $\theta_{12}, \theta_{13}, \theta_{23}$ and one phase δ_{13}:

$$
V = \begin{pmatrix} c_{12}c_{13} & s_{12}c_{13} & s_{13}e^{-i\delta_{13}} \\ -s_{12}c_{23} - c_{12}s_{23}s_{13}e^{i\delta_{13}} & c_{12}c_{23} - s_{12}s_{23}s_{13}e^{i\delta_{13}} & s_{23}c_{13} \\ s_{12}s_{23} - c_{12}c_{23}s_{13}e^{i\delta_{13}} & -c_{12}s_{23} - s_{12}c_{23}s_{13}e^{i\delta_{13}} & c_{23}c_{13} \end{pmatrix} \quad (1.2.4)
$$

where $s_{ij} = \sin\theta_{ij}$ and $c_{ij} = \cos\theta_{ij}$, while the Wolfenstein parameterisation uses the four parameters λ, A, ρ, η:

$$
V = \begin{pmatrix} 1 - \lambda^2/2 & \lambda & A\lambda^3(\rho - i\eta) \\ -\lambda & 1 - \lambda^2/2 & A\lambda^2 \\ A\lambda^3(1 - \rho - i\eta) & -A\lambda^2 & 1 \end{pmatrix} + \mathcal{O}\left(\lambda^4\right). \quad (1.2.5)
$$

The Wolfenstein parameterisation was originally conceived as an expansion in the small parameter $V_{us} \approx 0.2$, and nicely shows off several features of the CKM matrix that are seen numerically in Eq. 1.2.3:

1. It is close to the identity matrix—so transitions between same generation flavours are the least suppressed.
2. There is the most mixing between the first and second generation, and the least between the first and third.
3. The complex elements are $\mathcal{O}\left(\lambda^3\right)$, so CP violation is highly suppressed.

Since this parameterisation is not exact, it should not be used for detailed calculations.

Now that we have seen how flavour changing interactions arise, we make note of another interesting feature of the SM. Since FCNCs don't appear at tree level, they arise at loop level through loops involving the flavour changing W vertices. But they

---

[6]The UTfit collaboration [27, 28] also produces similar results, using a different statistical approach to the fit—in this work we use the CKMfitter results throughout.

are suppressed even beyond this. If we consider the amplitude for a loop level FCNC, changing a quark from flavour i to flavour j, we see it will (schematically at least) behave like

$$i\mathcal{M} \sim \sum_q V_{iq} V_{jq}^* \times f(m_q).$$

In the limit of equal quark masses, the unitarity of the CKM guarantees this amplitude vanishes since if f does not depend on q, we simply get $\sum_q V_{iq} V_{jq}^* = \sum_q (VV^\dagger)_{ij} = 0$ for $i \neq j$. This is known as the GIM mechanism [31][7]—loop flavour changing processes get suppressed as long as they have a weak dependence on the mass of the quarks in the loop.

The multiple sources of suppression in the flavour sector of the SM for FCNCs means that they can be ideal places to search for NP—any difference in flavour structure will likely lift the strong suppression, and greatly enhance the rate for these rare processes.

1.2.2   CP Violation

In the previous section, we made the point that while the CKM matrix has non-real elements, they are very small. Why are complex couplings interesting? Because the CP operator has the effect of replacing coupling constants with their complex conjugate, and so non-real couplings imply a theory is not CP invariant.

We make a quick aside here about the discrete symmetries of the SM. It has long been known that the CPT theorem holds for most quantum field theories [32–34]—which means that any Lorentz invariant local quantum field theory with a Hermitian Hamiltonian must obey CPT symmetry (i.e. symmetry under the combined effects of the charge conjugation C, parity inversion P, and time reversal T operators). For a long time, it was assumed that these each held individually as well, until the idea of a parity violating theory was proposed by Lee and Yang in [35] and discovered by Wu et al. [36] a year later, and CP violation was found in 1964 by Christenson et al. [37]. As a result of the CPT theorem, these results mean that C, P and T must all be violated individually in nature.

As we will discuss in Sect. 1.3.2, CP violation is required to reproduce various observed features of the universe. It is interesting to note that if we had just two generations of quarks, there would be no physical phases in the CKM matrix (as they could all be absorbed by rephasing of the quark fields), and so no CP violation could be present in the flavour sector. Hence the question of why there are (at least) three generations is intimately tied up with the origin of CP violation in our universe.

It can be shown that the amount of CP violation in the SM can be represented in a parameterisation independent way by the Jarlskog invariant, which comes from the

---

[7]In that work they predicted the existence of the charm quark through the non-observation of $K^0 \to \mu^+ \mu^-$.

commutator of the two quark Yukawa matrices. The original definition [38] (in our notation) is

$$\det\left(-i\left[\frac{M^u}{m_t}, \frac{M^d}{m_b}\right]\right) = -\frac{2}{m_t^3 m_b^3}(m_t - m_c)(m_t - m_u)(m_c - m_u)\times$$

$$(m_b - m_s)(m_b - m_d)(m_s - m_d)J \quad (1.2.6)$$

with

$$\left(\sum_{mn}\varepsilon_{ikm}\varepsilon_{jln}\right)J = \mathrm{Im}(V_{ij}V_{kl}V_{il}^*V_{kj}^*). \quad (1.2.7)$$

Nowadays, it is more common to discuss the modified invariant, first proposed in [39], defined by

$$\det\left(-i\left[(M^u)^2, (M^d)^2\right]\right) = -2(m_t^2 - m_c^2)(m_t^2 - m_u^2)(m_c^2 - m_u^2)\times$$

$$(m_b^2 - m_s^2)(m_b^2 - m_d^2)(m_s^2 - m_d^2)J, \quad (1.2.8)$$

which has the advantage of removing the unphysical sensitivity to the sign of the quark masses.

1.3 Beyond the Standard Model?

The Standard Model as detailed in Sect. 1.1 is, alongside being the most comprehensive theoretical description of the fundamental nature of reality, one of the most well tested and predictive physical theories ever devised. As an example of this, we take from ATLAS [40] and CMS [41] Figs. 1.1 and 1.2, which summarise cross section measurements for a variety of different production processes at the LHC and compare them to the corresponding theoretical prediction. In these measurements, which span many orders of magnitude, no significant (by which we mean $>5\sigma$) deviations been observed. This is true across almost the entirety of high energy particle physics.

However the SM is not (and cannot be) the final theory—there a few areas where it fails. Notably, the SM does not incorporate gravity and so must inevitably be superseded one day by a model that combines a full quantum theory of gravity with the strong and electroweak forces. But even beyond this fundamental weakness, there are a small number of well known questions in particle physics alone that cannot be understood in the SM—that of dark matter, the matter-antimatter asymmetry and neutrino masses.

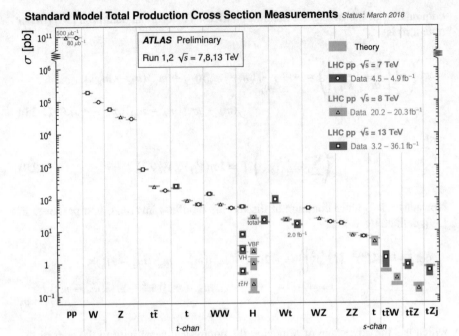

Fig. 1.1 ATLAS summary plot showing various theoretical predictions against experimental data [40]

1.3.1 Dark Matter

The existence of dark matter has a long history in physics, and was one of the earliest signs of new physics to be found—in fact the history is so long it predates the SM. One of the earliest suggestions of dark matter can be found in the work of Zwicky, who in the 1930s used the virial theorem to calculate the mass of the Coma Cluster [42, 43], and found a calculated mass of around 500 times that which was expected based on the luminosity. This difference he attributed to non-luminous matter or *dark matter*. A similar problem, of a discrepancy in the behaviour of astrophysical objects, appeared in the 1970s following the development of more advanced instruments. Observations of galactic rotation curves by Rubin and Ford [44] showed that stars orbiting far out from the centre of galaxies were moving much faster than would be expected from the observed distribution of matter. With the confirmation of these findings over the rest of that decade, and more detailed studies since then [45–47], dark matter became an established part of the scientific scenery.

Further astrophysical observations that reinforce the need for dark matter have not been slow to arise. On the largest scales, we have seen evidence for DM in the Cosmic Microwave Background (CMB) [48]. Studying the power spectrum allows us to infer the relative amounts of regular matter (which interacted with the photons that formed the CMB) and dark matter which only interacts gravitationally (at least to

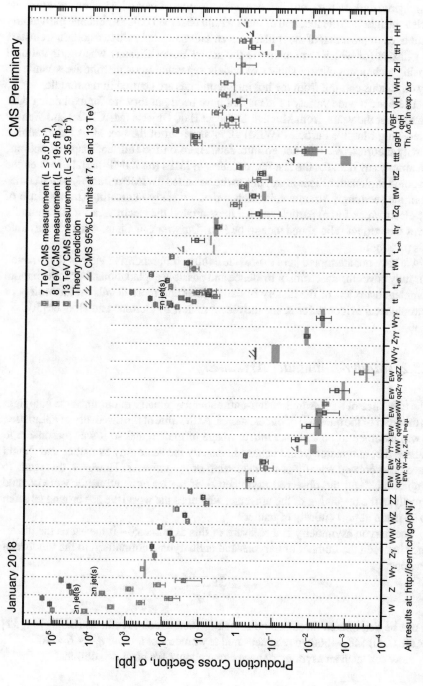

Fig. 1.2 CMS summary plot showing various theoretical predictions against experimental data [41]

a first approximation). Observations of gravitational lensing effects also provide an insight into dark matter (see [49] for a review)—these kinds of measurements provide a clear insight into the location of mass within observed objects, and one particularly famous example is that of the Bullet Cluster. This name refers to two colliding clusters of galaxies (of which the smaller is the actual Bullet Cluster) where gravitational lensing techniques clearly show two widely separated mass distributions, while X-ray emissions coming from the hot interacting gas are centred in the middle.

There are a wide variety of ideas that have been put forward for explaining dark matter over the years, from Massive Compact Halo Objects (MACHOs) and Weakly Interacting Massive Particles (WIMPs), to more out of the box ideas like modified gravity theories. We will not go into detail here on the full landscape of options, but simply say that, for the last few decades, WIMPs (or WIMP-like) have been the favoured explanation. This is part due to the so-called "WIMP miracle", which comes from the seemingly strange coincidence that a weakly interacting (in the sense of the electroweak force) particle, with a mass around the electroweak scale, naturally gives the correct relic abundance to be DM. Our work in Chap. 4 fits roughly into this paradigm.

We have not attempted here to provide a fully comprehensive view of dark matter, nor fully cover the spectrum of evidence and possible explanations. For a much more complete overview of the history of dark matter and more details on the variety of astrophysical evidence for it, see the recent review by Bertone and Hooper [50].

1.3.2 Matter-Antimatter Asymmetry

The existence of a measured matter-antimatter asymmetry in our universe cuts right to the heart of the question of our existence. At the time of the discovery of antimatter, it was assumed to be an exact mirror image of matter, and so it was reasonable to believe that antimatter would behave in exactly the same way as matter and should have been created equally at the beginning of the universe. However, this leads to a question about the observed asymmetry—if matter and antimatter were created equally at the beginning of the universe, why does the world we see around us seem to be made almost entirely of matter?

The baryon asymmetry is a measure of this asymmetry—it measures the difference between the number of baryons and antibaryons, normalised to the number of photons:

$$\eta_B = \frac{n_B - n_{\bar{B}}}{n_\gamma}.$$

It can be determined through studies of Big Bang Nucleosynthesis (BBN) [51, 52] or the CMB [48]—both give a measured asymmetry of around $\eta_B \sim 6 \times 10^{-10}$.

How can such an asymmetry have come about? We have two options:

1. Have a non-zero asymmetry as an initial condition of the universe, i.e. $\eta_B(t = 0) \neq 0$
2. Generate an asymmetry dynamically during the evolution of the universe, i.e. $\eta_B(t = 0) = 0 \implies \eta_B(t > 0) \neq 0$

At first sight, the first options seems rather unnatural, but without a full understanding of the physics of the Big Bang it is hard to rule it out. However, this option has bigger problems. All the evidence currently points towards a inflationary phase at very early times, where the size of the universe increased by a factor of around $e^{60} \approx 10^{26}$ [53]. This means that any initial asymmetry will be "washed out" to a much smaller value.

So we must find a way for the asymmetry to be generated dynamically (known as baryogenesis)—the three conditions that must be fulfilled are known as the Sakharov conditions [54]:

1. C and CP violation
2. Baryon number violation
3. An out of thermal equilibrium phase

Given our current knowledge, is it possible these conditions were fulfilled in our universe?

1. Currently we know that these are not good symmetries of the SM. As we explained earlier in Sect. 1.2.2, we have observed P and CP violation, and hence C must also be violated as well.
2. It has been shown that a baryon number violating process exists as a non-perturbative solution to the SM electroweak field equations [55] (this solution is generally known as a sphaleron).
3. In first order phase transitions, parameters can change discontinuously, and this allows an out of thermal equilibrium situation to arise—the breaking of the electroweak symmetry via the BEH mechanism constitutes a phase transition, so it seems possible such a situation can occur.

It seems we can fulfil all of the three conditions—however, it is not as good as it looks at first glance. If we assume that electroweak symmetry breaking is when baryogenesis occurs, we have a problem. As discussed in Sect. 1.2.2, the Jarlskog invariant characterises the amount of CP violation in the SM. Normalising it to the electroweak scale, we find that

$$\frac{\det\left(-i\left[(M^u)^2, (M^d)^2\right]\right)}{v^{12}} \sim \frac{7 \times 10^5 \text{GeV}^{12}}{(246\,\text{GeV})^{12}} \sim 10^{-23} \ll \eta_B \qquad (1.3.1)$$

and so there is not enough CP violation in SM to account for the measured asymmetry. On top of this, given the mass of the Higgs boson it is believed (e.g. [56]) that the phase transition associated with the breaking of electroweak symmetry is second rather than first order (meaning phase parameters change continuously) and so there would have been no out of equilibrium phase.

As you might suspect from the positioning of this section, we are left with the conclusion that there must be BSM physics in order to explain the observed matter-antimatter asymmetry.

1.3.3 Neutrino Masses

While the issue of neutrino masses will bear no further relevance for the work in this thesis, it is worth remarking briefly upon them as they may be considered the clearest sign of BSM physics that we have yet discovered.

Neutrinos were first posited as a solution to the different energies of electrons emitted in β decays—a two body decay of a neutron to a proton and an electron has fixed kinematics, and so the energy and momentum of the electron should always be the same. Observationally, electrons with a range of energies were detected, and Pauli proposed that it was in fact a three body decay—proton to neutron, electron and neutrino—such that the electron and neutrino could share the released energy in a variety of ways.

It was for a long time presumed that the neutrino was massless, until the discovery of neutrino oscillations through various experiments [57–60]. Neutrino oscillations can only occur if the three flavours of neutrino have different masses, as the oscillation probability is proportional to the mass differences. In order to give mass to neutrinos in the same way as the other fermions (through interactions with the Higgs field), a right-handed neutrino is needed. However, such a particle would be a singlet under all the SM gauge groups, and hence would have no interactions except through the Higgs mechanism—for this reason, the SM does not contain such a field, as it was unnecessary for massless neutrinos. There is an alternative way to give the neutrino mass—as the only massive neutral fundamental particle, it is possible that it is its own antiparticle. In QFT terms, the neutrino field is Majorana rather than Dirac.

Either solution involves extending the SM (albeit possibly in a very minimal way) and so the observed masses are the first sign of BSM physics.

1.3.4 Flavour Anomalies

While all direct searches for new physics (NP) at the LHC have so far drawn a blank, there is intriguing evidence of a serious anomaly emerging in rare B meson decays. The story started back in 2013 when the LHCb collaboration reported a discrepancy, with a local significance of $3.7\,\sigma$, in an analysis of angular observables in $B^0 \to K^* \mu^+ \mu^-$ decays [61], making use of less form factor dependent observables defined in [62] (including the famous P_5'). A complete theory analysis of these decays was done in [63]. That result was based on $1\,\mathrm{fb}^{-1}$ of data, and has since been replicated with more data by the LHCb [64, 65], as well as by CMS [66, 67], ATLAS [68], BaBar [69], Belle [70, 71] collaborations in various similar decay modes.

Since that initial result, there have been more measurements across a variety of channels and observables—the common thread is the underlying decay $b \rightarrow s\ell\ell$. The significance of the effect is still under discussion because of the difficulty of determining the exact size of the hadronic contributions (see e.g. [72–78]). Estimates of the combined significance of all these deviations range between three and almost six standard deviations.

The picture developed still further with the measurement of the ratios

$$R_{K^{(*)}} = \frac{\mathcal{B}(B \rightarrow K^{(*)}\mu^+\mu^-)}{\mathcal{B}(B \rightarrow K^{(*)}e^+e^-)}, \tag{1.3.2}$$

which test lepton flavour universality (LFU), by LHCb [79, 80].[8] Such ratios were first suggested as ideal testing grounds for NP in [82], and recent SM predictions can be found in [83]. The results are highly suggestive of some non-universal physics at play, i.e. that the universality of lepton couplings that exists in the SM is violated by BSM physics.

There have been many global fits to all the relevant data [84–95], and what has emerged is probably the most coherent sign yet from LHC results—a single NP effect, contributing to the operator $(\bar{s}\gamma_\mu P_L b)(\mu\gamma_\mu\mu)$, can reduce the tension in all the anomalies.

Since then there has been a flood of interest from theorists attempting to explain the results—see for example [96–125] for an arbitrary set of papers investigating Z' models alone. In Chap. 5 we examine how a lepton flavour universal effect could be generated at 1-loop by effective operators, while in Chap. 7 we take the measurements of LFU violation seriously and study how models that explain it are strongly constrained by B_s mixing measurements.

1.4 Remainder of the Thesis

The remainder of this thesis will concern the study of quark flavour physics, starting with Chap. 2 where we study methods that will form the basis of some of the calculations in the following chapters. Having worked on our theoretical foundations, we try and answer some of the questions raised in Sect. 1.3, namely is there anything beyond the Standard Model, and if so where might we see it and what is it like? In order to tackle this we take a three pronged approach: we assess some of the underlying assumptions of our calculational tools; we look for new physics in both model independent and model dependent ways; and attempt to improve the precision of vital input parameters to our calculations.

Starting in Chap. 3, we examine quark-hadron duality and how well it can be tested using the current precision of B mixing observables, and how a small violation of

---

[8]These ratios were also measured much earlier by Belle [81], but with a larger uncertainty.

the duality could improve the status of D mixing calculations. Following this, in Chap. 4 we study a specific model of dark matter that can produce interesting effects in the up type quark sector and attempt to see what parameter space can be ruled out by a mix of different constraints. Still looking at new physics, we take an effective theory approach in Chap. 5 and look at a full set of new four quark operators of the form $(\bar{s}b)(\bar{c}c)$ and how a small set of precision observables can rule out or confirm new effects in this sector. We then move on to improving the precision of our calculations—so-called "bag" parameters are an important input parameter in many of our calculations, and in Chap. 6 we provide an independent and state-of-the-art calculation of these parameters as a cross-check on other, older, calculations. Our final study, in Chap. 7, is of how a whole class of new physics model that aim to explain the anomalies we introduced in Sect. 1.3.4 can be very strongly constrained by B_s mixing observables given recent improvements to the calculation. Finally we wrap up by discussing our conclusions from this body of work in Chap. 8.

References

1. Dirac PAM (1928) The quantum theory of the electron. Proc R Soc Lond A 117:610–624. https://doi.org/10.1098/rspa.1928.0023
2. Tomonaga S (1946) On a relativistically invariant formulation of the quantum theory of wave fields. Prog Theor Phys 1:27–42. https://doi.org/10.1143/PTP.1.27
3. Schwinger JS (1948) On quantum electrodynamics and the magnetic moment of the electron. Phys Rev 73:416–417. https://doi.org/10.1103/PhysRev.73.416
4. Schwinger JS (1948) Quantum electrodynamics. I A covariant formulation. Phys Rev 74:1439 https://doi.org/10.1103/PhysRev.74.1439
5. Feynman RP (1949) Space - time approach to quantum electrodynamics. Phys Rev 76:769–789. https://doi.org/10.1103/PhysRev.76.769
6. Feynman RP (1949) The theory of positrons. Phys Rev 76:749–759. https://doi.org/10.1103/PhysRev.76.749
7. Feynman RP (1950) Mathematical formulation of the quantum theory of electromagnetic interaction. Phys Rev 80:440–457. https://doi.org/10.1103/PhysRev.80.440
8. Englert F, Brout R (1964) Broken symmetry and the mass of gauge vector mesons. Phys Rev Lett 13:321–323. https://doi.org/10.1103/PhysRevLett.13.321
9. Higgs PW (1964) Broken symmetries and the masses of gauge bosons. Phys Rev Lett 13:508–509. https://doi.org/10.1103/PhysRevLett.13.508
10. Guralnik GS, Hagen CR, Kibble TWB (1964) Global conservation laws and massless particles. Phys Rev Lett 13:585–587. https://doi.org/10.1103/PhysRevLett.13.585
11. Glashow SL (1961) Partial symmetries of weak interactions. Nucl Phys 22:579–588. https://doi.org/10.1016/0029-5582(61)90469-2
12. Weinberg S (1967) A model of leptons. Phys Rev Lett 19:1264–1266. https://doi.org/10.1103/PhysRevLett.19.1264
13. Salam A (1968) Weak and electromagnetic interactions. Conf Proc C 680519:367–377
14. Fritzsch H, Gell-Mann M, Leutwyler H (1973) Advantages of the color octet gluon picture. Phys Lett 47B:365–368. https://doi.org/10.1016/0370-2693(73)90625-4
15. Gross DJ, Wilczek F (1973) Ultraviolet behavior of nonabelian gauge theories. Phys Rev Lett 30:1343–1346. https://doi.org/10.1103/PhysRevLett.30.1343
16. Politzer HD (1973) Reliable perturbative results for strong interactions? Phys Rev Lett 30:1346–1349. https://doi.org/10.1103/PhysRevLett.30.1346

17. Fritzsch H (2001) The physics of flavor is the flavor of physics. In: Flavor physics. Proceedings, international conference, ICFP 2001, Zhang-Jia-Jie, China, May 31–June 6, 2001, pp 453–462. https://doi.org/10.1142/9789812777379_0050, arXiv:hep-ph/0111051

18. Fritzsch H (2004) The physics of flavor: challenge for the future. J Korean Phys Soc 45:S297–S300. arXiv:0407069

19. Yukawa H (1935) On the interaction of elementary particles I. Proc Phys Math Soc Jap 17:48–57. https://doi.org/10.1143/PTPS.1.1

20. Buras AJ, Gambino P, Gorbahn M, Jäger S, Silvestrini L (2001) Universal unitarity triangle and physics beyond the standard model. Phys Lett B 500:161–167. https://doi.org/10.1016/S0370-2693(01)00061-2, arXiv:hep-ph/0007085

21. D'Ambrosio G, Giudice GF, Isidori G, Strumia A (2002) Minimal flavor violation: an effective field theory approach. Nucl Phys B 645:155–187. https://doi.org/10.1016/S0550-3213(02)00836-2, arXiv:hep-ph/0207036

22. Cabibbo N (1963) Unitary symmetry and leptonic decays. Phys Rev Lett 10:531–533. https://doi.org/10.1103/PhysRevLett.10.531

23. Kobayashi M, Maskawa T (1973) CP violation in the renormalizable theory of weak interaction. Prog Theor Phys 49:652–657. https://doi.org/10.1143/PTP.49.652

24. CKMfitter collaboration, ICHEP 2016 results. http://ckmfitter.in2p3.fr/www/results/plots_ichep16/num/ckmEval_results_ichep16.html

25. CKMfitter collaboration, Charles J, Hocker A, Lacker H, Laplace S, Le Diberder FR, Malcles J et al (2005) CP violation and the CKM matrix: assessing the impact of the asymmetric B factories. Eur Phys J C41:1–131. https://doi.org/10.1140/epjc/s2005-02169-1, arXiv:hep-ph/0406184

26. CKMfitter collaboration. http://ckmfitter.in2p3.fr/

27. UTfit collaboration, Bona M et al (2006) The unitarity triangle fit in the standard model and hadronic parameters from lattice QCD: a reappraisal after the measurements of Δm_s and BR($B \rightarrow \tau\nu_\tau$), JHEP 10:081. https://doi.org/10.1088/1126-6708/2006/10/081, arXiv:hep-ph/0606167

28. UTfit collaboration. http://www.utfit.org/UTfit/WebHome

29. Chau L-L, Keung W-Y (1984) Comments on the parametrization of the Kobayashi-Maskawa matrix. Phys Rev Lett 53:1802. https://doi.org/10.1103/PhysRevLett.53.1802

30. Wolfenstein L (1983) Parametrization of the Kobayashi-Maskawa matrix. Phys Rev Lett 51:1945. https://doi.org/10.1103/PhysRevLett.51.1945

31. Glashow SL, Iliopoulos J, Maiani L (1970) Weak interactions with Lepton-Hadron symmetry. Phys Rev D 2:1285–1292. https://doi.org/10.1103/PhysRevD.2.1285

32. Luders G (1954) On the equivalence of invariance under time reversal and under particle-antiparticle conjugation for relativistic field theories. Kong Dan Vid Sel Mat Fys Med 28N5:1–17

33. Pauli W (1955) Niels Bohr and the development of physics: essays dedicated to Niels Bohr on the occasion of his seventieth birthday, W. Pauli edn. McGraw-Hill

34. Bell JS (1954) Ph.D. thesis, University of Birmingham

35. Lee TD, Yang C-N (1956) Question of parity conservation in weak interactions. Phys Rev 104:254–258. https://doi.org/10.1103/PhysRev.104.254

36. Wu CS, Ambler E, Hayward RW, Hoppes DD, Hudson RP (1957) Experimental test of parity conservation in beta decay. Phys Rev 105:1413–1414. https://doi.org/10.1103/PhysRev.105.1413

37. Christenson JH, Cronin JW, Fitch VL, Turlay R (1964) Evidence for the 2π decay of the K_2^0 meson. Phys Rev Lett 13:138–140. https://doi.org/10.1103/PhysRevLett.13.138

38. Jarlskog C (1985) Commutator of the quark mass matrices in the standard electroweak model and a measure of maximal CP violation. Phys Rev Lett 55:1039. https://doi.org/10.1103/PhysRevLett.55.1039

39. Jarlskog C (1985) A basis independent formulation of the connection between quark mass matrices, CP violation and experiment. Z Phys C 29:491–497. https://doi.org/10.1007/BF01565198

40. ATLAS collaboration, Summary plots from the ATLAS standard model physics group. https://atlas.web.cern.ch/Atlas/GROUPS/PHYSICS/CombinedSummaryPlots/SM/
41. CMS collaboration, Summaries of CMS cross section measurements. https://twiki.cern.ch/twiki/bin/view/CMSPublic/PhysicsResultsCombined
42. Zwicky F (1933) Die Rotverschiebung von extragalaktischen Nebeln. Helv Phys Acta 6:110–127. https://doi.org/10.1007/s10714-008-0707-4
43. Zwicky F (1937) On the masses of nebulae and of clusters of nebulae. Astrophys J 86:217–246. https://doi.org/10.1086/143864
44. Rubin VC, Ford WK Jr (1970) Rotation of the andromeda nebula from a spectroscopic survey of emission regions. Astrophys J 159:379–403. https://doi.org/10.1086/150317
45. Begeman KG, Broeils AH, Sanders RH (1991) Extended rotation curves of spiral galaxies: dark haloes and modified dynamics. Mon Not R Astron Soc 249:523. https://doi.org/10.1093/mnras/249.3.523
46. Davis DS, White RE III (1996) Rosat temperatures and abundances for a complete sample of elliptical galaxies. Astrophys J 470:L35. https://doi.org/10.1086/310289, arXiv:astro-ph/9607052
47. Allen SW, Rapetti DA, Schmidt RW, Ebeling H, Morris G, Fabian AC (2008) Improved constraints on dark energy from chandra X-ray observations of the largest relaxed galaxy clusters. Mon Not R Astron Soc 383:879–896. https://doi.org/10.1111/j.1365-2966.2007.12610.x, arXiv:0706.0033
48. Planck collaboration, Ade PAR et al (2016) Planck 2015 results. XIII. Cosmological parameters. Astron Astrophys 594:A13. https://doi.org/10.1051/0004-6361/201525830, arXiv:1502.01589
49. Massey R, Kitching T, Richard J (2010) The dark matter of gravitational lensing. Rept Prog Phys 73:086901. https://doi.org/10.1088/0034-4885/73/8/086901, arXiv:1001.1739
50. Bertone G, Hooper D (2018) History of dark matter. Rev Mod Phys 90:045002. https://doi.org/10.1103/RevModPhys.90.045002, arXiv:1605.04909
51. Particle Data Group collaboration, Big-Bang Nucleosynthesis. http://pdg.lbl.gov/2017/reviews/rpp2017-rev-bbang-nucleosynthesis.pdf
52. Cyburt RH, Fields BD, Olive KA, Yeh T-H (2016) Big bang nucleosynthesis: 2015. Rev Mod Phys 88:015004. https://doi.org/10.1103/RevModPhys.88.015004, arXiv:1505.01076
53. Particle Data Group collaboration, Inflation. http://pdg.lbl.gov/2018/reviews/rpp2018-rev-inflation.pdf
54. Sakharov AD (1967) Violation of CP invariance, C asymmetry, and baryon asymmetry of the universe. Pisma Zh Eksp Teor Fiz 5:32–35. https://doi.org/10.1070/PU1991v034n05ABEH002497
55. Klinkhamer FR, Manton NS (1984) A saddle point solution in the Weinberg-Salam theory. Phys Rev D 30:2212. https://doi.org/10.1103/PhysRevD.30.2212
56. Kajantie K, Laine M, Rummukainen K, Shaposhnikov ME (1996) Is there a hot electroweak phase transition at m(H) larger or equal to m(W)? Phys Rev Lett 77:2887–2890. https://doi.org/10.1103/PhysRevLett.77.2887, arXiv:hep-ph/9605288
57. Davis R Jr, Harmer DS, Hoffman KC (1968) Search for neutrinos from the sun. Phys Rev Lett 20:1205–1209. https://doi.org/10.1103/PhysRevLett.20.1205
58. Super-Kamiokande collaboration, Fukuda Y et al (1998) Evidence for oscillation of atmospheric neutrinos. Phys Rev Lett 81:1562–1567. https://doi.org/10.1103/PhysRevLett.81.1562, arXiv:hep-ex/9807003
59. Double Chooz collaboration, Abe Y et al (2012) Indication of reactor $\bar{\nu}_e$ disappearance in the double Chooz experiment. Phys Rev Lett 108:131801. https://doi.org/10.1103/PhysRevLett.108.131801, arXiv:1112.6353
60. OPERA collaboration, Agafonova N et al (2010) Observation of a first ν_τ candidate in the OPERA experiment in the CNGS beam. Phys Lett B691:138–145. https://doi.org/10.1016/j.physletb.2010.06.022, arXiv:1006.1623
61. LHCb collaboration, Aaij R et al (2013) Measurement of form-factor-independent observables in the decay $B^0 \rightarrow K^{*0}\mu^+\mu^-$. Phys Rev Lett 111:191801. https://doi.org/10.1103/PhysRevLett.111.191801, arXiv:1308.1707

62. Descotes-Genon S, Matias J, Ramon M, Virto J (2013) Implications from clean observables for the binned analysis of $B \to K*\mu^+\mu^-$ at large recoil. JHEP 01:048. https://doi.org/10.1007/JHEP01(2013)048, arXiv:1207.2753

63. Matias J, Mescia F, Ramon M, Virto J (2012) Complete anatomy of $\bar{B}_d \to \bar{K}^{*0}(\to K\pi)l^+l^-$ and its angular distribution. JHEP 04:104. https://doi.org/10.1007/JHEP04(2012)104, arXiv:1202.4266

64. LHCb collaboration, Aaij R et al (2015) Angular analysis and differential branching fraction of the decay $B_s^0 \to \phi\mu^+\mu^-$. JHEP 09:179. https://doi.org/10.1007/JHEP09(2015)179, arXiv:1506.08777

65. LHCb collaboration, Aaij R et al (2016) Angular analysis of the $B^0 \to K^{*0}\mu^+\mu^-$ decay using 3 fb^{-1} of integrated luminosity. JHEP 02:104. https://doi.org/10.1007/JHEP02(2016)104, arXiv:1512.04442

66. CMS collaboration, Khachatryan V et al (2016) Angular analysis of the decay $B^0 \to K^{*0}\mu^+\mu^-$ from pp collisions at $\sqrt{s} = 8$ TeV. Phys Lett B 753:424–448. https://doi.org/10.1016/j.physletb.2015.12.020, arXiv:1507.08126

67. CMS collaboration, Sirunyan AM et al (2018) Measurement of angular parameters from the decay $B^0 \to K^{*0}\mu^+\mu^-$ in proton-proton collisions at $\sqrt{s} = 8$ TeV. Phys Lett B781:517–541. https://doi.org/10.1016/j.physletb.2018.04.030, arXiv:1710.02846

68. ATLAS collaboration, Aaboud M et al (2018) Angular analysis of $B_d^0 \to K^*\mu^+\mu^-$ decays in pp collisions at $\sqrt{s} = 8$ TeV with the ATLAS detector. JHEP 10:047. https://doi.org/10.1007/JHEP10(2018)047, arXiv:1805.04000

69. BaBar collaboration, Lees JP et al (2016) Measurement of angular asymmetries in the decays $B \to K^*\ell^+\ell^-$. Phys Rev D 93:052015. https://doi.org/10.1103/PhysRevD.93.052015, arXiv:1508.07960

70. Belle collaboration, Abdesselam A et al (2016) Angular analysis of $B^0 \to K^*(892)^0\ell^+\ell^-$, in Proceedings, LHCSki 2016 - A first discussion of 13 TeV results: Obergurgl, Austria, April 10–15, 2016, arxiv:1604.04042, https://inspirehep.net/record/1446979/files/arXiv:1604.04042.pdf

71. Belle collaboration, Wehle S et al (2017) Lepton-flavor-dependent angular analysis of $B \to K^*\ell^+\ell^-$. Phys Rev Lett 118:111801. https://doi.org/10.1103/PhysRevLett.118.111801, arXiv:1612.05014

72. Jäger S, Camalich JM (2013) On $B \to V\ell\ell$ at small dilepton invariant mass, power corrections, and new physics. JHEP 05:043. https://doi.org/10.1007/JHEP05(2013)043, arXiv:1212.2263

73. Jäger S, Camalich JM (2016) Reassessing the discovery potential of the $B \to K^*\ell^+\ell^-$ decays in the large-recoil region: SM challenges and BSM opportunities. Phys Rev D 93:014028. https://doi.org/10.1103/PhysRevD.93.014028, arXiv:1412.3183

74. Descotes-Genon S, Hofer L, Matias J, Virto J (2014) On the impact of power corrections in the prediction of $B \to K^*\mu^+\mu^-$ observables. JHEP 12:125. https://doi.org/10.1007/JHEP12(2014)125, arXiv:1407.8526

75. Ciuchini M, Fedele M, Franco E, Mishima S, Paul A, Silvestrini L et al (2016) $B \to K^*\ell^+\ell^-$ decays at large recoil in the standard model: a theoretical reappraisal. JHEP 06:116. https://doi.org/10.1007/JHEP06(2016)116, arXiv:1512.07157

76. Chobanova VG, Hurth T, Mahmoudi F, Santos DM, Neshatpour S (2017) Large hadronic power corrections or new physics in the rare decay $B \to K^*\mu^+\mu^-$? JHEP 07:025. https://doi.org/10.1007/JHEP07(2017)025, arXiv:1702.02234

77. Capdevila B, Descotes-Genon S, Hofer L, Matias J (2017) Hadronic uncertainties in $B \to K^*\mu^+\mu^-$: a state-of-the-art analysis. JHEP 04:016. https://doi.org/10.1007/JHEP04(2017)016, arXiv:1701.08672

78. Bobeth C, Chrzaszcz M, van Dyk D, Virto J (2018) Long-distance effects in $B \to K^*\ell\ell$ from analyticity. Eur Phys J C 78:6. https://doi.org/10.1140/epjc/s10052-018-5918-6, arXiv:1707.07305

79. LHCb collaboration, Aaij R et al (2014) Test of lepton universality using $B^+ \to K^+ \ell^+ \ell^-$ decays. Phys Rev Lett 113:151601. https://doi.org/10.1103/PhysRevLett.113.151601, arXiv:1406.6482

80. LHCb collaboration, Aaij R et al (2017) Test of lepton universality with $B^0 \to K^{*0} \ell^+ \ell^-$ decays. JHEP 08:055. https://doi.org/10.1007/JHEP08(2017)055, arXiv:1705.05802

81. Belle collaboration, Wei JT et al (2009) Measurement of the differential branching fraction and forward-backword asymmetry for $B \to K^{(*)} \ell^+ \ell^-$. Phys Rev Lett 103:171801 https://doi.org/10.1103/PhysRevLett.103.171801, arXiv:0904.0770

82. Hiller G, Kruger F (2004) More model-independent analysis of $b \to s$ processes. Phys Rev D 69:074020. https://doi.org/10.1103/PhysRevD.69.074020, arXiv:hep-ph/0310219

83. Bordone M, Isidori G, Pattori A (2016) On the standard model predictions for R_K and R_{K^*}. Eur Phys J C 76:440. https://doi.org/10.1140/epjc/s10052-016-4274-7, arXiv:1605.07633

84. Descotes-Genon S, Matias J, Virto J (2013) Understanding the $B \to K^* \mu^+ \mu^-$ anomaly. Phys Rev D 88:074002. https://doi.org/10.1103/PhysRevD.88.074002, arXiv:1307.5683

85. Beaujean F, Bobeth C, van Dyk D (2014) Comprehensive Bayesian analysis of rare (semi)leptonic and radiative B decays. Eur Phys J C 74:2897. https://doi.org/10.1140/epjc/s10052-014-3179-6, arXiv:1310.2478

86. Altmannshofer W, Straub DM (2015) New physics in $b \to s$ transitions after LHC run 1. Eur Phys J C 75:382. https://doi.org/10.1140/epjc/s10052-015-3602-7, arXiv:1411.3161

87. Descotes-Genon S, Hofer L, Matias J, Virto J (2016) Global analysis of $b \to s\ell\ell$ anomalies. JHEP 06:092. https://doi.org/10.1007/JHEP06(2016)092, arXiv:1510.04239

88. Hurth T, Mahmoudi F, Neshatpour S (2016) On the anomalies in the latest LHCb data. Nucl Phys B 909:737–777. https://doi.org/10.1016/j.nuclphysb.2016.05.022, arXiv:1603.00865

89. Altmannshofer W, Niehoff C, Stangl P, Straub DM (2017) Status of the $B \to K^* \mu^+ \mu^-$ anomaly after Moriond 2017. Eur Phys J C 77:377. https://doi.org/10.1140/epjc/s10052-017-4952-0, arXiv:1703.09189

90. Ciuchini M, Coutinho AM, Fedele M, Franco E, Paul A, Silvestrini L et al (2017) On flavourful easter eggs for new physics hunger and lepton flavour universality violation. Eur Phys J C 77:688. https://doi.org/10.1140/epjc/s10052-017-5270-2, arXiv:1704.05447

91. Geng L-S, Grinstein B, Jäger S, Camalich JM, Ren X-L, Shi R-X (2017) Towards the discovery of new physics with lepton-universality ratios of $b \to s\ell\ell$ decays. Phys Rev D 96:093006. https://doi.org/10.1103/PhysRevD.96.093006, arXiv:1704.05446

92. Capdevila B, Crivellin A, Descotes-Genon S, Matias J, Virto J (2018) Patterns of new physics in $b \to s\ell^+ \ell^-$ transitions in the light of recent data. JHEP 01:093. https://doi.org/10.1007/JHEP01(2018)093, arXiv:1704.05340

93. Altmannshofer W, Stangl P, Straub DM (2017) Interpreting hints for lepton flavor universality violation. Phys Rev D 96:055008. https://doi.org/10.1103/PhysRevD.96.055008, arXiv:1704.05435

94. D'Amico G, Nardecchia M, Panci P, Sannino F, Strumia A, Torre R et al (2017) Flavour anomalies after the R_{K^*} measurement. JHEP 09:010. https://doi.org/10.1007/JHEP09(2017)010, arXiv:1704.05438

95. Alok AK, Bhattacharya B, Datta A, Kumar D, Kumar J, London D (2017) New physics in $b \to s \mu^+ \mu^-$ after the measurement of R_{K^*}. Phys Rev D 96:095009. https://doi.org/10.1103/PhysRevD.96.095009, arXiv:1704.07397

96. Buras AJ, Girrbach J (2013) Left-handed Z' and Z FCNC quark couplings facing new $b \to s \mu^+ \mu^-$ data. JHEP 12:009. https://doi.org/10.1007/JHEP12(2013)009, arXiv:1309.2466

97. Gauld R, Goertz F, Haisch U (2014) An explicit Z'-boson explanation of the $B \to K^* \mu^+ \mu^-$ anomaly. JHEP 01:069. https://doi.org/10.1007/JHEP01(2014)069, arXiv:1310.1082

98. Buras AJ, De Fazio F, Girrbach J (2014) 331 models facing new $b \to s \mu^+ \mu^-$ data. JHEP 02:112. https://doi.org/10.1007/JHEP02(2014)112, arXiv:1311.6729

99. Altmannshofer W, Gori S, Pospelov M, Yavin I (2014) Quark flavor transitions in $L_\mu - L_\tau$ models. Phys Rev D 89:095033. https://doi.org/10.1103/PhysRevD.89.095033, arXiv:1403.1269

100. Crivellin A, D'Ambrosio G, Heeck J (2015) Explaining $h \to \mu^\pm \tau^\mp$, $B \to K^* \mu^+ \mu^-$ and $B \to K \mu^+ \mu^- / B \to K e^+ e^-$ in a two-Higgs-doublet model with gauged $L_\mu - L_\tau$. Phys Rev Lett 114:151801. https://doi.org/10.1103/PhysRevLett.114.151801, arXiv:1501.00993

101. Crivellin A, D'Ambrosio G, Heeck J (2015) Addressing the LHC flavor anomalies with horizontal gauge symmetries. Phys Rev D 91:075006. https://doi.org/10.1103/PhysRevD.91. 075006, arXiv:1503.03477

102. Celis A, Fuentes-Martin J, Jung M, Serôdio H (2015) Family nonuniversal Z' models with protected flavor-changing interactions. Phys Rev D 92:015007. https://doi.org/10.1103/ PhysRevD.92.015007, arXiv:1505.03079

103. Belanger G, Delaunay C, Westhoff S (2015) A dark matter relic from muon anomalies. Phys Rev D 92:055021. https://doi.org/10.1103/PhysRevD.92.055021, arXiv:1507.06660

104. Falkowski A, Nardecchia M, Ziegler R (2015) Lepton flavor non-universality in B-meson decays from a U(2) flavor model. JHEP 11:173. https://doi.org/10.1007/JHEP11(2015)173, arXiv:1509.01249

105. Carmona A, Goertz F (2016) Lepton flavor and nonuniversality from minimal composite higgs setups. Phys Rev Lett 116:251801. https://doi.org/10.1103/PhysRevLett.116.251801, arXiv:1510.07658

106. Allanach B, Queiroz FS, Strumia A, Sun S (2016) Z? models for the LHCb and $g - 2$ muon anomalies. Phys Rev D 93:055045. https://doi.org/10.1103/PhysRevD.95.119902, arXiv:1511.07447

107. Chiang C-W, He X-G, Valencia G (2016) Z? model for b?s?$\bar{?}$ flavor anomalies. Phys Rev D 93:074003. https://doi.org/10.1103/PhysRevD.93.074003, arXiv:1601.07328

108. Boucenna SM, Celis A, Fuentes-Martin J, Vicente A, Virto J (2016) Non-abelian gauge extensions for B-decay anomalies. Phys Lett B 760:214–219. https://doi.org/10.1016/j.physletb. 2016.06.067, arXiv:1604.03088

109. Megias E, Panico G, Pujolas O, Quiros M (2016) A natural origin for the LHCb anomalies. JHEP 09:118. https://doi.org/10.1007/JHEP09(2016)118, arXiv:1608.02362

110. Boucenna SM, Celis A, Fuentes-Martin J, Vicente A, Virto J (2016) Phenomenology of an $SU(2) \times SU(2) \times U(1)$ model with lepton-flavour non-universality. JHEP 12:059. https:// doi.org/10.1007/JHEP12(2016)059, arXiv:1608.01349

111. Altmannshofer W, Gori S, Profumo S, Queiroz FS (2016) Explaining dark matter and B decay anomalies with an $L_\mu - L_\tau$ model. JHEP 12:106. https://doi.org/10.1007/JHEP12(2016)106, arXiv:1609.04026

112. Crivellin A, Fuentes-Martin J, Greljo A, Isidori G (2017) Lepton flavor non-universality in B decays from dynamical yukawas. Phys Lett B 766:77–85. https://doi.org/10.1016/j.physletb. 2016.12.057, arXiv:1611.02703

113. Garcia IG (2017) LHCb anomalies from a natural perspective. JHEP 03:040. https://doi.org/ 10.1007/JHEP03(2017)040, arXiv:1611.03507

114. Bhatia D, Chakraborty S, Dighe A (2017) Neutrino mixing and R_K anomaly in U(1)$_X$ models: a bottom-up approach. JHEP 03:117. https://doi.org/10.1007/JHEP03(2017)117, arXiv:1701.05825

115. Cline JM, Cornell JM, London D, Watanabe R (2017) Hidden sector explanation of B-decay and cosmic ray anomalies. Phys Rev D 95:095015. https://doi.org/10.1103/PhysRevD.95. 095015, arXiv:1702.00395

116. Baek S (2018) Dark matter contribution to $b \to s\mu^+\mu^-$ anomaly in local $U(1)_{L_\mu-L_\tau}$ model. Phys Lett B 781:376–382. https://doi.org/10.1016/j.physletb.2018.04.012, arXiv:1707.04573

117. Cline JM, Camalich JM (2017) B decay anomalies from nonabelian local horizontal symmetry. Phys Rev D 96:055036. https://doi.org/10.1103/PhysRevD.96.055036, arXiv:1706.08510

118. Di Chiara S, Fowlie A, Fraser S, Marzo C, Marzola L, Raidal M et al (2017) Minimal flavor-changing Z' models and muon $g - 2$ after the R_{K^*} measurement. Nucl Phys B 923:245–257. https://doi.org/10.1016/j.nuclphysb.2017.08.003, arXiv:1704.06200

119. Kamenik JF, Soreq Y, Zupan J (2018) Lepton flavor universality violation without new sources of quark flavor violation. Phys Rev D 97:035002. https://doi.org/10.1103/PhysRevD. 97.035002, arXiv:1704.06005

120. Ko P, Omura Y, Shigekami Y, Yu C (2017) LHCb anomaly and B physics in flavored Z' models with flavored Higgs doublets. Phys Rev D 95:115040. https://doi.org/10.1103/PhysRevD.95. 115040, arXiv:1702.08666

121. Ko P, Nomura T, Okada H (2017) Explaining $B \rightarrow K^{(*)}\ell^+\ell^-$ anomaly by radiatively induced coupling in $U(1)_{\mu-\tau}$ gauge symmetry. Phys Rev D 95:111701. https://doi.org/10.1103/PhysRevD.95.111701, arXiv:1702.02699

122. Alonso R, Cox P, Han C, Yanagida TT (2017) Anomaly-free local horizontal symmetry and anomaly-full rare B-decays. Phys Rev D 96:071701. https://doi.org/10.1103/PhysRevD.96.071701, arXiv:1704.08158

123. Ellis J, Fairbairn M, Tunney P (2018) Anomaly-free models for flavour anomalies. Eur Phys J C 78:238. https://doi.org/10.1140/epjc/s10052-018-5725-0, arXiv:1705.03447

124. Alonso R, Cox P, Han C, Yanagida TT (2017) Flavoured $B - L$ local symmetry and anomalous rare B decays. Phys Lett B 774:643–648. https://doi.org/10.1016/j.physletb.2017.10.027, arXiv:1705.03858

125. Carmona A, Goertz F (2018) Recent B physics anomalies: a first hint for compositeness? Eur Phys J C 78:979. https://doi.org/10.1140/epjc/s10052-018-6437-1, arXiv:1712.02536

Chapter 2
Theoretical Tools

In this chapter we explore some of the concepts, tools and methods which will be used in the rest of the thesis. The idea of effective field theories (EFTs) is one of the most powerful in physics, and we will explain them in Sect. 2.1, along with a specific example of an EFT in Sect. 2.3. Another omnipresent tool is the Heavy Quark Expansion, which we see in Sect. 2.2. Finally, we will explore two simple examples of operator matching and renormalisation—the construction of the Weak Effective Theory (WET) and the calculation of a B_s mixing observable—in Sect. 2.4.

2.1 Effective Field Theories

The basic idea of an effective field theory is simple—that a problem can be successfully analysed only making reference to the physics that is most relevant. To talk about a basic mechanics problem of e.g. balls colliding, we don't need to know the details of the individual atoms in the ball, only that the total effect is to make the balls bounce off each other elastically. We have chosen a description with only the relevant degrees of freedom—in our example, Newtonian mechanics can do the job most effectively. An EFT is simply a formalised version of this principle, applied to field theories as our computational tool. If we have a theory with some high energy degrees of freedom, but want to calculate at low energies (or equivalently larger scales), we can produce a new theory with the irrelevant parts removed, and do it in a consistent way such that we understand how to include corrections in a systematic fashion.

There are two approaches to constructing an EFT—generally called *top down* and *bottom up*. In the bottom up approach, you take a model that is known to work well at some energy scale and build up all the new higher dimensional operators from the fields you know while respecting your low energy symmetries—an example of such an EFT is the SMEFT (Standard Model Effective Field Theory) [1, 2]. Alternatively, you can take the top down approach. Start with a theory you know (or believe) to be valid at a high scale, and integrate out the irrelevant degrees of freedom such that

© Springer Nature Switzerland AG 2019
M. J. Kirk, *Charming New Physics in Beautiful Processes?*,
Springer Theses, https://doi.org/10.1007/978-3-030-19197-9_2

you end up with a simpler theory that is more useful for certain classes of problems. Two examples are shown in the remainder of this chapter—Heavy Quark Effective Theory, which we describe in Sect. 2.3, and the Weak Effective Theory, the basic derivation of which we show in Sect. 2.4.1.

The principle of the EFT formalism (separating out the scales of a problem) provides another benefit for our computations. One central aspect of QCD is the running behaviour of the coupling constant – α_s becomes larger at lower energies, diverging at the scale Λ_{QCD}. As such, QCD corrections become increasingly large. An EFT allows us to separate out the matrix elements (which are determined by long distance, low energy behaviour) from the effective coupling constants (which come from high energy interactions). As we will see, the low energy matrix elements can be determined through lattice QCD calculations, while we determine the effective coupling constants using perturbation theory. This separation can be seen in the Lagrangian, which takes the form

$$\mathcal{L} \sim C(\mu)\mathcal{O}(\mu),$$

where μ is the scale of our calculation (which we will see more of later in this chapter). Both the effective coupling constant C and the operator $\mathcal{O}$ are individually scale dependent, but their product (which is directly related to a physical observable) is not—the scale variation cancels between them.

2.2 Heavy Quark Expansion

The Heavy Quark Expansion (HQE) is a tool for studying inclusive decays of heavy hadrons, particularly the decay of b mesons. In an inclusive decay we calculate a quark level process, but don't worry about the details of how the final state quarks can hadronise into many different hadrons. The corresponding experimental result is found by measuring and summing over all the possible hadronic end products. (The opposite would be exclusive decays, where a single hadronic decay mode is studied and the theoretical calculation is challenging.)

The HQE is an operator product expansion (OPE), in inverse powers of m_b.[1] (See [3–10] for the pioneering papers and [11] for a recent review.) The reason for such an expansion can be seen schematically in the following way. The decay width of a B meson can be written in the form

$$\Gamma(B \to X) = \frac{1}{2M_B} \sum_X \int_{\text{p.s.}} (2\pi)^4 \delta^4(p_B - p_X)|i\mathcal{M}(B \to X)|^2 \qquad (2.2.1)$$

---

[1]In the case of the total lifetime and Γ_{12}, you also find that an overall factor of $\sqrt{1 - M_f^2/m_b^2}$, where M_f is the total mass of the final state quarks, appears in the calculations.

where $\int_{\text{p.s.}}$ stands for an integration over the phase space of the final state particles. Making use of the optical theorem,[2] we can rewrite this as

$$\Gamma(B) = \frac{1}{M_B} \, \text{Im}(\mathcal{M}(B \to B)) \,. \tag{2.2.2}$$

An alternative formulation is more common, using the Taylor expansion of the S-matrix in terms of time-ordered products of the Hamiltonian:

$$\Gamma(B) = \frac{1}{M_B} \, \text{Im}\langle B|\mathcal{T}|B\rangle \text{ where } \mathcal{T} = i \int d^4x \, T(\mathcal{H}(x)\mathcal{H}(0)) \,. \tag{2.2.3}$$

The OPE tells us that since m_b is large in this context, the integral is dominated by small distances $x \sim m_b^{-1}$, and we can expand $\mathcal{T}$ as a series of local operators.

If we consider operators that can mediate the $B \to B$ process, we can see what the power series looks like.

- The smallest dimension operator we can write down is $\bar{b}b$, with mass dimension three.
- At first glance, it would seem we could write down $\bar{b}\slashed{D}b$ as a dimension-four operator (where $\slashed{D}$ is the QCD + QED covariant derivative), but this can be reduced to $\bar{b}b$ using the corresponding equations of motion, and so is not independent.
- The next independent operator has dimension five: $\bar{b}\sigma_{\mu\nu}G^{\mu\nu}b$.
- At dimension six, there is $(\bar{b}\Gamma q)(\bar{b}\Gamma' q)$—notice this is the first time we have a coupling to the second quark that makes up the meson, whereas for the lower dimensional operators this other quark is a *spectator* to the decay.

We can expand the width in terms of these operators and their (dimensionless) coefficients as

$$\Gamma(B) \sim C_{\bar{b}b} \frac{\langle B|\bar{b}b|B\rangle}{M_B} + \frac{C_{\bar{b}Gb}}{m_b^2} \frac{\langle B|\bar{b}\sigma_{\mu\nu}G^{\mu\nu}b|B\rangle}{M_B} + \frac{C_{\bar{b}q\bar{b}q}}{m_b^3} \frac{\langle B|(\bar{b}\Gamma q)(\bar{b}\Gamma' q)|B\rangle}{M_B} + \cdots, \tag{2.2.4}$$

which is the *Heavy Quark Expansion*. It is interesting to note two properties we expect from the HQE—the corrections to free b quark decay arise at $\mathcal{O}\left(1/m_b^2\right)$, and the spectator quark only gets involved at $\mathcal{O}\left(1/m_b^3\right)$ (although some of the $\mathcal{O}\left(1/m_b^3\right)$ terms get a numerical enhancement of the order $16\pi^2$ as they arise from 1-loop rather than 2-loop diagrams). This is realised in the experimental data, where we can see from Table 2.1 that many B mesons and baryons have a lifetime of approximately 1.5 ps, and e.g. the lifetime ratio $\tau(B_s)/\tau(B_d)$ deviates from 1 at the sub-percent level.

---

[2]In the language of amplitudes, the optical theorem can be written in the form

$$i\mathcal{M}(i \to f) + [i\mathcal{M}(f \to i)]^* = -(2\pi)^4 \sum_X \int_{\text{p.s.}} [i\mathcal{M}(f \to X)]^* \, i\mathcal{M}(i \to X) \,.$$

Table 2.1 Lifetime data for B mesons and baryons, taken from the HFLAV [12, 13] 2018 results [14]. Those marked with a † are not directly given by HFLAV, and are my own calculation

Particle	Lifetime/ps	Lifetime ratio ($\tau(X)/\tau(B_d)$)
$B_d(=\bar{b}d)$	1.520 ± 0.004	1
$B_s(=\bar{b}s)$	1.509 ± 0.004	0.993 ± 0.004
$B^+(=\bar{b}u)$	1.638 ± 0.004	1.076 ± 0.004
$B_c^+(=\bar{b}c)$	0.507 ± 0.009	0.334 ± 0.006 †
$\Lambda_b(=bdu)$	1.470 ± 0.010	0.967 ± 0.007
$\Xi_b^0(=bsu)$	1.479 ± 0.031	0.97 ± 0.02 †
$\Xi_b^-(=bsd)$	1.571 ± 0.040	1.03 ± 0.26 †
$\Omega_b^-(=bss)$	$1.64^{+0.18}_{-0.17}$	$1.08^{+0.12}_{-0.11}$ †

We have focused on the HQE for B mesons here, but it is worth a small discussion about the charm sector. Since $m_c \approx m_b/3$, it is not clear that the HQE will converge as well for D mesons as it does for B. The simplest way to test this is to make predictions for D lifetimes and see how they compare to data. However, this approach has been hindered by the lack of availability of non-perturbative matrix element determinations, leaving us reliant on the Vacuum Saturation Approximation (VSA). Our work in Chap. 6 provides some of the first results in this area, and is encouraging in terms of the reliability of the HQE for these calculations.

2.3 Heavy Quark Effective Theory

As we discussed in Sect. 2.1 the basic idea of an EFT is that given a scale that is widely separated from all others in a problem, we can remove the isolated scale to construct a simpler theoretical description. For the Heavy Quark Effective Theory, the heavy quark mass is the isolated scale. (See [10, 15–24] for some of the early development of the HQET, or [25, 26] for in depth reviews.) What do we mean by a "heavy" quark? In many processes with a single bottom or charm quark, the other relevant scales are that of the QCD scale and/or the lighter (i.e. s, d, u) quark masses and momenta, which are both on the order of 100MeV. Since $\Lambda_{QCD}/m_{b,c} \ll 1$, we have the basis for our EFT.

The formulation of HQET can be thought of in the following way: we work in a frame where the heavy quark is almost at rest,

$$p = m_Q v + \tilde{p}, \quad (\text{with } v \cdot v = 1) \tag{2.3.1}$$

where the residual momentum is small ($\tilde{p} \sim \Lambda_{QCD}$). Although choosing such a frame breaks Lorentz invariance, it is possible to write the theory in a Lorentz covariant fashion. In this frame, the heavy quark acts as a static source of the QCD potential. A more physical understanding can be seen by analogy to the case of the electron cloud surrounding a proton in a hydrogen atom. At first approximation, we model the proton at rest and talk about the electrons moving around in a fixed electromagnetic (QED) potential—this gives rise to the measurable spectra of hydrogen atoms. Corrections to this picture arise at $\mathcal{O}\left(1/m_Q\right)$, in the same way that the spectrum of different isotopes of hydrogen differ only in their hyperfine structure. A final point to note about HQET is that the heavy quark and antiquark fields are separate—we see that in the frame where the heavy quark is at rest, there is not enough energy to create an antiquark where none existed before.

Feynman Rules

The Feynman rules for HQET can be found in Table B.1 of Appendix B, but we briefly outline here how their form can be derived. Since HQET is based on the large quark mass, we can get an idea of the behaviour by considering $m_Q \to \infty$. By taking this limit of the QCD Feynman rules, we can see, for example, that the HQET heavy quark propagator will look like:

$$\frac{i(\not{p} + m)}{p^2 - m^2} = \frac{im(\not{v} + \mathbb{1}) + i\not{\tilde{p}}}{m^2 v \cdot v + 2mv \cdot \tilde{p} + \tilde{p}^2 - m^2}$$
$$= \frac{im(\not{v} + \mathbb{1}) + i\not{\tilde{p}}}{2mv \cdot \tilde{p} + \tilde{p}^2}$$
$$= \frac{i(\not{v} + \mathbb{1})}{2v \cdot \tilde{p}} + \mathcal{O}\left(\frac{\tilde{p}}{m}\right).$$

2.4 Example Calculations

As a round off to this section, we aim to show in detail two particular calculations, including some points that are often less well elaborated on. To start, we will show the calculation of the Wilson coefficients for the first two operators of the WET and how the renormalisation and renormalisation group running works. As a follow up, we detail the calculation of the mass difference arising from B_s mixing, hopefully illuminating the origin of many of the more obscure factors.

2.4.1 Matching and RG Running

Matching is the procedure by which we relate two field theories with different regions of applicability (generally a more fundamental theory to its low energy effective version) to each other—we calculate some process in both the EFTs, and set the Wilson coefficients by requiring that we get the same answer in both (in general many processes may be necessary to specify all the unknown coefficients). Once we

Table 2.2 Coefficients in a perturbative expansion

	LL	NLL	$NNLL$	N^3LL
Tree-level	1	–	–	–
1-loop	$\alpha_s \ln$	α_s	–	–
2-loop	$\alpha_s^2 \ln^2$	$\alpha_s^2 \ln$	α_s^2	–
3-loop	$\alpha_s^3 \ln^3$	$\alpha_s^3 \ln^2$	$\alpha_s^3 \ln$	α_s^3

go beyond tree level, a scale dependence will appear in our calculation—this is the "matching scale" and there is a certain amount of freedom in our choice of this scale as it is unphysical (by which we mean that any observable quantity cannot depend on it). As we have said, often the idea of EFTs is to allow us to calculate at an energy scale that is very far removed from some fundamental scale in our "full" theory. Once we have our Wilson coefficients as a function of our matching scale, μ_m, can we just set this scale to whatever low scale we want to work at? Mathematically, yes perhaps, but if we do that we see that large logarithms arise in our results—generally of the form ln(scale of removed physics/scale of calculation), which break the validity of our perturbation series. Naively this might be solved by going to higher orders in the calculation, as this is "well-known" as the technique to reduce scale variation (which amounts to calculating more rows in Table 2.2). A better solution is to use renormalisation group (RG) improved perturbation theory to sum up these large logs to *all* orders—this corresponds to calculating column by column in Table 2.2.

The process we study in this example is $b^i u^j \rightarrow c^k d^l$, which can arise from two $\Delta F = 1$ operators in the WET:

$$Q_1 = (\bar{c}^i \gamma^\mu P_L b^j)(\bar{d}^j \gamma_\mu P_L u^i), \qquad Q_2 = (\bar{c}^i \gamma^\mu P_L b^i)(\bar{d}^j \gamma_\mu P_L u^j) \qquad (2.4.1)$$

(there is an alternative way of defining the two operators in terms of colour singlets and octets, which can be related to our definition using the Fierz relation Eq. A.0.1). The numbering of these two operators is convention dependent—our choice matches that of [27]. At tree level only one operator would be needed (Q_2), but QCD corrections generate the other colour structure.

In this example we match onto the WET at LO + LL accuracy—this means calculating the matching at 1-loop as this is where the leading order QCD corrections come in. We keep terms that are $\mathcal{O}(\alpha_s \ln)$ since these will turn out to be large, while discarding terms that are simply $\mathcal{O}(\alpha_s)$—looking at Table 2.2 we see that these terms count at NLL accuracy. We can choose our external quarks to all have equal, off-shell momentum p (with $p^2 < 0$), despite this obviously being unphysical, as well as further approximating our external states as massless. The previous two choices have no effect on the resulting Wilson coefficients (as only the infrared (IR) behaviour is affected, and our Wilson coefficients come from the different ultraviolet (UV) behaviour of the two theories) but greatly simplify the calculation.

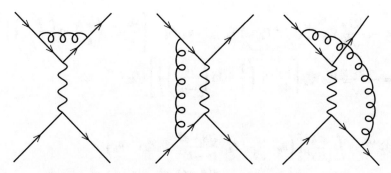

Fig. 2.1 QCD correction diagrams in the SM (symmetric diagrams not shown)

SM Calculation

In addition to the tree-level diagram the QCD corrections give us another six diagrams, some of which are shown in Fig. 2.1. The amplitude for the tree diagram is simple:

$$i\mathcal{M}^{SM,(0)} = -i\frac{4G_F}{\sqrt{2}} V_{cb} V_{ud}^* (\bar{u}_c \gamma^\mu P_L u_b)(\bar{u}_d \gamma_\mu P_L u_u)\delta_{ki}\delta_{lj} \equiv -i\frac{4G_F}{\sqrt{2}} V_{cb} V_{ud}^* \langle Q_2 \rangle^{\text{tree}} ,$$

where we have defined the tree-level insertion in what will be a sensible way when we come to the WET calculation, and the Fermi constant G_F is given by

$$\frac{G_F}{\sqrt{2}} \equiv \frac{g^2}{8M_W^2} . \tag{2.4.2}$$

The 1-loop diagrams are where the interesting results come in—we will see that different colour structures are generated by the exchange of gluons between the two sets of fermion lines. (In the following, the subscripts on $i\mathcal{M}$ refer to the left, middle and right diagrams in Fig. 2.1 respectively.)

$$i\mathcal{M}_1^{SM,(1)} = \int \frac{d^d k}{(2\pi)^d} \left[\left(\bar{u}_c \cdot i g_s t_{km}^a \gamma^\alpha \cdot \frac{i(\not{p} - \not{k})}{(p-k)^2} \cdot \frac{ig}{\sqrt{2}} \gamma^\mu P_L \cdot \frac{i(\not{p} - \not{k})}{(p-k)^2} \cdot i g_s t_{mi}^a \gamma^\beta \cdot u_b \right) \right.$$
$$\left(\bar{u}_d \cdot \frac{ig}{\sqrt{2}} \delta_{jl} \gamma^\nu P_L \cdot u_u \right) \times \frac{-i g_{\mu\nu}}{-M_W^2} \times \frac{-i g_{\alpha\beta}}{k^2} \times V_{cb} V_{ud}^* \right]$$
$$= -i\frac{4G_F}{\sqrt{2}} V_{cb} V_{ud}^* \delta_{ki}\delta_{lj} \frac{\alpha_s}{4\pi} C_F \left[(\bar{u}_c \gamma^\mu P_L u_b)(\bar{u}_d \gamma_\mu P_L u_u) \left(\frac{1}{\epsilon} + \ln\left(\frac{\mu^2}{-p^2}\right) + 1 \right) \right.$$
$$\left. - \frac{2}{p^2}(\bar{u}_c \not{p} P_L u_b)(\bar{u}_d \not{p} P_L u_u) \right]$$

and now dropping terms that don't contribute at leading log accuracy we find

$$= -i\frac{4G_F}{\sqrt{2}} V_{cb} V_{ud}^* (\bar{u}_c \gamma^\mu P_L u_b)(\bar{u}_d \gamma_\mu P_L u_u)\delta_{ki}\delta_{lj}\left[\frac{\alpha_s}{4\pi}C_F\left(\frac{1}{\epsilon}+\ln\left(\frac{\mu^2}{-p^2}\right)\right)\right]$$

$$= -i\frac{4G_F}{\sqrt{2}} V_{cb} V_{ud}^* \left[\frac{\alpha_s}{4\pi}C_F\left(\frac{1}{\epsilon}+\ln\left(\frac{\mu^2}{-p^2}\right)\right)\right]\langle Q_2\rangle^{\text{tree}}.$$

$$i\mathcal{M}_2^{\text{SM},(1)} = \int \frac{d^d k}{(2\pi)^d}\left[\left(\bar{u}_c \cdot \frac{ig}{\sqrt{2}}\gamma^\mu P_L \cdot \frac{i(\slashed{p}-\slashed{k})}{(p-k)^2}\cdot ig_s t_{ki}^a \gamma^\alpha \cdot u_b\right)\right.$$
$$\left(\bar{u}_d \cdot \frac{ig}{\sqrt{2}}\gamma^\nu P_L \cdot \frac{i(\slashed{p}+\slashed{k})}{(p+k)^2}\cdot ig_s t_{lj}^a \gamma^\beta \cdot u_u\right)$$
$$\left.\times \frac{-ig_{\mu\nu}}{k^2-M_W^2}\times\frac{-ig_{\alpha\beta}}{k^2}\times V_{cb} V_{ud}^*\right]$$

$$= -i\frac{8G_F}{\sqrt{2}} V_{cb} V_{ud}^* (\bar{u}_c \gamma^\mu P_L u_b)(\bar{u}_d \gamma_\mu P_L u_u)\left(\frac{1}{N_c}\delta_{ki}\delta_{lj}-\delta_{kj}\delta_{li}\right)\left[\frac{\alpha_s}{4\pi}\ln\left(\frac{M_W^2}{-p^2}\right)\right]$$

$$= -i\frac{8G_F}{\sqrt{2}} V_{cb} V_{ud}^* \left[\frac{\alpha_s}{4\pi}\ln\left(\frac{M_W^2}{-p^2}\right)\right]\left(\frac{1}{N_c}\langle Q_2\rangle^{\text{tree}}-\langle Q_1\rangle^{\text{tree}}\right).$$

$$i\mathcal{M}_3^{\text{SM},(1)} = \int \frac{d^d k}{(2\pi)^d}\left[\left(\bar{u}_c \cdot \frac{ig}{\sqrt{2}}\gamma^\mu P_L \cdot \frac{i(\slashed{p}+\slashed{k})}{(p+k)^2}\cdot ig_s t_{ki}^a \gamma^\alpha \cdot u_b\right)\right.$$
$$\left(\bar{u}_d \cdot ig_s t_{lj}^a \gamma^\beta \cdot \frac{i(\slashed{p}+\slashed{k})}{(p+k)^2}\cdot \frac{ig}{\sqrt{2}}\gamma^\nu P_L \cdot u_u\right)$$
$$\left.\times \frac{-ig_{\mu\nu}}{k^2-M_W^2}\times\frac{-ig_{\alpha\beta}}{k^2}\times V_{cb} V_{ud}^*\right]$$

$$= i\frac{2G_F}{\sqrt{2}} V_{cb} V_{ud}^* (\bar{u}_c \gamma^\mu P_L u_b)(\bar{u}_d \gamma_\mu P_L u_u)\left(\frac{1}{N_c}\delta_{ki}\delta_{lj}-\delta_{kj}\delta_{li}\right)\left[\frac{\alpha_s}{4\pi}\ln\left(\frac{M_W^2}{-p^2}\right)\right]$$

$$= -i\frac{2G_F}{\sqrt{2}} V_{cb} V_{ud}^* \left[\frac{\alpha_s}{4\pi}\ln\left(\frac{M_W^2}{-p^2}\right)\right]\left(\langle Q_1\rangle^{\text{tree}}-\frac{1}{N_c}\langle Q_2\rangle^{\text{tree}}\right).$$

The symmetric diagrams give the same results as their corresponding diagrams, and so we find

$$i\mathcal{M}^{\text{SM}} = i\mathcal{M}^{\text{SM},(0)} + 2(i\mathcal{M}_1^{\text{SM},(1)}+i\mathcal{M}_2^{\text{SM},(1)}+i\mathcal{M}_3^{\text{SM},(1)})$$
$$= -i\frac{4G_F}{\sqrt{2}} V_{cb} V_{ud}^* \left[\left(1+\frac{\alpha_s}{4\pi}\left[2C_F\left\{\frac{1}{\epsilon}+\ln\left(\frac{\mu^2}{-p^2}\right)\right\}+\frac{3}{N_c}\ln\left(\frac{M_W^2}{-p^2}\right)\right]\right)\langle Q_2\rangle^{\text{tree}}\right.$$
$$\left.+\left(\frac{\alpha_s}{4\pi}\left[-3\ln\left(\frac{M_W^2}{-p^2}\right)\right]\right)\langle Q_1\rangle^{\text{tree}}\right]$$

WET Calculation

We define our effective Hamiltonian as

$$\mathcal{H}_{\text{eff}} = \frac{4G_F}{\sqrt{2}} V_{cb} V_{ud}^* (C_1 Q_1 + C_2 Q_2) + \text{h.c.} \qquad (2.4.3)$$

where we have already included some prefactors for convenience.

The diagrams which contribute in the effective theory are just those of the SM (see Fig. 2.1) with the W propagator contracted to a point. At tree level, the amplitude is just

$$
i\mathcal{M}^{\text{WET},(0)} = -i\frac{4G_F}{\sqrt{2}} V_{cb} V_{ud}^* (\bar{u}_c \gamma^\mu P_L u_b)(\bar{u}_d \gamma_\mu P_L u_u)(C_1 \delta_{ki}\delta_{lj} + C_2 \delta_{kj}\delta_{li})
$$

$$
\equiv -i\frac{4G_F}{\sqrt{2}} V_{cb} V_{ud}^* (C_1 \langle Q_1 \rangle^{\text{tree}} + C_2 \langle Q_2 \rangle^{\text{tree}})
$$

which justifies our earlier definition of the tree-level insertion.

The 1-loop QCD corrections are:

$$
i\mathcal{M}_1^{\text{WET},(1)} = \int \frac{d^d k}{(2\pi)^d} \left[\left(\bar{u}_c \cdot ig_s t_{km}^a \gamma^\alpha \cdot \frac{i(\not{p}-\not{k})}{(p-k)^2} \cdot \gamma^\mu P_L \cdot \frac{i(\not{p}-\not{k})}{(p-k)^2} \cdot ig_s t_{ni}^a \gamma^\beta \cdot u_b \right) \right.
$$

$$
\times \left(\bar{u}_d \cdot \gamma^\mu P_L \cdot u_u \right)
$$

$$
\left. \times \frac{-ig_{\alpha\beta}}{k^2} \times -i\frac{4G_F}{\sqrt{2}} V_{cb} V_{ud}^* (C_1 \delta_{mj}\delta_{ln} + C_2 \delta_{mn}\delta_{lj}) \right]
$$

$$
= -i\frac{4G_F}{\sqrt{2}} V_{cb} V_{ud}^* \left[\frac{\alpha_s}{4\pi} \left(\frac{1}{\epsilon} + \ln\left(\frac{\mu^2}{-p^2}\right) \right) \right] \times
$$

$$
\left[\left(\frac{C_1}{2} + C_F C_2 \right) \langle Q_2 \rangle^{\text{tree}} - \frac{C_1}{2N_c} \langle Q_1 \rangle^{\text{tree}} \right].
$$

$$
i\mathcal{M}_2^{\text{WET},(1)} = \int \frac{d^d k}{(2\pi)^d} \left[\left(\bar{u}_c \cdot \gamma^\mu P_L \cdot \frac{i(\not{p}-\not{k})}{(p-k)^2} \cdot ig_s t_{mi}^a \gamma^\alpha \cdot u_b \right) \right.
$$

$$
\times \left(\bar{u}_d \cdot \gamma^\mu P_L \cdot \frac{i(\not{p}+\not{k})}{(p+k)^2} \cdot ig_s t_{nj}^a \gamma^\beta \cdot u_u \right)
$$

$$
\left. \times \frac{-ig_{\alpha\beta}}{k^2} \times -i\frac{4G_F}{\sqrt{2}} V_{cb} V_{ud}^* (C_1 \delta_{km}\delta_{ln} + C_2 \delta_{kn}\delta_{lm}) \right]
$$

$$
= -i\frac{8G_F}{\sqrt{2}} V_{cb} V_{ud}^* \left[\frac{\alpha_s}{4\pi} \left(\frac{1}{\epsilon} + \ln\left(\frac{\mu^2}{-p^2}\right) \right) \right] \times
$$

$$
\left[\left(\frac{C_1}{N_c} - C_2 \right) \langle Q_1 \rangle^{\text{tree}} + \left(\frac{C_2}{N_c} - C_1 \right) \langle Q_2 \rangle^{\text{tree}} \right].
$$

$$i\mathcal{M}_3^{\text{WET},(1)} = \int \frac{d^d k}{(2\pi)^d} \left[\left(\bar{u}_c \cdot \gamma^\mu P_L \cdot \frac{i(\not{p} - \not{k})}{(p-k)^2} \cdot i g_s t^a_{mi} \gamma^\beta \cdot u_b \right) \right.$$

$$\times \left(\bar{u}_d \cdot i g_s t^a_{ln} \gamma^\alpha \cdot \frac{i(\not{p} - \not{k})}{(p-k)^2} \gamma^\mu P_L \cdot \cdot u_u \right)$$

$$\left. \times \frac{-i g_{\alpha\beta}}{k^2} \times -i \frac{4G_F}{\sqrt{2}} V_{cb} V_{ud}^* (C_1 \delta_{kj} \delta_{nm} + C_2 \delta_{km} \delta_{nj}) \right]$$

$$= -i \frac{4G_F}{\sqrt{2}} V_{cb} V_{ud}^* \left[\frac{\alpha_s}{4\pi} \left(\frac{1}{\epsilon} + \ln \left(\frac{\mu^2}{-p^2} \right) \right) \right] \times$$

$$\left[\left(\frac{C_2}{2} + C_F C_1 \right) \langle Q_1 \rangle^{\text{tree}} - \frac{C_2}{2N_c} \langle Q_2 \rangle^{\text{tree}} \right].$$

As in the SM, the symmetric diagrams give the same result, and hence the total result for our EFT calculation is

$$i\mathcal{M}^{\text{WET}} = i\mathcal{M}^{\text{WET},(0)} + 2(i\mathcal{M}_1^{\text{WET},(1)} + i\mathcal{M}_2^{\text{WET},(1)} + i\mathcal{M}_3^{\text{WET},(1)})$$

$$= -i \frac{4G_F}{\sqrt{2}} V_{cb} V_{ud}^* \left[\langle Q_1 \rangle^{\text{tree}} \left(C_1 \left\{ 1 + \frac{\alpha_s}{4\pi} \left(\frac{1}{\epsilon} + \ln \left(\frac{\mu^2}{-p^2} \right) \right) \left(2C_F + \frac{3}{N_c} \right) \right\} \right. \right.$$

$$\left. - 3C_2 \left\{ \frac{\alpha_s}{4\pi} \left(\frac{1}{\epsilon} + \ln \left(\frac{\mu^2}{-p^2} \right) \right) \right\} \right)$$

$$+ \langle Q_2 \rangle^{\text{tree}} \left(+ C_2 \left\{ 1 + \frac{\alpha_s}{4\pi} \left(\frac{1}{\epsilon} + \ln \left(\frac{\mu^2}{-p^2} \right) \right) \left(2C_F + \frac{3}{N_c} \right) \right\} \right.$$

$$\left. \left. - 3C_1 \left\{ \frac{\alpha_s}{4\pi} \left(\frac{1}{\epsilon} + \ln \left(\frac{\mu^2}{-p^2} \right) \right) \right\} \right) \right].$$

Comparing $i\mathcal{M}^{\text{SM}}$ with $i\mathcal{M}^{\text{WET}}$, we see that the effective amplitude has extra divergences compared to the full theory. This is related to the fact that our effective theory is non-renormalisable—both have the same IR behaviour but the UV behaviour is worse in the WET. By setting $i\mathcal{M}^{\text{SM}} = i\mathcal{M}^{\text{WET}}$, we can find the Wilson coefficients needed such that the two theories give the same result as the calculated order[3]:

$$C_1 = -3 \frac{\alpha_s}{4\pi} \left(\frac{1}{\epsilon} + \ln \left(\frac{\mu^2}{M_W^2} \right) \right),$$

$$C_2 = 1 - \frac{3}{N_c} \frac{\alpha_s}{4\pi} \left(\frac{1}{\epsilon} + \ln \left(\frac{\mu^2}{M_W^2} \right) \right).$$

We see that the result still has divergences—as expected they did not all cancel in the matching, only those common to both theories. The Wilson coefficients can be

---

[3] Note that the explicit μ dependence in this result cancels with that of α_s (from Eq. 2.4.8) to give a μ independent result for C_1 and C_2, as expected since these are bare coefficients.

renormalised by a two by two matrix, which in our case will be non-diagonal since the QCD corrections generate one operator from the other.

$$C_i^{\text{bare}} = Z_{ij}(\mu) C_j^{\text{ren}}(\mu) \qquad (2.4.4)$$

The renormalisation constants Z_{ij} can be expanded perturbatively in α_s, and it is easy to read off that the matrix

$$Z = 1 + \frac{\alpha_s}{4\pi} \frac{1}{\epsilon} \begin{pmatrix} 0 & 3 \\ 0 & -\frac{3}{N_c} \end{pmatrix}$$

renders our Wilson coefficients finite, with the following form:

$$C_1(\mu) = -3 \frac{\alpha_s}{4\pi} \ln\left(\frac{\mu^2}{M_W^2}\right),$$

$$C_2(\mu) = 1 - \frac{3}{N_c} \frac{\alpha_s}{4\pi} \ln\left(\frac{\mu^2}{M_W^2}\right).$$

There is a slight problem, however—the Z_{i1} elements of our renormalisation matrix drop out of the calculation at this order since C_1 only appears at $\mathcal{O}(\alpha_s)$. Hence we have an ambiguity in the result—any constant times α_s could be added here without affecting the renormalisation property. This ambiguity can be resolved by a different choice of operator basis, e.g. if we use $Q_\pm$ (as defined later in this section). Doing so, we can then transform back into the $Q_{1,2}$ basis and find the renormalisation matrix for the Wilson coefficients $C_{1,2}$ is

$$Z = 1 + \frac{\alpha_s}{4\pi} \frac{1}{\epsilon} \begin{pmatrix} -\frac{3}{N_c} & 3 \\ 3 & -\frac{3}{N_c} \end{pmatrix}. \qquad (2.4.5)$$

It is also worth noting that we can instead consider renormalising the operators $Q_{1,2}$ themselves to remove the new UV divergence, and this process gives us a non-ambiguous renormalisation matrix, which is related to the one we have found as follows: let

$$C_i^{\text{bare}} = Z_{ij} C_j^{\text{ren}} \text{ and } Q_i^{\text{bare}} = Z_{ij}^Q Q_j^{\text{ren}}$$

such that

$$(C_i^{\text{bare}})^T Q_i^{\text{bare}} = (C_j^{\text{ren}})^T Z_{ji} Z_{ik}^Q Q_k^{\text{ren}} = (C_i^{\text{ren}})^T Q_i^{\text{ren}}$$

where the second equality holds since we can equally well formulate our theory in terms of renormalised or bare constituents. As such, we must have

$$Z_{ji} Z_{ik}^Q = \delta_{jk} \Rightarrow Z_{ij}^Q = Z_{ji}^{-1}. \qquad (2.4.6)$$

Renormalisation Group Evolution

In principle we can now use our EFT to make calculations at low ($\mu \sim m_b$) scales, as we have a perturbative expression for our coupling constants as a function of the scale. But looking at the numbers, it is not quite so simple—if we set our scale to the b quark mass, what do we find? Our expansion parameter, rather than being $\alpha_s(m_b) \sim 0.2$, is instead $\alpha_s \ln m_b^2/M_W^2 \sim 1$—the entire basis of perturbation theory, that successive corrections get smaller, seems to have broken down. Including all the factors we find our Wilson coefficients are then $C_1 \sim -0.3$, $C_2 \sim 1.1$—not quite as worrying, but still a relatively large change from tree level.

How do we resolve this crisis? Going to higher orders will not help—looking at Table 2.2 we see at n-loops we have terms that look like $(\alpha_s \ln)^n$ which still break the convergence of the perturbative series. We have to sum up these terms to all orders—i.e. calculate column by column rather than row by row in Table 2.2. The tool to do this is renormalisation group improved perturbation theory, which we will now demonstrate.

If we recall our renormalised Wilson coefficients are defined as $\vec{C}^{\text{bare}} = Z \cdot \vec{C}^{\text{ren}}$, and that the bare coefficients should be independent of the scale μ, we can write

$$\frac{d\vec{C}^{\text{bare}}}{d\mu} = 0 = \frac{dZ}{d\mu} \cdot \vec{C}^{\text{ren}} + Z \cdot \frac{d\vec{C}^{\text{ren}}}{d\mu}.$$

Rearranging this equation, we get

$$\mu \frac{d\vec{C}^{\text{ren}}}{d\mu} = -Z^{-1} \cdot \mu \frac{dZ}{d\mu} \cdot \vec{C}^{\text{ren}}$$

$$\frac{d\vec{C}^{\text{ren}}}{d\ln\mu} \equiv \gamma^T \cdot \vec{C}^{\text{ren}} \qquad (2.4.7)$$

which defines the *anomalous dimension* matrix γ (the transpose in the definition is conventional).

The simplest way to solve this differential equation is to move to a basis of Wilson coefficients (or equivalently operators) where the renormalisation matrix is diagonal. Defining new operators $Q_\pm = \frac{1}{2}(Q_2 \pm Q_1)$, we find the following result

$$C_\pm^{\text{bare}} = Z_\pm C_\pm^{\text{ren}} \text{ with } Z_\pm = 1 \pm \frac{3}{N_c} \frac{\alpha_s}{4\pi} \frac{1}{\epsilon}(N_c \mp 1)$$

for which our now diagonal anomalous dimension matrix is

$$\gamma_\pm = -\frac{1}{Z_\pm} \frac{dZ_\pm}{d\ln\mu}.$$

The running of α_s is given by the QCD beta function, defined in the following way:

$$\frac{d\alpha_s}{d\ln\mu} \equiv \beta(\alpha_s) = -2\alpha_s\left(\epsilon + \frac{\alpha_s}{4\pi}\beta_0 + \left(\frac{\alpha_s}{4\pi}\right)^2\beta_1 + \dots\right), \qquad (2.4.8)$$

where we have included the $\mathcal{O}(\epsilon)$ term as well. The β_i coefficients are determined by an i-loop calculation—in our notation, $\beta_0 = 11 - 2n_f/3$. Using this, we can expand our anomalous dimension matrix as

$$
\begin{aligned}
\gamma_\pm &= -\frac{1}{Z_\pm}\frac{dZ_\pm}{d\alpha_s}\frac{d\alpha_s}{d\ln\mu}\\
&= -\left(1 \pm \frac{3}{N_c}\frac{\alpha_s}{4\pi}\frac{1}{\epsilon}(N_c \mp 1)\right)^{-1} \times \frac{d}{d\alpha_s}(1 \pm \frac{3}{N_c}\frac{\alpha_s}{4\pi}\frac{1}{\epsilon}(N_c \mp 1)) \times \beta(\alpha_s)\\
&= -\left(1 \mp \frac{3}{N_c}\frac{\alpha_s}{4\pi}\frac{1}{\epsilon}(N_c \mp 1)\right)\left(\pm\frac{3}{N_c}\frac{1}{4\pi}\frac{1}{\epsilon}(N_c \mp 1)\right)\left(-2\alpha_s\left(\epsilon + \frac{\alpha_s}{4\pi}\beta_0\right)\right) + \mathcal{O}\left(\alpha_s^2\right)\\
&= -\frac{\alpha_s}{4\pi}\frac{6}{N_c}(1 \mp N_c)\\
&\equiv \frac{\alpha_s}{4\pi}\gamma_\pm^{(0)}.
\end{aligned}
$$

We can now solve the RG equation for the evolution of our Wilson coefficients, in the following way:

$$\frac{dC_\pm}{d\ln\mu} = \gamma_\pm C_\pm$$

$$\frac{1}{C_\pm}\frac{dC_\pm}{d\alpha_s} = \frac{\gamma_\pm}{\beta(\alpha_s)}$$

$$\int_{C_\pm(\mu_0)}^{C_\pm(\mu)}\frac{dC_\pm}{C_\pm} = \int_{\alpha_s(\mu_0)}^{\alpha_s(\mu)}\frac{\frac{\alpha_s}{4\pi}\gamma_\pm^{(0)}}{-2\alpha_s(\frac{\alpha_s}{4\pi}\beta_0)}d\alpha_s$$

$$\ln\left(\frac{C_\pm(\mu)}{C_\pm(\mu_0)}\right) = -\frac{\gamma_\pm^{(0)}}{2\beta_0}\int_{\alpha_s(\mu_0)}^{\alpha_s(\mu)}\frac{d\alpha_s}{\alpha_s}$$

$$\ln\left(\frac{C_\pm(\mu)}{C_\pm(\mu_0)}\right) = -\frac{\gamma_\pm^{(0)}}{2\beta_0}\ln\left(\frac{\alpha_s(\mu)}{\alpha_s(\mu_0)}\right)$$

$$\Rightarrow C_\pm(\mu) = C_\pm(\mu_0)\left[\frac{\alpha_s(\mu)}{\alpha_s(\mu_0)}\right]^{-\frac{\gamma_\pm^{(0)}}{2\beta_0}}$$

where μ_0 is some scale close to the matching scale of our EFT. From the start of our calculation, and our definition $C_\pm = C_2 \pm C_1$, we have

$$C_\pm(\mu) = 1 - \frac{3}{N_c}(1 \pm N_c)\frac{\alpha_s}{4\pi}\ln\left(\frac{\mu^2}{M_W^2}\right)$$

and so our full result is

$$C_{\pm}(\mu) = \left(1 - \frac{3}{N_c}(1 \pm N_c)\frac{\alpha_s}{4\pi} \ln\left(\frac{\mu_0^2}{M_W^2}\right)\right)\left[\frac{\alpha_s(\mu)}{\alpha_s(\mu_0)}\right]^{-\frac{\gamma_{\pm}^{(0)}}{2\beta_0}} \qquad (2.4.9)$$

Before continuing, a brief interlude about the above result. The first term (in round brackets) is a fixed order calculation at some high scale μ_0, which displays the poor convergence behaviour we discussed earlier. The second term (in square brackets) is the result of summing up all large logs, which arise when we take μ very different from μ_0.

Our final step is to revert back to the original basis of $C_{1,2}$ coefficients. We have the freedom to choose the matching scale μ_0 (which acts as a boundary condition on our RGE equations) to be anything of the order of M_W, such that large logs don't appear. The simplest solution is to set $\mu_0 = M_W$, as at this scale the logs vanish at this order, and we are left with a simpler result. Plugging in all the numbers, we find as our final result

$$C_1(\mu) = \frac{1}{2}\left(\left[\frac{\alpha_s(\mu)}{\alpha_s(M_W)}\right]^{-\frac{6}{23}} - \left[\frac{\alpha_s(\mu)}{\alpha_s(M_W)}\right]^{\frac{12}{23}}\right), \qquad (2.4.10)$$

$$C_2(\mu) = \frac{1}{2}\left(\left[\frac{\alpha_s(\mu)}{\alpha_s(M_W)}\right]^{-\frac{6}{23}} + \left[\frac{\alpha_s(\mu)}{\alpha_s(M_W)}\right]^{\frac{12}{23}}\right). \qquad (2.4.11)$$

2.4.2 B_s Mixing

The formalism of meson mixing can be understood as a "simple"[4] application of time-dependent perturbation theory with two discrete eigenstates of the unperturbed Hamiltonian, plus a continuum of lighter states. They are unstable particles, which can transition into other quantum states through the decay process. Treating the particles as quantum mechanical states, and applying the Wigner-Weisskopf method (see Appendix I of [28] for the details of this), we can show that the time evolution can be written as

$$\frac{\partial}{\partial t}\begin{pmatrix} B_s \\ \bar{B}_s \end{pmatrix} = \left(\hat{M} - \frac{i}{2}\hat{\Gamma}\right)\begin{pmatrix} B_s \\ \bar{B}_s \end{pmatrix} \qquad (2.4.12)$$

where $\hat{M}$ and $\hat{\Gamma}$ are both Hermitian matrices. Because of the weak interaction these two matrices are not diagonal, and the eigenstates of the Hamiltonian can be found by diagonalising $\hat{M}$ and $\hat{\Gamma}$. This gives

---

[4]I.e. not at all.

$$(\Delta M)^2 - \frac{1}{4}(\Delta \Gamma)^2 = 4|M_{12}|^2 - |\Gamma_{12}|^2, \qquad (2.4.13)$$

$$\Delta M \cdot \Delta \Gamma = -4 \operatorname{Re}(M_{12}\Gamma_{12}^*), \qquad (2.4.14)$$

where M_{12} and Γ_{12} are the off-diagonal elements of $\hat{M}$ and $\hat{\Gamma}$, and ΔM and $\Delta \Gamma$ correspond to the mass and width difference of the two mass eigenstates, conventionally labelled B_H, B_L (heavy and light) such that

$$\Delta M \equiv M_{B_H} - M_{B_L}, \quad \Delta \Gamma \equiv \Gamma_{B_L} - \Gamma_{B_H}. \qquad (2.4.15)$$

It has been experimentally measured that the ratio $\Delta \Gamma / \Delta M$ is small, and it is also true that in the SM we have $|\Gamma_{12}/M_{12}| \ll 1$.[5] Expanding in either of these small parameters, we find

$$\Delta M \approx 2|M_{12}|, \quad \Delta \Gamma \approx 2|\Gamma_{12}|\cos\phi_{12}, \quad \text{with } \phi_{12} = \arg(-M_{12}/\Gamma_{12}), \quad (2.4.16)$$

where we have neglected terms of $\mathcal{O}\left(|\Gamma_{12}/M_{12}|^2\right)$.

In order to predict ΔM, we see that we must calculate M_{12}. The quantum mechanical calculation[6] tells us that the off diagonal element M_{12} is given by

$$M_{12} = \langle B_s | H' | \bar{B}_s \rangle,$$

where H' is the weak Hamiltonian (note that this is actually the Hamiltonian here, not the density) that acts as the perturbation. Since in QFT states are normalised such that $\langle i|i\rangle = 2E_i V$ (where V is the volume of a hypothetical finite spacetime), compared to the unit normalisation $\langle B_s | \bar{B}_s \rangle = 1$ in QM, this expression becomes

$$\begin{aligned}
M_{12} &= \frac{\langle B_s | H' | \bar{B}_s \rangle}{2M_{B_s} V} \\
&= \frac{\langle B_s | \int d^4 x \mathcal{H}(x) | \bar{B}_s \rangle}{2M_{B_s} V} \\
&= \frac{\langle B_s | \mathcal{H} | \bar{B}_s \rangle}{2M_{B_s}}, \qquad (2.4.17)
\end{aligned}$$

where we have used the translational invariance of the Hamiltonian density to cancel off the volume factor.

In our above expression, what is the Hamiltonian we must use? Since we want to calculate the overlap between two hadronic states, it seems reasonable that the

---

[5] Note that while this hierarchy is true for B and B_s mesons, they do not hold for D mixing.

[6] A full account of this calculation can be found in Chap. 1.2.1 of [29].

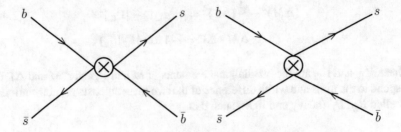

Fig. 2.2 Leading order diagrams for B_s mixing in the weak effective theory

Hamiltonian must be one that is most appropriate for that scale, i.e. $\sim m_b$—so we should use an effective Hamiltonian.

So a rough outline of the calculation is

1. Match the SM onto the WET (which we will do using the process $b^i \bar{s}^j \to \bar{b}^k s^l$)
2. (RG run the Wilson coefficients down to $\sim m_b$)
3. Find a way to evaluate the matrix element of our low scale Hamiltonian.

The matching process proceeds similarly to that done in the previous section.

WET Calculation

It can be shown that the appropriate basis of operators for $\Delta B = 2$ processes is the so-called "SUSY" basis [30], which contains eight operators (see Appendix C for a list). For our purposes, we need only one however (as it is the only one that arises in the SM):

$$\mathcal{O}_1 = (\bar{s}^i \gamma^\mu P_L b^i)(\bar{s}^j \gamma_\mu P_L b^j)$$

which means our effective Hamiltonian is

$$\mathcal{H}_{\text{eff}} = C_1 \mathcal{O}_1 + \text{h.c.}$$

(note that there is a factor of four and a hermitian conjugation relative to e.g. FLAG [31, 32]).

There are two diagrams that contribute in the WET (at leading order), shown in Fig. 2.2.

The amplitude in the effective theory is

$$
\begin{aligned}
i\mathcal{M}^{\text{WET}} &= (-iC) \cdot 2(\bar{v}\gamma^\mu P_L u)(\bar{u}\gamma_\mu P_L v)\delta_{ij}\delta_{kl} - (-iC) \cdot 2(\bar{v}\gamma^\mu P_L v)(\bar{u}\gamma_\mu P_L u)\delta_{il}\delta_{kj} \\
&= -2iC\left[(\bar{v}\gamma^\mu P_L u)(\bar{u}\gamma_\mu P_L v)\delta_{ij}\delta_{kl} - (\bar{v}\gamma^\mu P_L v)(\bar{u}\gamma_\mu P_L u)\delta_{il}\delta_{kj}\right] .
\end{aligned}
$$

(2.4.18)

The factor of two comes from being able to contract the external legs multiple ways in the diagram (this would normally be cancelled out by a normalisation factor in the Lagrangian, like e.g. the 4! factor in scalar ϕ^4 theory), and the minus sign from the

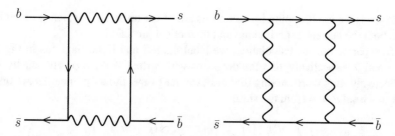

Fig. 2.3 Box diagrams that contribute to B_s mixing in the SM

different orientation of the fermion lines. Both these factors can be most easily seen by doing the Wick contraction by hand.

SM Calculation

In the SM there are many more diagrams, two of which are shown in Fig. 2.3. We have to sum over the possible internal quarks (u, c, t), and in Feynman gauge you can also substitute either of the W bosons for a charged Goldstone boson. In total, this gives us $2 \times 4 \times 3 \times 3 = 72$ diagrams. However, what seems like a large computation can be simplified by examining the structure of the amplitudes.

For each diagram, we get a result like $i\mathcal{M} \sim \sum_{q,q'} \lambda_q \lambda_{q'} F(x_q, x_{q'})$ where $\lambda_q \equiv V_{qb} V_{qs}^*$, $x_q \equiv m_q/M_W$ and F is the function resulting from the loop integral, which we have written as a function of the dimensionless quark mass ratios rather than the masses alone. Using the unitarity of the CKM matrix, the above sum can be written in the following form

$$
\begin{aligned}
i\mathcal{M} = \; & \lambda_u^2 \left[F(x_u, x_u) - 2F(x_c, x_u) + F(x_c, x_c) \right] \\
& + 2\lambda_u \lambda_c \left[F(x_c, x_c) - F(x_c, x_u) - F(x_t, x_c) + F(x_t, x_u) \right] \qquad (2.4.19) \\
& + \lambda_t^2 \left[F(x_t, x_t) - 2F(x_t, x_u) + F(x_u, x_u) \right].
\end{aligned}
$$

There are two sources of suppression in this expression, GIM and CKM. GIM suppression (see the end of Sect. 1.2.1) comes from the fact that if the quark masses were equal, each set of terms in the square brackets would vanish, while CKM suppression can be seen from examining size of the prefactors:

$$
\lambda_t^2 \sim 10^{-3}, \quad \lambda_t \lambda_u \sim 10^{-5}, \quad \lambda_u^2 \sim 10^{-7}. \qquad (2.4.20)
$$

While obviously the approximation $m_t = m_c = m_u$ is not a good one, setting $m_c = m_u = 0$ is much better, and so we will drop the first two terms in Eq. 2.4.19, leaving us with

$$
i\mathcal{M} = \lambda_t^2 \left[F(x_t, x_t) - 2F(x_t, 0) + F(0, 0) \right]. \qquad (2.4.21)
$$

(The terms we are dropping by doing this are of the order 10^{-4} which is much less than both the current experimental and theoretical precision.)

To separate out the calculation, we label the left and right diagrams in Fig. 2.3 by a and b respectively, and the diagrams with virtual WW, $W\phi$, ϕW, $\phi\phi$ by 1–4 respectively. It is worth noting that in Feynman gauge, the loop integrals are finite, and so we set $d = 4$ from the start.

$$i\mathcal{M}_{1a}^{\text{SM}} = \lambda_t^2 \int \frac{d^4k}{(2\pi)^4} \left[\left(\bar{v} \cdot \frac{ig}{\sqrt{2}} \gamma^\mu P_L \cdot \frac{i(\slashed{k} + m_q)}{k^2 - m_q^2} \cdot \frac{ig}{\sqrt{2}} \gamma^\nu P_L \cdot u \right) \right.$$

$$\left(\bar{u} \cdot \frac{ig}{\sqrt{2}} \gamma^\alpha P_L \cdot \frac{i(\slashed{k} + m_{q'})}{k^2 - m_{q'}^2} \cdot \frac{ig}{\sqrt{2}} \gamma^\beta P_L \cdot v \right)$$

$$\left. \times \frac{-ig_{\mu\nu}}{k^2 - M_W^2} \times \frac{-ig_{\alpha\beta}}{k^2 - M_W^2} \times \delta_{ij}\delta_{kl} \right]$$

$$= \delta_{ij}\delta_{kl} \lambda_t^2 \frac{g^4}{16} (\bar{v}\gamma^\mu\gamma^\alpha\gamma^\nu P_L u)(\bar{u}\gamma_\nu\gamma_\alpha\gamma_\mu P_L v) \times$$

$$\int \frac{d^4k}{(2\pi)^4} \frac{k^2}{(k^2 - m_q^2)(k^2 - m_{q'}^2)(k^2 - M_W^2)^2}.$$

For the diagrams with virtual Goldstone bosons, since the Goldstone coupling is proportional to the quark mass, we need only calculate the result for the case where both virtual quarks are tops, as the others give zero contribution in our approximation $m_c = m_u = 0$.

$$i\mathcal{M}_{2a}^{\text{SM}} = i\mathcal{M}_{3a}^{\text{SM}} = \lambda_t^2 \int \frac{d^4k}{(2\pi)^4} \left[\left(\bar{v} \cdot \frac{ig}{\sqrt{2}} \frac{m_t}{M_W} P_R \cdot \frac{i(\slashed{k} + m_t)}{k^2 - m_t^2} \cdot \frac{ig}{\sqrt{2}} \gamma^\mu P_L \cdot u \right) \right.$$

$$\left(\bar{u} \cdot \frac{ig}{\sqrt{2}} \gamma^\nu P_L \cdot \frac{i(\slashed{k} + m_t)}{k^2 - m_t^2} \cdot \frac{ig}{\sqrt{2}} \frac{m_t}{M_W} P_L \cdot v \right)$$

$$\left. \times \frac{-ig_{\mu\nu}}{k^2 - M_W^2} \times \frac{i}{k^2 - M_W^2} \times \delta_{ij}\delta_{kl} \right]$$

$$= -\delta_{ij}\delta_{kl} \lambda_t^2 \frac{g^4}{4} \frac{m_t^4}{M_W^2} (\bar{v}\gamma^\mu P_L u)(\bar{u}\gamma_\mu P_L v) \times$$

$$\int \frac{d^4k}{(2\pi)^4} \frac{1}{(k^2 - m_t^2)^2(k^2 - M_W^2)^2},$$

$$i\mathcal{M}_{4a}^{SM} = \lambda_t^2 \int \frac{d^4k}{(2\pi)^4} \left[\left(\bar{v} \cdot \frac{ig}{\sqrt{2}} \frac{m_t}{M_W} P_R \cdot \frac{i(\slashed{k}+m_t)}{k^2 - m_t^2} \cdot \frac{ig}{\sqrt{2}} \frac{m_t}{M_W} P_L \cdot u \right) \right.$$

$$\left(\bar{u} \cdot \frac{ig}{\sqrt{2}} \frac{m_t}{M_W} P_R \cdot \frac{i(\slashed{k}+m_t)}{k^2 - m_t^2} \cdot \frac{ig}{\sqrt{2}} \frac{m_t}{M_W} P_L \cdot v \right)$$

$$\left. \times \frac{i}{k^2 - M_W^2} \times \frac{i}{k^2 - M_W^2} \times \delta_{ij}\delta_{kl} \right]$$

$$= -\delta_{ij}\delta_{kl}\lambda_t^2 \frac{g^4}{16} \frac{m_t^4}{M_W^4} (\bar{v}\gamma^\mu P_L u)(\bar{u}\gamma_\mu P_L v) \int \frac{d^4k}{(2\pi)^4} \frac{k^2}{(k^2-m_t^2)^2(k^2-M_W^2)^2}.$$

The b diagrams give the same result, up to a relative minus sign[7] (which will match the sign difference in the effective theory), and having the opposite spinor structure $((\bar{v}v)(\bar{u}u)$ instead of $(\bar{v}u)(\bar{u}v))$ and colour structure ($\delta_{il}\delta_{kj}$ instead of $\delta_{ij}\delta_{kl}$).

Putting our results together, and using the Fierz identity Eq. A.0.3 (which allows us to relate the two different Dirac structures), we find

$$i\mathcal{M}^{SM} = -i\frac{G_F^2}{2\pi^2} M_W^4 \lambda_t^2 S_0(x_t) \left[(\bar{v}\gamma^\mu P_L u)(\bar{u}\gamma_\mu P_L v)\delta_{ij}\delta_{kl} - (\bar{v}\gamma^\mu P_L v)(\bar{u}\gamma_\mu P_L u)\delta_{i,l}\delta_{k,j} \right]$$

where

$$S_0(x_t) = \frac{4x_t - 11x_t^2 + x_t^3}{4(x_t^2 - 1)} + \frac{3x_t^3 \ln x_t}{2(x_t^2 - 1)} \tag{2.4.22}$$

is the Inami-Lim function [34].

$\Delta B = 2$ Matching

Setting $i\mathcal{M}^{WET} = i\mathcal{M}^{SM}$, we can read off the coefficient of our effective Hamiltonian to be

$$C = \frac{G_F^2}{4\pi^2} M_W^2 \lambda_t^2 S_0(x_t).$$

Hence,

$$M_{12} = \frac{G_F^2}{4\pi^2} \lambda_t^2 M_W^2 S_0(x_t) \times \frac{1}{2M_{B_s}} \langle B_s | \mathcal{O}_1 | \bar{B}_s \rangle$$

$$= \frac{G_F^2}{12\pi^2} \lambda_t^2 M_W^2 S_0(x_t) f_{B_s}^2 M_{B_s} B$$

where B, known as a bag parameter, parameterises the deviation from the VSA (see Equation Appendix C for details).

---

[7] See for example [33] for a set of Feynman rules which make this sign clear.

In principle we must also consider the RG group running of our operator and Wilson coefficient—we will not go into detail here, but simply state the effects. Having matched at the scale $\mu_t \sim M_W, m_t$, we run down to the scale $\mu_b \sim m_b$. It turns out that QCD corrections, in contrast to our other EFT case study, do not bring in any other operators, and so we simply find

$$M_{12}(\mu_t, \mu_b) = \frac{G_F^2}{12\pi^2} \lambda_t^2 M_W^2 S_0\left(\frac{m_t^2(\mu_t)}{M_W^2}\right) \bar{\eta}_{2B}(\mu_t, \mu_b) f_{B_s}^2 M_{B_s} B(\mu_b)$$

where at leading order the correction factor is

$$\bar{\eta}_{2B} = \left(\frac{\alpha_s(\mu_t)}{\alpha_s(\mu_b)}\right)^{6/23} .$$

(Note that in e.g. [35], this quantity is denoted $\hat{\eta}_B$.) At higher orders, it is seen that $\bar{\eta}_{2B}$ is only very weakly dependent on μ_t if we set our matching scale to be close to the top mass, as the logarithmic terms are of the form $\ln \mu_t/m_t$ [36]. Traditionally, the μ_b dependence is absorbed into the bag parameter, which then formally cancels its scale and scheme dependence, and this parameter then gets called $\hat{B}$—this notation is widely used by lattice groups. We then can write

$$M_{12} = \frac{G_F^2}{12\pi^2} \lambda_t^2 M_W^2 S_0\left(\frac{m_t^2(m_t)}{M_W^2}\right) \eta_{2B}(m_t) f_{B_s}^2 M_{B_s} \hat{B} , \qquad (2.4.23)$$

with $\eta_{2B} \approx 0.55$. (In the literature, since the m_t dependence of η_{2B} is so small, the m_t dependence is often dropped.) It is useful to convert between the "hatted" and "unhatted" bag parameters, so we quote here the NLO conversion factor [36]:

$$\hat{B} = \alpha_s(\mu_b)^{-\frac{6}{23}} \left(1 + \frac{\alpha_s(\mu_b)}{4\pi} \frac{5165}{3174}\right) B(\mu_b) \approx 1.5 B(\mu_b) \qquad (2.4.24)$$

from which we can find $\bar{\eta}_{2B} \approx 0.84$.

With all this in place, we can finally make a prediction for $\Delta M-$ using a recent set of input (circa 2015) parameters, we find

$$\Delta M_s = 18.3 \pm 2.7 \text{ps}^{-1}$$

as the SM prediction [35]. This has been the standard result for the last few years, but in Chap. 7 we update this in light of new inputs, and examine the consequences for certain classes of NP model.

2.5 *B* **Mixing Observables**

Following on from our in depth discussion of the calculation of ΔM_s, now seems an appropriate time to give an overview of the other observables that arise from meson mixing.

The width difference $\Delta \Gamma$ is calculated in a similar way to ΔM, but with the difference that we only include diagrams where the intermediate particles can go on shell. A simple way to understand this is by considering $\Delta \Gamma$ written in the form

$$\Delta \Gamma = \Gamma_{B_L} - \Gamma_{B_H} = \sum_f \left| \langle f | \mathcal{H} | B_L \rangle \right|^2 - \sum_f \left| \langle f | \mathcal{H} | B_H \rangle \right|^2 ,$$

where f denotes final states common to B_s and $\bar{B}_s$—the lifetime calculation can only involve on shell final states.

In the HQE Γ_{12} can be expanded in powers of Λ / m_b

$$\Gamma_{12} = \frac{\Lambda^3}{m_b^3} \left(\Gamma_3^{(0)} + \frac{\alpha_s}{4\pi} \Gamma_3^{(1)} + \dots \right) + \frac{\Lambda^4}{m_b^4} \left(\Gamma_4^{(0)} + \dots \right) + \dots . \tag{2.5.1}$$

The leading term $\Gamma_3^{(0)}$ was calculated quite some time ago by [37–42], the NLO-QCD corrections $\Gamma_3^{(1)}$ were determined in [43–45] and subleading mass corrections were done in [46–48]. Lattice values for the corresponding operators were determined by [49–52].

Alongside ΔM and $\Delta \Gamma$, there is another main observable associated with mixing—the semileptonic asymmetry or a_{sl}. This is defined by

$$a_{\text{sl}} \equiv \frac{\Gamma(\bar{B}(t) \to f) - \Gamma(B(t) \to \bar{f})}{\Gamma(\bar{B}(t) \to f) + \Gamma(B(t) \to \bar{f})} , \tag{2.5.2}$$

where f is a flavour-specific final state, i.e. $\bar{B} \to f$ and $B \to \bar{f}$ are forbidden without mixing.[8] Expanding in our small parameters as we did in Eq. 2.4.16 we find

$$a_{\text{sl}} \approx \text{Im} \left(\frac{\Gamma_{12}}{M_{12}} \right) , \tag{2.5.3}$$

where as before we have dropped terms of $\mathcal{O} \left(|\Gamma_{12}/M_{12}|^2 \right)$. This ratio expression is nice from a calculational point of view, as various factors which are less precisely

---

[8]For this reason, a_{sl} is often referred to as the flavour specific asymmetry, or a_{fs}.

known cancel out when we take the ratio, allowing us to make a more precise prediction. Something similar can be done for the mass and width differences, where we find that at the same level of accuracy the real part of the ratio is

$$\text{Re}\left(\frac{\Gamma_{12}}{M_{12}}\right) \approx -\frac{\Delta\Gamma}{\Delta M}.$$ (2.5.4)

References

1. Buchmuller W, Wyler D (1986) Effective lagrangian analysis of new interactions and flavor conservation. Nucl Phys B 268:621–653. https://doi.org/10.1016/0550-3213(86)90262-2
2. Grzadkowski B, Iskrzynski M, Misiak M, Rosiek J (2010) Dimension-six terms in the standard model lagrangian. JHEP 10:085. https://doi.org/10.1007/JHEP10(2010)085, arXiv:1008.4884
3. Khoze VA, Shifman MA (1983) HEAVY QUARKS. Sov Phys Usp 26:387. https://doi.org/10.1070/PU1983v026n05ABEH004398
4. Shifman MA, Voloshin MB (1985) Preasymptotic effects in inclusive weak decays of charmed particles. Sov J Nucl Phys 41:120
5. Bigi IIY, Uraltsev NG (1992) Gluonic enhancements in non-spectator beauty decays: an Inclusive mirage though an exclusive possibility. Phys Lett B 280:271–280. https://doi.org/10.1016/0370-2693(92)90066-D
6. Bigi IIY, Uraltsev NG, Vainshtein AI (1992) Nonperturbative corrections to inclusive beauty and charm decays: QCD versus phenomenological models. Phys Lett B 293:430–436. https://doi.org/10.1016/0370-2693(92)90908-M, https://doi.org/10.1016/0370-2693(92)91287-J, arXiv:hep-ph/9207214
7. Blok B, Shifman MA (1993) The Rule of discarding 1/N(c) in inclusive weak decays. 1. Nucl Phys B399: 441–458. https://doi.org/10.1016/0550-3213(93)90504-I, arXiv:hep-ph/9207236
8. Blok B, Shifman MA (1993) The Rule of discarding 1/N(c) in inclusive weak decays. 2. Nucl Phys B399:459–476.https://doi.org/10.1016/0550-3213(93)90505-J, arXiv:hep-ph/9209289
9. Chay J, Georgi H, Grinstein B (1990) Lepton energy distributions in heavy meson decays from QCD. Phys Lett B 247:399–405. https://doi.org/10.1016/0370-2693(90)90916-T
10. Luke ME (1990) Effects of subleading operators in the heavy quark effective theory. Phys Lett B 252:447–455. https://doi.org/10.1016/0370-2693(90)90568-Q
11. Lenz A (2015) Lifetimes and heavy quark expansion. Int J Mod Phys A 30:1543005. https://doi.org/10.1142/S0217751X15430058, arXiv:1405.3601
12. HFLAV collaboration, Amhis Y et al (2017) Averages of b-hadron, c-hadron, and τ-lepton properties as of summer 2016. Eur Phys J C77:895, https://doi.org/10.1140/epjc/s10052-017-5058-4, arXiv:1612.07233
13. HFLAV collaboration. https://hflav.web.cern.ch
14. HFLAV collaboration (2018) B lifetime and oscillation parameters, PDG. http://www.slac.stanford.edu/xorg/hflav/osc/PDG_2018/
15. Eichten E, Feinberg F (1981) Spin dependent forces in QCD. Phys Rev D 23:2724. https://doi.org/10.1103/PhysRevD.23.2724
16. Caswell WE, Lepage GP (1986) Effective lagrangians for bound state problems in QED, QCD, and other field theories. Phys Lett 167B:437–442. https://doi.org/10.1016/0370-2693(86)91297-9
17. Politzer HD, Wise MB (1988) Leading logarithms of heavy quark masses in processes with light and heavy quarks. Phys Lett B206:681–684, https://doi.org/10.1016/0370-2693(88)90718-6
18. Politzer HD, Wise MB (1988) Effective field theory approach to processes involving both light and heavy fields. Phys Lett B 208:504–507. https://doi.org/10.1016/0370-2693(88)90656-9

19. Eichten E, Hill BR (1990) An effective field theory for the calculation of matrix elements involving heavy quarks. Phys Lett B 234:511–516. https://doi.org/10.1016/0370-2693(90)92049-O
20. Eichten E, Hill BR (1990) Static effective field theory: 1/m corrections. Phys Lett B 243:427–431. https://doi.org/10.1016/0370-2693(90)91408-4
21. Grinstein B (1990) The static quark effective theory. Nucl Phys B 339:253–268. https://doi.org/10.1016/0550-3213(90)90349-I
22. Georgi H (1990) An effective field theory for heavy quarks at low-energies. Phys Lett B 240:447–450. https://doi.org/10.1016/0370-2693(90)91128-X
23. Falk AF, Georgi H, Grinstein B, Wise MB (1990) Heavy meson form-factors from QCD. Nucl Phys B 343:1–13. https://doi.org/10.1016/0550-3213(90)90591-Z
24. Falk AF, Grinstein B, Luke ME (1991) Leading mass corrections to the heavy quark effective theory. Nucl Phys B 357:185–207. https://doi.org/10.1016/0550-3213(91)90464-9
25. Georgi H (1991) Heavy quark effective field theory. In: Theoretical advanced study institute in elementary particle physics (TASI 91): perspectives in the standard model boulder, Colorado, June 2–28, pp. 0589–630. http://www.people.fas.harvard.edu/~hgeorgi/tasi.pdf
26. Neubert M (1994) Heavy quark symmetry. Phys Rept 245:259–396. https://doi.org/10.1016/0370-1573(94)90091-4, arXiv:hep-ph/9306320
27. Buras AJ (1998) Weak Hamiltonian, CP violation and rare decays. In: Probing the standard model of particle interactions. Proceedings, summer School in theoretical physics, NATO advanced study Institute, 68th session, Les Houches, France, July 28–September 5, 1997. Pt. 1, 2, pp. 281–539, arXiv:hep-ph/9806471
28. Nachtmann O (1990) Elementary particle physics: concepts and phenomena
29. Bóna M (2001) Analysis of two-body charmless decays for branching ratio and CP-violating asymmetry measurements with the BaBar experiment, PhD thesis, Universita' di Torino, 2001. http://inspirehep.net/record/923685/files/cer-002642923.pdf
30. Gabbiani F, Gabrielli E, Masiero A, Silvestrini L (1996) A Complete analysis of FCNC and CP constraints in general SUSY extensions of the standard model. Nucl Phys B 477:321–352. https://doi.org/10.1016/0550-3213(96)00390-2, arXiv:hep-ph/9604387
31. Aoki S et al (2017) Review of lattice results concerning low-energy particle physics. Eur Phys J C 77:112. https://doi.org/10.1140/epjc/s10052-016-4509-7, arXiv:1607.00299
32. FLAG collaboration. http://flag.unibe.ch/MainPage
33. Srednicki M (2004) Quantum field theory. Part 2. Spin one half, arXiv:hep-th/0409036
34. Inami T, Lim CS (1981) Effects of superheavy quarks and leptons in low-energy weak processes $K_L \to \mu\bar{\mu}$, $K^+ \to \pi^+\nu\bar{\nu}$ and $K^0 \longleftrightarrow \bar{K}^0$. Prog Theor Phys 65:297. https://doi.org/10.1143/PTP.65.297
35. Artuso M, Borissov G, Lenz A (2016) CP violation in the B_s^0 system. Rev Mod Phys 88:045002. https://doi.org/10.1103/RevModPhys.88.045002, arXiv:1511.09466
36. Buras AJ, Jamin M, Weisz PH (1990) Leading and next-to-leading QCD corrections to ϵ parameter and $B^0 - \bar{B}^0$ mixing in the presence of a heavy top quark. Nucl Phys B 347:491–536. https://doi.org/10.1016/0550-3213(90)90373-L
37. Ellis JR, Gaillard MK, Nanopoulos DV, Rudaz S (1977) The phenomenology of the next left-handed quarks. Nucl Phys B 131:285. https://doi.org/10.1016/0550-3213(77)90374-1
38. Hagelin JS (1981) Mass mixing and CP violation in the $B^0 - \bar{B}^0$ system. Nucl Phys B 193:123–149. https://doi.org/10.1016/0550-3213(81)90521-6
39. Franco E, Lusignoli M, Pugliese A (1982) Strong interaction corrections to CP violation in B0 anti-b0 mixing. Nucl Phys B 194:403. https://doi.org/10.1016/0550-3213(82)90018-9
40. Chau L-L (1983) Quark mixing in weak interactions. Phys Rept 95:1–94. https://doi.org/10.1016/0370-1573(83)90043-1
41. Buras AJ, Slominski W, Steger H (1984) B^0-$\bar{B}^0$ mixing, C violation and the B-meson decay. Nucl Phys B 245:369–398. https://doi.org/10.1016/0550-3213(84)90437-1
42. Khoze VA, Shifman MA, Uraltsev NG, Voloshin MB (1987) On inclusive hadronic widths of beautiful particles. Sov J Nucl Phys 46:112
43. Beneke M, Buchalla G, Greub C, Lenz A, Nierste U (1999) Next-to-leading order QCD corrections to the lifetime difference of B_s mesons. Phys Lett B 459:631–640. https://doi.org/10.1016/S0370-2693(99)00684-X, arXiv:hep-ph/9808385

44. Ciuchini M, Franco E, Lubicz V, Mescia F, Tarantino C (2003) Lifetime differences and CP violation parameters of neutral B mesons at the next-to-leading order in QCD. JHEP 08:031. https://doi.org/10.1088/1126-6708/2003/08/031, arXiv:hep-ph/0308029
45. Beneke M, Buchalla G, Lenz A, Nierste U (2003) CP asymmetry in flavor specific B decays beyond leading logarithms. Phys Lett B 576:173–183. https://doi.org/10.1016/j.physletb.2003.09.089, arXiv:hep-ph/0307344
46. Beneke M, Buchalla G, Dunietz I (1996) Width difference in the $B_s - \bar{B}_s$ system. Phys Rev D 54:4419–4431. https://doi.org/10.1103/PhysRevD.54.4419, https://doi.org/10.1103/PhysRevD.83.119902, arXiv:hep-ph/9605259
47. Dighe AS, Hurth T, Kim CS, Yoshikawa T (2002) Measurement of the lifetime difference of B_d mesons: possible and worthwhile? Nucl Phys B 624:377–404. https://doi.org/10.1016/S0550-3213(01)00655-1, arXiv:hep-ph/0109088
48. Badin A, Gabbiani F, Petrov AA (2007) Lifetime difference in B_s mixing: Standard model and beyond. Phys Lett B 653:230–240. https://doi.org/10.1016/j.physletb.2007.07.049, arXiv:0707.0294
49. Becirevic D, Gimenez V, Martinelli G, Papinutto M, Reyes J (2002) B-parameters of the complete set of matrix elements of $\Delta B = 2$ operators from the lattice. JHEP 04:025. https://doi.org/10.1088/1126-6708/2002/04/025, arXiv:hep-lat/0110091
50. Bouchard CM, Freeland ED, Bernard C, El-Khadra AX, Gamiz E, Kronfeld AS et al (2011) Neutral B mixing from 2+1 flavor lattice-QCD: the standard model and beyond, PoS LATTICE2011 274, https://doi.org/10.22323/1.139.0274, arXiv:1112.5642
51. ETM collaboration, Carrasco N et al (2014) B-physics from $N_f = 2$ tmQCD: the standard model and beyond. JHEP 03:016. https://doi.org/10.1007/JHEP03(2014)016, arXiv:1308.1851
52. Dowdall RJ, Davies CTH, Horgan RR, Lepage GP, Monahan CJ, Shigemitsu J (2014) B-meson mixing from full lattice QCD with physical u, d, s and c quarks, arXiv:1411.6989

Chapter 3
Quark-Hadron Duality

3.1 Introduction

In Chap. 1 we discussed how flavour physics is an ideal testing ground for searches for new physics, as there are many observables which are highly suppressed in the SM but could easily be enhanced by new physics, as well as many observables for which a high experimental precision is available—meson mixing observables satisfy both these criteria. In order to make use of this fact, we must be sure that hadronic uncertainties are under control in our theoretical calculations, as otherwise we cannot tell for certain if we are seeing anything new. In this chapter we aim to tackle one part of this question.

Many current theory predictions rely on the HQE and we will examine how the idea of quark-hadron duality—which is a basic assumption of the HQE—can be tested. In order to do this, we use current data from B mixing and B meson lifetimes to constrain violations of quark-hadron duality, and then see how this affects the predicted values of other observables. As part of this, we discuss how future improvements in theory and experiment could further constraint duality violations, and what level of precision could allow us to properly distinguish genuine new physics from some non-perturbative contribution to the SM calculation. We also investigate how the current trouble with inclusive predictions of mixing in the charm sector can be explained through a mild violation of quark-hadron duality.

Our work in this chapter is organised as follows: in Sect. 3.2 we explain the basic ideas of duality violation in the HQE, then we introduce in Sect. 3.2.1 a simple parameterisation for duality violation in B mixing and we derive bounds on its possible size. The lifetime ratio $\tau(B_s)/\tau(B_d)$ can provide complementary bounds on duality violation, and so we discuss this in Sect. 3.2.2. The bounds in the B system depend strongly on the theory uncertainties, hence we present in Sect. 3.3 a numerical update of the mixing observables with an aggressive error estimate for the input parameters. In Sect. 3.4 we study possible effects of duality violation in D mixing. We summarise our findings in Sect. 3.5 and discuss the outlook for the future.

© Springer Nature Switzerland AG 2019
M. J. Kirk, *Charming New Physics in Beautiful Processes?*,
Springer Theses, https://doi.org/10.1007/978-3-030-19197-9_3

3.2 Duality Violation

In 1979 the notion of duality was introduced by Poggio, Quinn and Weinberg [1]
for the process $e^+e^- \to$ hadrons.[1] The basic assumption is that this process can
be well approximated by a quark level calculation of $e^+e^- \to q\bar{q}$. In this work we
will investigate duality in the case of decays of heavy hadrons, which are described
by the HQE (see Sect. 2.2 for an overview and references to the early works). The
HQE assumes quark-hadron duality, i.e. that the hadron decays can be described at
the quark level. A violation of duality could correspond to non-perturbative terms
like $\exp(-m_b/\Lambda)$ (where Λ is the numerator in the HQE) which give vanishing
contributions when Taylor expanded around $\Lambda/m_b = 0$ (see e.g. [4] and also [5]
for a detailed discussion of duality, its violations and some possible models for
duality violations). To estimate the possible size of these non-perturbative terms
we note that in the case of lifetimes and mixing, the expansion parameter of the
HQE is not Λ/m_b but the hadronic scale Λ normalised to the momentum release
$\sqrt{M_i^2 - M_f^2}$, where M_i is the mass of the initial state and M_f the sum of the final
state masses. As such, the expansion parameter for the quark-level decay $b \to c\bar{c}s$,
$\Lambda/\sqrt{m_b^2 - 4m_c^2}$, is considerably larger than for the decay $b \to u\bar{u}u$, where it is Λ/m_b.
Put another way, the less phase space that is accessible in the final state, the worse
the convergence property of the HQE for this class of decays, and the larger might
be the hypothetical duality violating terms. The remaining phase space for B_s decay
into light mesons (e.g. $B_s \to K^-\pi^+$, via $b \to u\bar{u}d$), due to the dominant quark level
decay (e.g. $B_s \to D_s^-\pi^+$, via $b \to c\bar{u}d$) and into the leading contribution to $\Delta\Gamma_s$
(e.g. $B_s \to D_s^+D_s^-$, via $b \to c\bar{c}s$) reads

$$M_{B_s} - M_K - M_\pi = 4.73 \,\text{GeV}, \qquad (3.2.1)$$

$$M_{B_s} - M_{D_s^+} - M_\pi = 3.26 \,\text{GeV}, \qquad (3.2.2)$$

$$M_{B_s} - 2M_{D_s^{(*)+}} = 1.43(1.15) \,\text{GeV}. \qquad (3.2.3)$$

The crucial question is whether the phase space in $B_s \to D_s^{(*)+}D_s^{(*)-}$ is still large
enough to ensure quark-hadron duality.

To get some idea of the possible values of the expansion parameter and the non-
perturbative terms in inclusive b quark decays, we vary Λ within 0.2 and 2 GeV,[2]
m_b within 4.18 and 4.78 GeV, and m_c within 0.975 and 1.67 GeV—the resulting
numerics are shown in Table 3.1. From this simple numerical exercise we find that
duality violating terms could easily be of a similar size as the expansion parameter of
the HQE. Moreover decay channels like $b \to c\bar{c}s$ might be more strongly affected by

---

[1]The concept of duality was already used in 1970 for electron proton scattering by Bloom and
Gilman [2, 3].
[2]This is twice the scale one finds in $\Delta\Gamma_s$, where $\Lambda/m_b \approx 1/5$ [6].

Table 3.1 Value of the HQE expansion parameter for a range of values of Λ, m_b and m_c

Channel	Expansion parameter ($= x$)	Value of x	$\exp(-1/x)$
$b \to c\bar{c}s$	$\dfrac{\Lambda}{\sqrt{m_b^2 - 4m_c^2}} \approx \dfrac{\Lambda}{m_b}\left(1 + 2\dfrac{m_c^2}{m_b^2}\right)$	0.054–0.58	$9.4 \times 10^{-9} - 0.18$
$b \to c\bar{u}s$	$\dfrac{\Lambda}{\sqrt{m_b^2 - m_c^2}} \approx \dfrac{\Lambda}{m_b}\left(1 + \tfrac{1}{2}\dfrac{m_c^2}{m_b^2}\right)$	0.045–0.49	$1.9 \times 10^{-10} - 0.13$
$b \to u\bar{u}s$	$\dfrac{\Lambda}{\sqrt{m_b^2 - 4m_u^2}} = \dfrac{\Lambda}{m_b}$	0.042–0.48	$4.2 \times 10^{-11} - 0.12$

duality violations compared to e.g. $b \to u\bar{u}s$. This agrees with the naive expectation that decays with a smaller final state phase space might be more sensitive to duality violation.

Obviously duality cannot be proved directly because this would require a complete solution of QCD and a subsequent comparison with the HQE expectations, which is clearly not possible. To make statements about duality violation, in principle there are two strategies that could be performed:

1. Study simplified models for QCD, e.g. the 't Hooft model (a two-dimensional model for QCD, see e.g. [4, 5, 7–10]) and develop models for duality violations, like instanton-based and resonance-based models (see e.g. [4, 5]).
2. Use a pure phenomenological approach, by comparing experiment with HQE predictions.

In this chapter we will follow the second strategy and use a simple parameterisation for duality violation in mixing observables and lifetime ratios, which will be most pronounced for the $b \to c\bar{c}s$ channel.

At this stage it is interesting to note that for many years there have been problems related to applications of the HQE for inclusive b hadron decays and most of them seemed to be related to the $b \to c\bar{c}s$ channel:

- For quite some time the experimental value for the Λ_b lifetime was considerably lower than early theory predictions [11], which indicated a value quite close to the B_d lifetime (see [12] for a detailed review). These results triggered theoretical attempts to explain the discrepancy with a failure of the HQE, see e.g. the discussion in [13], where a simple model for a modification of the HQE was suggested in order to explain experiment, see also [14, 15]. The dominant contribution to the Λ_b lifetime is given by the $b \to c\bar{u}d$ and $b \to c\bar{c}s$ transitions. To a large extent the Λ_b lifetime problem has now been solved experimentally, see the detailed discussion in [12], mostly by new measurements from LHCb [16–18] and the new data confirm the early theory estimates [11]. However, there is still a large theory uncertainty remaining due to unknown non-perturbative matrix elements that could be calculated with current lattice QCD techniques.

- For quite some time the values of the inclusive semi-leptonic branching ratio of B mesons as well as the average number of charm quarks per b decay (missing charm puzzle) disagreed between experiment and theory, see e.g. [19–22]. Modifications of the decay $b \to c\bar{c}s$ were considered as a potential candidate for solving this problem. This issue has been improved considerably by new data and new calculations [23]. Again, there still a considerable uncertainty remains due to unknown NNLO QCD corrections. First estimates suggest that such corrections could be large [24].
- Because of a cancellation of weak annihilation contributions it is theoretically expected (based on the HQE) that the B_s lifetime is more or less equal to the B_d lifetime, see e.g. an early estimate from 1986 [11] or the review [12] for updated values. For quite some time experiment found deviations of $\tau(B_s)/\tau(B_d)$ from one—we have plotted the experimental averages from HFLAV [25] from 2003 onwards in Fig. 3.1. Currently there is still a small difference between data and the HQE prediction, which will be discussed further in Sect. 3.2.2. Here again a modification of the $b \to c\bar{u}d$ and/or the $b \to c\bar{c}s$ transitions might solve the problem.

All of these problems have softened considerably since being recognised and huge duality violations are now ruled out by experiment [26], in particular by the measurement of the decay rate difference of neutral B_s mesons, $\Delta\Gamma_s$, which is to a good approximation a $b \to c\bar{c}s$ transition. But there is still space for a small amount of duality violation—which will be quantified in this chapter. We will thus investigate the decay rate difference $\Delta\Gamma_s$ in more detail.

We show in Table 3.2 the theory and experimental values we use in the rest of this chapter. The most recent numerical update for the mixing quantities is given in [27] (superseding the numerical predictions in [28, 29]) which is where we take our SM predictions from, and can be compared to the experimental values from HFLAV

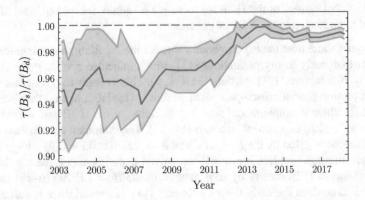

Fig. 3.1 Historical values of the lifetime ratio $\tau(B_s)/\tau(B_d)$ as reported by HFLAV [25] since 2003. The solid line shows the central value and the shaded line indicates the $1\,\sigma$ region, the dotted line corresponds to the theory prediction, which is essentially one, with a tiny uncertainty

Table 3.2 SM predictions and experimental measurements that will be used for the work in this chapter

Observable	SM	Experiment
$\Delta M_s/\mathrm{ps}^{-1}$	18.3 ± 2.7	17.757 ± 0.021
$\Delta \Gamma_s/\mathrm{ps}^{-1}$	0.088 ± 0.020	0.082 ± 0.006
$a_{\mathrm{sl}}^s/10^{-5}$	2.22 ± 0.27	170 ± 300
$\Delta \Gamma_s/\Delta M_s/10^{-4}$	48.1 ± 8.32	46.2 ± 3.4
$\Delta M_d/\mathrm{ps}^{-1}$	0.528 ± 0.078	0.5055 ± 0.0020
$\Delta \Gamma_d/10^{-3}\mathrm{ps}^{-1}$	2.61 ± 0.59	0.66 ± 6.6
$a_{\mathrm{sl}}^d/10^{-4}$	-4.7 ± 0.6	-15 ± 17
$\Delta \Gamma_d/\Delta M_d/10^{-4}$	49.4 ± 8.5	13 ± 130

[30, 31] (except the experimental average for a_{sl}^s which has been taken from [32]). The theory prediction uses conservative ranges for the input parameters—we will present a more aggressive estimate in Sect. 3.3. Experiment and theory agree very well for the quantities ΔM_q and $\Delta \Gamma_s$. The semileptonic asymmetries and the decay rate difference in the B_d system have not been observed yet. More profound statements about the validity of the theory can be made by comparing the ratio of $\Delta \Gamma_s$ and ΔM_s, where some theoretical uncertainties cancel and we get

$$\frac{\left(\frac{\Delta \Gamma_s}{\Delta M_s}\right)^{\mathrm{exp}}}{\left(\frac{\Delta \Gamma_s}{\Delta M_s}\right)^{\mathrm{SM}}} = 0.96 \pm 0.22 \,. \tag{3.2.4}$$

The central value shows a very good agreement of experiment and HQE predictions. The remaining uncertainty leaves some space for new physics effects or for violations of duality. We have taken here the 2σ range of the experimental value, while we consider the theory range to cover all allowed values. Thus we conclude that in the most sensitive decay channel, $b \to c\bar{c}s$, duality seems to be violated by at most 22%. In the next section we try to investigate these possibilities in more detail.

3.2.1 B Mixing

For the discussion in the rest of this section, it will be useful to separate out the CKM dependence of M_{12} and Γ_{12}. We can write

$$\Gamma_{12}^s = -\sum_{x=u,c}\sum_{y=u,c} \lambda_x \lambda_y \Gamma_{12}^{s,xy}, \qquad M_{12}^s = \lambda_t^2 \tilde{M}_{12}^s \,. \tag{3.2.5}$$

where $\lambda_q = V_{qs}^* V_{qb}$. For simplicity we give only the expressions for B_s mesons when modifications for B_d mesons are obvious, and we will explicitly present expressions for the B_d sector only when they are non-trivial. As we saw in Sect. 2.5, the physical observables ΔM_s, $\Delta \Gamma_s$ and a_{sl}^s are related to the ratio Γ_{12}^s / M_{12}^s via Eqs. 2.5.3 and 2.5.4. Using the unitarity of the CKM matrix we can write

$$-\frac{\Gamma_{12}^s}{M_{12}^s} = \frac{\Gamma_{12}^{s,cc}}{\tilde{M}_{12}^s} + 2\frac{\lambda_u}{\lambda_t} \frac{\Gamma_{12}^{s,cc} - \Gamma_{12}^{s,uc}}{\tilde{M}_{12}^s} + \left(\frac{\lambda_u}{\lambda_t}\right)^2 \frac{\Gamma_{12}^{s,cc} - 2\Gamma_{12}^{s,uc} + \Gamma_{12}^{s,uu}}{\tilde{M}_{12}^s} \qquad (3.2.6)$$

$$= -10^{-4} \left[c + a\frac{\lambda_u}{\lambda_t} + b \left(\frac{\lambda_u}{\lambda_t}\right)^2 \right]. \qquad (3.2.7)$$

The a, b and c notation of Eq. 3.2.7 comes from [33]. The way of writing Γ_{12}^s / M_{12}^s in Eqs. 3.2.6 and 3.2.7 can be viewed as a Taylor expansion in the small CKM parameter λ_u / λ_t, for which we get (using the input parameters of [27])

$$B_s : \frac{\lambda_u}{\lambda_t} = (-8.05 + 18.1i) \times 10^{-3}, \qquad \left(\frac{\lambda_u}{\lambda_t}\right)^2 = (-2.63 - 2.91i) \times 10^{-4}.$$

$$B_d : \frac{\lambda_u}{\lambda_t} = (7.55 - 405i) \times 10^{-3}, \qquad \left(\frac{\lambda_u}{\lambda_t}\right)^2 = (-164 - 6.11i) \times 10^{-3}.$$
$$\qquad (3.2.8)$$

In addition to the CKM suppression a pronounced GIM cancellation arises in the coefficients a and b in Eq. 3.2.7. With the input parameters described in [27] we get for the numerical values of a, b and c:

$$B_s : c = -48.0 \pm 8.3, \quad a = 12.3 \pm 1.4, \qquad b = 0.79 \pm 0.12.$$
$$B_d : c = -49.5 \pm 8.5, \quad a = 11.7 \pm 1.3, \qquad b = 0.24 \pm 0.06. \qquad (3.2.9)$$

From this hierarchy we see that (to a very good approximation) $\Delta\Gamma_q / \Delta M_q$ is given by $-c \times 10^{-4}$ and a_{sl}^q by $\mathrm{Im}(\lambda_u / \lambda_t) \times a \times 10^{-4}$.

We now introduce our simple phenomenological model for duality violation. As such effects are expected to be inversely proportional to the phase space of a decay, as we observed in Sect. 3.2, we write to a first approximation[3]:

$$\Gamma_{12}^{s,cc} \rightarrow \Gamma_{12}^{s,cc}(1 + 4\delta), \qquad (3.2.10)$$
$$\Gamma_{12}^{s,uc} \rightarrow \Gamma_{12}^{s,uc}(1 + \delta), \qquad (3.2.11)$$
$$\Gamma_{12}^{s,uu} \rightarrow \Gamma_{12}^{s,uu}(1 + 0\delta). \qquad (3.2.12)$$

It is already obvious at this stage that such a model will soften the GIM cancellations in the ratio Γ_{12}^s / M_{12}^s—we get

---

[3] Similar models have been used in [34–36] for penguin insertions with a $c\bar{c}$-loop.

Table 3.3 Dependence on δ of the mixing observables. The expressions for $\Delta\Gamma_q$ were obtained by simply multiplying the theory ratio $\Delta\Gamma_q/\Delta M_q$ with the theoretical values of the mass difference, as given in Table 3.2

Observable	B_s	B_d
$\frac{\Delta\Gamma}{\Delta M}$	$48.1(1 + 3.95\delta) \times 10^{-4}$	$49.5(1 + 3.76\delta) \times 10^{-4}$
$\Delta\Gamma$	$0.0880(1 + 3.95\delta)\text{ps}^{-1}$	$2.61(1 + 3.76\delta) \times 10^{-3}\text{ps}^{-1}$
a_{sl}	$2.225(1 - 22.3\delta) \times 10^{-5}$	$-4.74(1 - 24.5\delta) \times 10^{-4}$

$$\frac{\Gamma_{12}^s}{M_{12}^s} = 10^{-4}\left[c(1 + 4\delta) + \frac{\lambda_u}{\lambda_t}(a + \delta(6c + a)) + \frac{\lambda_u^2}{\lambda_t^2}(b + \delta(2c + a))\right].$$

(3.2.13)

From our earlier observations we expect $\Delta\Gamma_s/\Delta M_s \approx -c(1 + 4\delta) \times 10^{-4}$, but more interestingly the semi-leptonic CP asymmetries will be approximately given by $a_{\text{sl}}^s \approx \text{Im}(\lambda_u/\lambda_t)[a + \delta(6c + a)] \times 10^{-4}$—the duality violating coefficient δ is GIM *enhanced* by $(6c + a)$ compared to the leading term a. As such, an agreement between experiment and theory for the semileptonic CP asymmetries could provide very strong constraints on duality violation. Using the values of a, b and c from Eq. 3.2.9 and the CKM elements from Eq. 3.2.8, we get the dependence on the duality violating parameter δ of the mixing observables that is shown in Table 3.3. As expected we find that the duality violating parameter δ has a decent leverage on $\Delta\Gamma_q$ and a sizeable one on a_{sl}^q.

Comparing experiment and theory for the ratio of the decay rate difference $\Delta\Gamma_s$ and the mass difference ΔM_s we found in Eq. 3.2.4 an agreement with a deviation of at most 22%. Thus the duality violation—i.e. the factor $1 + 3.95\delta$ in Table 3.3—has to be smaller than this uncertainty:

$$1 + 3.95\delta \leq 0.96 \pm 0.22 \Rightarrow \delta \in [-0.066, 0.046],$$

(3.2.14)

Equivalently this bound tells us that the duality violation in the cc-channel is at most $+18.2\%$, or -26.3% if the effect turns out to be negative. If experiment and theory would also agree to within 22% for the semileptonic asymmetry a_{sl}^s, then we could shrink the bound to δ down to 0.01, but for the moment experiment is still far away from the SM prediction (see Table 3.2). However, we can turn around the argument: even in the most pessimistic scenario—i.e. having a duality violation that lifts GIM suppression—the theory prediction of a_{sl}^s can at most be in the range

$$a_{\text{sl}}^s \in [-0.06, 5.50] \times 10^{-5}.$$

(3.2.15)

In the B_d system, making a similar comparison of experiment and theory for $\Delta\Gamma/\Delta M$ turns out to be tricky, since $\Delta\Gamma_d$ is not yet measured. Because of this large uncertainty in the current experimental bound on $\Delta\Gamma_d$, we would get artificially large bounds on δ. If we look at the structure of the loop contributions necessary to calculate

Γ_{12}^d and Γ_{12}^s, we find very similar $c\bar{c}$-, $u\bar{c}$-, $c\bar{u}$- and $u\bar{u}$-contributions. Our duality violation model is based on the phase space differences of decays like $B_s \to D_s D_s$ $(c\bar{c})$, $B_s \to D_s K$ $(u\bar{c})$, $(c\bar{u})$ and $B_s \to \pi K$ $(u\bar{u})$, which are very pronounced. On the other hand we find that the phase space differences of B_s and B_d decays are not very pronounced, i.e. the difference between e.g. $B_s \to D_s D_s$ versus $B_d \to D_s D$ is small, compared to the above differences due to different internal quarks. As such, we conclude that applying the duality violation bounds from the B_s system to the B_d system is a good approximation. With the B_s bound we find that the theory prediction of a_{sl}^d and $\Delta\Gamma_d$ can be altered due to duality violations to at most

$$a_{sl}^d \in [-12.4, -0.6] \times 10^{-4}, \tag{3.2.16}$$

$$\Delta\Gamma_d \in [1.96, 3.06] \times 10^{-3} \text{ps}^{-1}. \tag{3.2.17}$$

These numbers can be compared to the SM values shown in Table 3.2. In principle any measurement of these observables outside the ranges in Eqs. 3.2.15–3.2.17 would be a clear indication of new physics.

Since our conclusions (new physics or unknown hadronic effects) are quite far-reaching, we try to be as conservative as possible and from now on use a likelihood ratio test to set a 95% confidence limit on our duality violating parameters. Our more conservative bound for δ is now given by

$$\delta \in [-0.12, 0.10]. \tag{3.2.18}$$

This more conservative statistical method doubles the allowed region for δ. Inserting these values into the predictions for $a_{sl}^{d,s}$ and $\Delta\Gamma_d$ we see that duality violation can give at most the following ranges for the mixing observables:

$$a_{sl}^s \in [-2.8, 8.2] \times 10^{-5}, \tag{3.2.19}$$

$$a_{sl}^d \in [-18.7, 6.9] \times 10^{-4}, \tag{3.2.20}$$

$$\Delta\Gamma_d \in [1.4, 3.6] \times 10^{-3} \text{ps}^{-1}. \tag{3.2.21}$$

The second modification to ensure that our estimates are conservative concerns our ad-hoc ansatz in Eqs. 3.2.10–3.2.12, where we assumed that the cc-part is affected by duality violations four times as much as the cu-part and the uu-part is not affected at all; we can obtain more general results with the following modification

$$\Gamma_{12}^{s,cc} \to \Gamma_{12}^{s,cc}(1 + \delta^{cc}), \tag{3.2.22}$$

$$\Gamma_{12}^{s,uc} \to \Gamma_{12}^{s,uc}(1 + \delta^{uc}), \tag{3.2.23}$$

$$\Gamma_{12}^{s,uu} \to \Gamma_{12}^{s,uu}(1 + \delta^{uu}), \tag{3.2.24}$$

with $\delta^{cc} \geq \delta^{uc} \geq \delta^{uu}$ and the requirement that all δs must have the same sign. Now we get for the observables

$$\frac{\Delta\Gamma_s}{\Delta M_s} = 48.1(1 + 0.982\delta^{cc} + 0.0187\delta^{uc} - 0.000326\delta^{uu}) \times 10^{-4}, \qquad (3.2.25)$$

$$\Delta\Gamma_d = 26.1(1 + 0.852\delta^{cc} + 0.350\delta^{uc} - 0.202\delta^{uu}) \times 10^{-4}\,\mathrm{ps}^{-1}, \qquad (3.2.26)$$

$$a_{sl}^s = 2.225(1 - 7.75\delta^{cc} + 8.67\delta^{uc} + 0.0780\delta^{uu}) \times 10^{-5}, \qquad (3.2.27)$$

$$a_{sl}^d = -4.74(1 - 8.52\delta^{cc} + 9.60\delta^{uc} - 0.0787\delta^{uu}) \times 10^{-4}. \qquad (3.2.28)$$

In the case of $\Delta\Gamma_s/\Delta M_s$, which will be used to determine the size of the duality violating δs, the coefficients of the uu component are suppressed by more than three orders of magnitude compared to the rest and therefore neglected. For the semileptonic CP asymmetries the uu duality violating component is about two orders of magnitude lower than the rest, and again we neglect the uu component in the following. This might lead to an uncertainty of about 20% in the duality bounds for $\Delta\Gamma_d$, which we will keep in mind.

Considering only δ^{cc} and δ^{uc} we get with the likelihood ratio test the bounds depicted in Fig. 3.2. In the figure we see that a duality violation of no more than 60% is allowed in either Γ_{cc}^s or in Γ_{uc}^s. We also see that it is in principle possible to see duality violation in $\Delta\Gamma_s$ but not in a_{sl}^s and vice versa. Moreover we find from the functional form of a_{sl}^s, that this quantity achieves a maximum (minimum) when $\delta^{uc} = 0$ and $\delta^{cc} < 0 \,(> 0)$. Our generalised parameterisation of duality violation gives now the most conservative bounds on the mixing observables

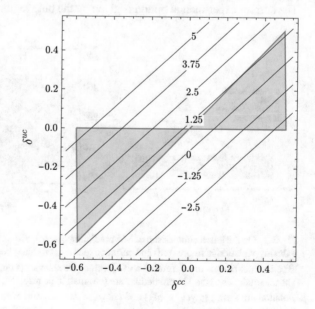

Fig. 3.2 95% confidence limits on δ^{cc} and δ^{uc} for the B_s system from a comparison of the experimentally allowed region of $\Delta\Gamma_s/\Delta M_s$ with the theory expression in Eq. 3.2.25. The allowed regions for the δs are shaded blue. A deviation of the δs from zero will also affect the theory prediction of a_{sl}^s in Eq. 3.2.27. The modification factors of $a_{sl}^s/a_{sl}^{s,SM}$ are denoted by the black lines

$$a_{\rm sl}^s \in [-6.7, 12.5] \times 10^{-5}, \tag{3.2.29}$$

$$a_{\rm sl}^d \in [-29, 16] \times 10^{-4}, \tag{3.2.30}$$

$$\Delta\Gamma_d \in [0.7, 4.2] \times 10^{-3}{\rm ps}^{-1}. \tag{3.2.31}$$

The duality bound on $a_{\rm sl}^d$ overlaps largely with the current experimental bound on this observable—in this case a future improvement in the measurement of $a_{\rm sl}^d$ will give an additional bound on duality violation.

We are now in a position to make a strong statement: any measurement outside this range is very unlikely to be due to duality violation, and so must be considered as a strong signal for new physics.

Since the ranges in Eqs. 3.2.29–3.2.31 are considerably larger than the uncertainties of the corresponding SM prediction given in Table 3.2, we now discuss the question of how to further shrink our bounds on duality violation. Currently the bound comes entirely from $\Delta\Gamma_s/\Delta M_s$, where the combined experimental and theoretical uncertainty comes to 22%. Any improvement on this uncertainty will shrink the allowed regions on δ. In Sect. 3.3 we will discuss a more aggressive estimate of the theory predictions for the mixing observable, indicating that a theory uncertainty of about 10% or even 5% in $\Delta\Gamma_s/\Delta M_s$ might come into sight. Also including possible improvements in experiment, this indicates a region for δ could be found that is considerably smaller than the ones given in Eqs. 3.2.29–3.2.31. The current (and a possible future) situation is summarised in Fig. 3.3. On the left $\Delta\Gamma_d$ is investigated. The current experimental bound is given by the blue region, which can be compared

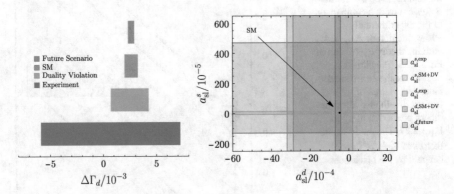

Fig. 3.3 On the left, a comparison of SM prediction (green), SM + duality violation (yellow), SM + duality violation in future (red) and current experimental (blue) bound for $\Delta\Gamma_d$. On the right, the experimental bounds on $a_{\rm sl}^d$ (green) and $a_{\rm sl}^s$ (blue) are shown in comparison to their theory values. The uncertainties of the SM predictions are too small to be resolved, the regions allowed by duality violation are shown in yellow ($a_{\rm sl}^s$) and red ($a_{\rm sl}^d$). Any measurement outside these duality allowed theory regions will be a clear indication for new physics. For $a_{\rm sl}^d$ the duality allowed region (red) has a pronounced overlap with the experimental one, while in the future this region could be shrink to the dark blue region. The theory uncertainties for the future duality region of $a_{\rm sl}^s$ are so small, that they cannot be resolved in the plot

to the SM prediction (green). As we have seen above, because of still sizeable uncertainties in $\Delta\Gamma_s$ duality violation of up to 60% cannot currently be excluded—this leads to an extended region (yellow) for the SM prediction including the effects of duality violation. If in future $\Delta\Gamma_s$ were known with a precision of about 5% in both theory and experiment, then the yellow region will shrink to the red one—obviously in this scenario the intrinsic precision of the SM value will be reduced. In other words: currently any measurement of $\Delta\Gamma_d$ outside the yellow region will be a clear signal of new physics, in the future any measurement outside the red region will be a signal of new physics. The same logic is applied for the right of Fig. 3.3, where a_{sl}^d and a_{sl}^s are investigated simultaneously. For a_{sl}^s still any measurement outside the bounds in Eq. 3.2.29 would be a clear indication of new physics. This bound is denoted in Fig. 3.3 by the tiny yellow region. For a_{sl}^d the current experimental region is given by the green area, which is slightly smaller than the red region, which is indicating the theory prediction including duality violation. Future improvement in experiment and theory for $\Delta\Gamma_s$ will reduce the red region to the purple one and then any measurement outside the dark blue region will be a clear signal of new physics.

Following this discussion, we now turn to the question of if there are more observables that will be affected by the duality violations. An obvious candidate is the lifetime ratio $\tau(B_s)/\tau(B_d)$, where the dominant diagrams are very similar to the mixing ones, and this observable will be studied further in Sect. 3.2.2.

3.2.2 Duality Bounds from Lifetime Ratios

Very similar diagrams to the ones in Γ_{12}^q arise in the lifetime ratio $\tau(B_s)/\tau(B_d)$, as we see from Fig. 3.4. The obvious difference between the two diagrams is the trivial exchange of b and q lines at the right end of the diagrams. A more subtle and more important difference lies in the possible intermediate states, which comes from cutting the diagrams down the middle. In the case of lifetimes all possible intermediate states that can originate from a $x\bar{y}$ quark pair can appear, while in the case of mixing, we have only the subset of all intermediate states into which both B_q and $\bar{B}_q$ can decay. Because of the larger set of intermediate states, one would expect that duality works better in the lifetimes than in mixing. Independent of this observation, our initial argument that the phase space for intermediate $c\bar{c}$-states is smaller than the one for intermediate $u\bar{c}$-states, which is again smaller that the $u\bar{u}$-case, still holds. Hence we assume that the $x\bar{y}$-loop for the lifetime ratio, has the same duality violating factor δ^{xy} as the $x\bar{y}$-loop for Γ_{12}^q. It turns out that the largest weak annihilation contribution to the B_s lifetime is given by a cc-loop, while for the B_d lifetime a uc-loop is dominating. This tells us that duality will not drop out in the lifetime ratio, because the dominating contributions for B_s and B_d are affected differently. Using our previously described model and modifying the cc-loop with a factor $1 + 4\delta$ and the uc-loop with a factor $1 + \delta$, we get with the expressions in [12, 14, 15, 37]

$$\frac{\tau(B_s)}{\tau(B_d)} = 1.00050 \pm 0.00108 - 0.0225\,\delta\,. \tag{3.2.32}$$

A detailed estimate of the theoretical error is given in Table D.1, since the previous SM prediction from [12] had an error twice as large. Unfortunately the SM prediction relies strongly on lattice calculations that are 17 years old [38], and the only update since then is our work in Chap. 6 (a more detailed discussion of the status of lifetime predictions can be found in [12]). Despite this, we find that the duality violating factor δ can have a large impact on the final result; a value of $\delta = 0.1$ would give corrections of the same size as the SM uncertainty.

Our theory prediction can be compared to the current experimental value for the lifetime ratio [31]

$$\frac{\tau(B_s)}{\tau(B_d)} = 0.990 \pm 0.004\,. \tag{3.2.33}$$

If the deviation between theory and experiment is attributed to duality violation, then we get an allowed range for δ of

$$\delta \in [0.13, 0.80]\,. \tag{3.2.34}$$

There is currently a discrepancy of about $2.5\,\sigma$ between experiment (Eq. 3.2.33) and theory (Eq. 3.2.32) and this difference could stem from new physics or a sizeable duality violation of $\delta \approx 0.5$ in lifetimes. Of course it is also worth bearing in mind the historical evolution of the experimental values, as seen in Fig. 3.1, when considering this deviation. The allowed region of the duality violating parameter δ canbe read off Fig. 3.5, where the current experimental bound from Eq. 3.2.33 is given by the blue region and theory prediction including hypothetical duality violation by the red region. Since $2.5\,\sigma$ is not enough to justify profound statements, we consider next future scenarios where the experimental uncertainty of the lifetime ratio will be reduced to 0.1%.

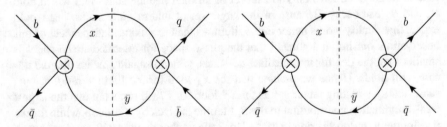

Fig. 3.4 Diagrams contributing to the Γ_{12}^q (left) and diagrams contributing to the lifetime of a B_q meson (right)

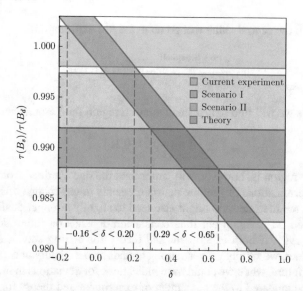

Fig. 3.5 Duality bounds extracted from the lifetime ratio $\tau(B_s)/\tau(B_d)$. The red band shows the theoretical expected value, with the δ dependence given in Eq. 3.2.32. The current experimental bound is given by the blue region and the overlap of both gives the current allowed region δ, indicated in Eq. 3.2.34. The future scenarios are indicated by the purple band (Scenario I) and the green band (Scenario II). Again the overlap of the future scenarios with the theory prediction gives future allowed regions for δ—in this figure the naive overlap of both regions is shown, this corresponds to a linear addition of uncertainties and leads thus to slightly bigger ranges of δ compared to the main text (Eqs. 3.2.36 and 3.2.38)

- Scenario I: the central value will stay at the current slight deviation from one:

$$\frac{\tau(B_s)}{\tau(B_d)}^{\text{Scenario I}} = 0.990 \pm 0.001 \,. \tag{3.2.35}$$

This scenario corresponds to a clear sign of duality violation or new physics in the lifetime ratio. Assuming the first one, we get a range of δ of

$$\delta \in [0.34, 0.60] \,. \tag{3.2.36}$$

Thus the lifetime ratio requires large values of δ. Our final conclusions depend now on the future developments of $\Delta\Gamma_s$. Currently $\Delta\Gamma_s$ requires small values of δ, which is in contrast to scenario I. Thus we have to assume additional new physics effects—either in mixing or in lifetimes—that might solve the discrepancy. Alternatively, if in the future the theory value of $\Delta\Gamma_s$ goes up or the experimental value goes down, then mixing might also require a big value of δ and we then would have duality violation as a simple solution for explaining discrepancies in both lifetimes and B_s mixing.

• Scenario II: the central value will go up to the SM expectation:

$$\frac{\tau(B_s)}{\tau(B_d)}^{\text{Scenario II}} = 1.000 \pm 0.001 \,, \qquad (3.2.37)$$

In that case we will find only a small allowed region for δ around zero

$$\delta \in [-0.11, 0.15] \,. \qquad (3.2.38)$$

The above region is, however, still larger than the one obtained from $\Delta\Gamma_s$. New lattice determinations of lifetime matrix elements might change this picture and in the end the lifetime ratio might also lead to slightly stronger duality violating bounds than $\Delta\Gamma_s$. Again our final conclusion depends on future developments related to $\Delta\Gamma_s$. If both experiment and theory for mixing stay at their current central values, we simply get very strong bounds on δ. If theory or experiment will change in future, when we could have indications for deviations in mixing, which have to be compared to the agreement of experiment and theory for lifetimes in Scenario II.

In Sect. 3.3 we will discuss a possible future development of future theory predictions for mixing observables.

Before we proceed let us make a comment about our duality model. In principle we also could generalise our duality ansatz in the same way as we did for Γ_{12}. This would leave to the following expression

$$\frac{\tau(B_s)}{\tau(B_d)} = 1 + 0.0005(1 - 13.4\delta^{cc} + 8.92\delta^{uc}) \qquad (3.2.39)$$

Here we see a pronounced cancellation of the cc and the uc contribution, if we allow δ^{cc} to be close in size to δ^{uc}. This is not what we expect from our phase space estimates for duality violation, and would allow an unphysically large upper limit on duality violation, and so we use for the lifetime ratio only our simple model and the corresponding result given in Eq. 3.2.32.

3.3   Numerical Updates of Standard Model Predictions

We have already pointed out that more precise values of $\Delta\Gamma_s$ are needed to derive more stringent bounds on duality violation in the B system. Very recently the Fermilab MILC collaboration presented a comprehensive study of the non-perturbative parameters that enter B mixing [39].[4] A brief summary of their results reads:

---

[4] A numerical analysis with these new inputs was already performed in [40], but the authors put emphasis on the implications for the correlation between $\Delta M_{s,d}$ and ε_K in models with constrained MFV and the implications for $\Delta\Gamma_{s,d}$ were not been analysed.

- Improved numerical values for the matrix elements of the operators $\mathcal{O}_{1-5}$ and R_0 that are necessary for $\Delta\Gamma_q$ and ΔM_q.[5] With these results, we have numerical values for all operators that arise up to dimension seven in the HQE, except $\langle R_2 \rangle$ and $\langle R_3 \rangle$ which are still unknown and can only be estimated by assuming the vacuum saturation approximation.
- The numerical values of $f_{B_q}^2 B$ are larger than in most previous lattice calculations.
- We have a very strong confirmation of the vacuum saturation approximation, since all their bag parameters turn out to be in the range 0.8–1.2.

Based on these new results we perform a more aggressive numerical analysis of the SM predictions, where we try to push the current theory uncertainties to the limits. (These results also form part of the basis of our work in Chap. 7.) In particular we will modify the predictions in [27] by doing the following:

1. We use the most recent (at time of writing) values of the CKM parameter from CKMfitter [42].
2. We take the new Fermilab MILC results for the bag parameters of $\langle \mathcal{O}_1 \rangle$, $\langle \mathcal{O}_3 \rangle$, $\langle \mathcal{O}_4 \rangle$, $\langle \mathcal{O}_5 \rangle$ and $\langle R_0 \rangle$. (Note that they quote values for the bag parameters of the operators R_1, $\tilde{R}_1$ which are proportional to those for $\mathcal{O}_4$, $\mathcal{O}_5$, and it is these we use these in our calculation.) We do not try to average with other lattice results, e.g. the values given by FLAG [43].
3. We assume the vacuum saturation approximation for R_2 and R_3 with a small uncertainty of $B = 1 \pm 0.2$. We note that this is the most aggressive change as no calculation of these bag parameters has yet been done.
4. We use results derived from equations of motion $\tilde{B}_{R_3} = 7/5 B_{R_3} - 2/5 B_{R_2}$ and $\tilde{B}_{R_2} = -B_{R_2}$ [41].

All our inputs are listed in Table D.2. We first note a few things: that the overall normalisation due to $f_{B_q}^2 B$ seems to be considerably enhanced now, so we expect enhancements in ΔM_q and $\Delta\Gamma_q$ individually (but that will cancel in the ratio); that the uncertainty in the bag parameter ratio B_3/B_1 is larger than e.g. in [27]; and that the dominant uncertainty due to R_2 and R_3 will now be dramatically reduced.

Putting everything together, we show in Table 3.4 our updated predictions with the new parameters for the two neutral B systems, which are compared with the more conservative theory predictions [27] and the experimental values from HFLAV [31] which were already given in Table 3.2.

The new theory values for ΔM_q and $\Delta\Gamma_q$ are larger than the ones presented in [27] and they are further from experiment. For the ratios $\Delta\Gamma_q/\Delta M_q$ and a_{sl}^q the central values are only slightly enhanced. The overall error shrinks by about a factor of two for ΔM_s and also sizeably for ΔM_d, $\Delta\Gamma_q$ and the ratios $\Delta\Gamma_q/\Delta M_q$. For the semileptonic asymmetries the effect is less pronounced. A detailed breakdown of the errors is given in Tables D.3, D.4, D.5, D.6, D.7, D.8, D.9 and D.10.

It is now interesting to consider the ratios of the new SM predictions normalised to the experimental numbers.

---

[5] R_0 is a $1/m_b$ suppressed operator which is a linear combination of $\mathcal{O}_{1,2,3}$ [28, 41].

Table 3.4 New predictions for mixing observables with "aggressive" input choices, with a comparison to the usual SM predictions and experimental measurements

Observable	SM (conservative)	SM (agressive)	Experiment
$\Delta M_s/\mathrm{ps}^{-1}$	18.3 ± 2.7	20.11 ± 1.37	17.757 ± 0.021
$\Delta \Gamma_s/\mathrm{ps}^{-1}$	0.088 ± 0.020	0.098 ± 0.014	0.082 ± 0.006
$a_{sl}^s/10^{-5}$	2.22 ± 0.27	2.27 ± 0.25	170 ± 300
$\frac{\Delta \Gamma_s}{\Delta M_s}/10^{-4}$	$48.1(1 \pm 0.173)$	$48.8(1 \pm 0.125)$	$46.2(1 \pm 0.073)$
$\Delta M_d/\mathrm{ps}^{-1}$	0.528 ± 0.078	0.606 ± 0.056	0.5055 ± 0.0020
$\Delta \Gamma_d/10^{-3}\mathrm{ps}^{-1}$	2.61 ± 0.59	2.99 ± 0.52	0.66 ± 6.6
$a_{sl}^d/10^{-4}$	-4.7 ± 0.6	-4.9 ± 0.54	-15 ± 17
$\frac{\Delta \Gamma_d}{\Delta M_d}/10^{-4}$	$49.4(1 \pm 0.172)$	$49.3(1 \pm 0.149)$	$13(1 \pm 10)$

$$\frac{\Delta M_s^{\mathrm{exp}}}{\Delta M_s^{\mathrm{SM\,(agr.)}}} = 0.88(1 \pm 0.060(\text{theory}) \pm 0.001(\text{exp})) \qquad (3.3.1)$$

$$= 0.88(1 \pm 0.06)\,, \qquad (3.3.2)$$

$$\frac{\Delta \Gamma_s^{\mathrm{exp}}}{\Delta \Gamma_s^{\mathrm{SM\,(agr.)}}} = 0.836(1 \pm 0.12(\text{theory}) \pm 0.06(\text{exp})) \qquad (3.3.3)$$

$$= 0.84(1 \pm 0.13)\,, \qquad (3.3.4)$$

$$\frac{\left(\frac{\Delta \Gamma_s}{\Delta M_s}\right)^{\mathrm{exp}}}{\left(\frac{\Delta \Gamma_s}{\Delta M_s}\right)^{\mathrm{SM\,(agr.)}}} = 0.947(1 \pm 0.12(\text{theory}) \pm 0.07(\text{exp})) \qquad (3.3.5)$$

$$= 0.95(1 \pm 0.14)\,, \qquad (3.3.6)$$

$$\frac{\Delta M_d^{\mathrm{exp}}}{\Delta M_d^{\mathrm{SM\,(agr.)}}} = 0.83(1 \pm 0.08(\text{theory}) \pm 0.003(\text{exp})) \qquad (3.3.7)$$

$$= 0.83(1 \pm 0.08)\,. \qquad (3.3.8)$$

Here one clearly sees the enhancements of the mass differences, which are up to 20% or more than two standard deviations above the experimental value. The decay rate difference $\Delta \Gamma_s$ is also enhanced by about 20% above the measured value; due to larger uncertainties, this is statistically less significant. The dominant source for this enhancement is the new value of $\langle \mathcal{O}_1 \rangle$. The ratio $\Delta \Gamma_s / \Delta M_s$ is slightly lower than before, but still consistent with the corresponding experimental number.

Taking the deviations above seriously, we can think about several possible interpretations:

1. Since duality violations do not affect ΔM, they alone cannot explain the deviations. Statistical fluctuations in the experimental results of the order of three standard deviations might explain the deviation in $\Delta \Gamma_s$, while the deviation in ΔM_s cannot be explained by a fluctuation in the experiment.

2. The lattice normalisation for $f_B^2 B$ is simply too high, future investigations will bring down the value and there is no NP in mixing. Currently there is no reason to favour this possibility, but we try to leave no stone unturned. Since $f_B^2 B$ cancels in the ratio of mass and decay rate difference, we can use the new values to give the most precise SM prediction of $\Delta\Gamma_s$ via

$$\frac{\Delta\Gamma_s}{\Delta M_s} \cdot 17.757\,\text{ps}^{-1} (\equiv \Delta M_s^{\text{exp}}) = 0.087(10)\,\text{ps}^{-1}. \tag{3.3.9}$$

Now the theory error is very close to the experimental one and it would be desirable to have more precise values in both theory and experiment. In that case we also get an indication of the short-term perspectives for duality violating bounds. The above numbers indicate an uncertainty of ± 0.145 for the ratio $\Delta\Gamma_s/\Delta M_s$, which corresponds (if we had perfect agreement of experiment and theory) to a bound on δ of ± 0.037. This would already be a considerable improvement compared to the current situation.

3. Finally the slight deviation might be a first hint for NP effects.

   (a) To explain the deviation in the decay rate difference we need new physics effects in tree-level decays, while the deviation in M_{12} could be solved by new physics effects in loop contributions.
   (b) In principle there is also the possibility of new tree-level effects that modify both $\Delta\Gamma_s$ and ΔM_s, but which cancels in the ratio. (ΔM_s is affected by a double insertion of the new tree-level operators.) Following the strategy described in e.g. [44], we found, however, that the possible effects on the mass difference are much too small.
   (c) Finally there is also the possibility of a duality violation of about 20% in $\Delta\Gamma_s$, while the deviation in ΔM_s is due to new physics at loop level. This possibility can be tested in the future by more precise investigations of the lifetime ratio $\tau(B_s)/\tau(B_d)$.

In order to draw any definite conclusions about these interesting possibilities, we need improvements in several sectors: from experiment we need more precise values for $\Delta\Gamma_s$ and $\tau(B_s)/\tau(B_d)$, as well as a first measurement of $\Delta\Gamma_d$ which would also be very helpful. A measurement of the semileptonic asymmetries outside the duality-allowed regions would already be a clear manifestation of new physics in the mixing system. From the theory side we need (in ranked order):

1. A first principle determination of the dimension-seven operators $B_{R_{2,3}}$ and the corresponding colour-rearranged ones.
2. Independent non-perturbative determinations (lattice, sum rules) of the matrix elements $\langle \mathcal{O}_1 \rangle$, $\langle \mathcal{O}_2 \rangle$, $\langle \mathcal{O}_3 \rangle$, $\langle \mathcal{O}_4 \rangle$, $\langle \mathcal{O}_5 \rangle$ and $\langle R_0 \rangle$. We provide a sum rule determination of all these operators except R_0 in Chap. 6.
3. NNLO QCD calculations for the perturbative part of Γ_{12}. First steps in this direction have been done in [45].

4. An updated result for the dimension-six operators for meson lifetimes. We use sum rules to calculate these in Chap. 6, but a lattice calculation would still be desirable.

These improvements seem possible in the next few years and they might lead to a reduction of the theory error as low as 5% (see Chap. 7 where we achieve a precision of around 6% in ΔM_s) and thus might be the path to a detection of new physics effects in meson mixing.

3.4   *D* Mixing

Mixing in D mesons is by now experimentally well established and the values of the mixing parameters are quite well measured [46]:

$$x = (0.37 \pm 0.16) \times 10^{-2}, \tag{3.4.1}$$

$$y = (0.66^{+0.07}_{-0.10}) \times 10^{-2}. \tag{3.4.2}$$

Using $\tau(D^0) = 0.4101$ ps [47], this can be translated into

$$\Delta M_D = \frac{x}{\tau(D^0)} = 0.0090 \,\text{ps}^{-1}, \tag{3.4.3}$$

$$\Delta \Gamma_D = 2\frac{y}{\tau(D^0)} = 0.032 \,\text{ps}^{-1}. \tag{3.4.4}$$

When trying to compare these numbers with theory predictions, we face the problem that it is not obvious if our theory tools also work in the D system (see also the discussion in Sect. 2.2). The mixing quantities have been estimated via both inclusive and exclusive approaches. Inclusive HQE calculations work very well in the B system, but their naive application to the D system gives results that are several orders of magnitude lower than the experimental result [48, 49]. The exclusive approach is mostly based on phase space and $SU(3)_F$-symmetry arguments, see for example [50, 51]. Within this approach values for x and y of the order of 1% can be obtained. So while it is not a real first principles approach, this method seems to be our best currently available tool to describe D mixing. Given this status, it seems we are left with some of the following options:

• The HQE is simply not valid in the charm system. This obvious solution might however, be challenged by the fact that the tiny theoretical D mixing result is solely caused by an extremely effective GIM cancellation [52] (see e.g. the discussion in [53]), and not by the smallness of the first terms of the HQE expansion. A breakdown of the HQE in the charm system could best be tested by investigating the lifetime ratio of D mesons. From the theory side, the NLO QCD corrections for the lifetime ratio have been determined in [54] and it seems that the experimental measured values can be reproduced. To draw a definite conclusion about the agreement of experiment and theory for lifetimes and thus about the convergence of the HQE in the charm system, lattice evaluations of the unknown charm lifetime matrix elements are urgently needed. We make an initial calculation of these matrix elements in Chap. 6 using sum rules and at first sight it seems that the HQE is working.

- Bigi and Uraltsev pointed out in 2000 [55] that the extreme GIM cancellation in *D* mixing might be lifted by higher terms in HQE. There are indications for such an effect (see [53, 56]) but it is not yet clear whether the effect is large enough to explain the experimental mixing values. To make further progress in that direction we need the perturbative calculation of the higher order terms of the OPE and an idea of how to estimate the matrix elements of the operators that arise.
- The deviation of theory and experiment could of course also be due to new physics effects. Bounds on new physics models have been studied in e.g. [57], where they determined the NP contribution to *D* mixing, while more or less neglecting the SM contributions.

In this work we will investigate the related question of whether relatively small duality violating effects in inclusive charm decays could explain the deviation between experiment and the inclusive approach. We consider the decay rate difference $\Delta\Gamma_D$ for this task. According to the relation

$$\Delta\Gamma_D \leq 2|\Gamma_{12}|, \tag{3.4.5}$$

(see Appendix D.2 for a derivation) we will only study $|\Gamma_{12}|$ and test whether it can be enhanced close to the experimental value of the decay rate difference. This is of course a necessary, but not sufficient condition for an agreement of experiment and theory. A complete answer would also require a calculation of M_{12}, which is beyond the scope of this work.

Γ_{12} consists again of three CKM contributions

$$\Gamma_{12} = -\left(\lambda_s^2 \Gamma_{12}^{ss} + 2\lambda_s\lambda_d \Gamma_{12}^{sd} + \lambda_d^2 \Gamma_{12}^{dd}\right), \tag{3.4.6}$$

with the CKM elements $\lambda_d = V_{cd}V_{ud}^*$ and $\lambda_s = V_{cs}V_{us}^*$. Using as before the unitarity of the CKM matrix ($\lambda_d + \lambda_s + \lambda_b = 0$) we get

$$\Gamma_{12} = -\lambda_s^2 \left(\Gamma_{12}^{ss} - 2\Gamma_{12}^{sd} + \Gamma_{12}^{dd}\right) + 2\lambda_s\lambda_b \left(\Gamma_{12}^{sd} - \Gamma_{12}^{dd}\right) - \lambda_b^2 \Gamma_{12}^{dd}. \tag{3.4.7}$$

The CKM-factor have now a very pronounced hierarchy, they read

$$\lambda_s^2 = (4.82 - .0003i) \times 10^{-2}, \tag{3.4.8}$$

$$2\lambda_s\lambda_b = (2.50 + 5.91i) \times 10^{-5}, \tag{3.4.9}$$

$$\lambda_b^2 = (-1.49 + 1.53i) \times 10^{-8}. \tag{3.4.10}$$

The expressions for Γ_{12}^{xy} can be expanded in powers of $\bar{z}_s = (\bar{m}_s(\bar{m}_c)/\bar{m}_c(\bar{m}_c))^2 \approx$ 0.0092.

$$\Gamma_{12}^{ss} = 1.8696 - 5.5231\bar{z}_s - 13.8143\bar{z}^2 + \mathcal{O}\left(\bar{z}^3\right), \tag{3.4.11}$$

$$\Gamma_{12}^{sd} = 1.8696 - 2.7616\bar{z}_s - 7.4906\bar{z}^2 + \mathcal{O}\left(\bar{z}^3\right), \tag{3.4.12}$$

$$\Gamma_{12}^{dd} = 1.8696. \tag{3.4.13}$$

Looking at the expressions in Eq. 3.4.7 we see an extreme GIM cancellation in the CKM-leading term, while the last term without any GIM cancellation is strongly CKM suppressed. We get

$$\Gamma_{12}^{ss} - 2\Gamma_{12}^{sd} + \Gamma_{12}^{dd} = 1.17\bar{z}^2 - 59.5\bar{z}^3 + \dots, \tag{3.4.14}$$

$$\Gamma_{12}^{sd} - \Gamma_{12}^{dd} = -2.76\bar{z} + \dots. \tag{3.4.15}$$

Using our simplest duality violating model

$$\Gamma_{12}^{ss} \to \Gamma_{12}^{ss}(1 + 4\delta), \tag{3.4.16}$$

$$\Gamma_{12}^{sd} \to \Gamma_{12}^{sd}(1 + \delta), \tag{3.4.17}$$

$$\Gamma_{12}^{dd} \to \Gamma_{12}^{dd}(1 + 0\delta), \tag{3.4.18}$$

we find

$$\Gamma_{12}^{ss} - 2\Gamma_{12}^{sd} + \Gamma_{12}^{dd} = 1.17\bar{z}^2 - 59.5\bar{z}^3 + \dots \tag{3.4.19}$$
$$+ \delta(3.7392 - 16.5692\bar{z} - 40.276\bar{z}^2 + \cdots),$$

$$\Gamma_{12}^{sd} - \Gamma_{12}^{dd} = -2.76\bar{z} + \cdots + \delta\left(1.8696 - 2.7616\bar{z} - 7.4906\bar{z}^2 + \dots\right). \tag{3.4.20}$$

Equation 3.4.20 shows that our duality violating model completely lifts the GIM cancellation and that even tiny values of δ will lead to an overall result that is much bigger than the usual SM predictions within the inclusive approach. For our final conclusions we will use the generalised duality violating model

$$\Gamma_{12}^{ss} \to \Gamma_{12}^{ss}(1 + \delta^{ss}), \tag{3.4.21}$$

$$\Gamma_{12}^{sd} \to \Gamma_{12}^{sd}(1 + \delta^{sd}), \tag{3.4.22}$$

$$\Gamma_{12}^{dd} \to \Gamma_{12}^{dd}(1 + \delta^{dd}), \tag{3.4.23}$$

with $\delta^{ss} \geq \delta^{sd} \geq \delta^{dd}$. Next we test for what values of δ the inclusive approach can reproduce the experimental results for $\Delta\Gamma_D$. The corresponding allowed regions for $\delta^{ss,sd,dd}$ are given as shaded areas in Fig. 3.6. As expected, very small values

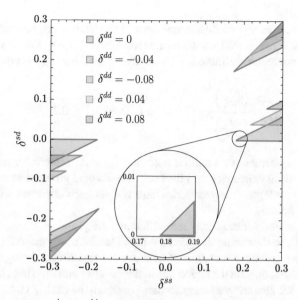

Fig. 3.6 Limits on δ^{ss}, δ^{sd} and δ^{dd} for the *D* system from a comparison of the experimentally allowed region of $\Delta\Gamma_D$ with the theory prediction based on the HQE. The allowed regions for the δs are shaded. Depending on the values of δ^{dd}, different colours are used. As expected, for small values of δ the experimental value of $\Delta\Gamma_D$ can not be reproduced and thus the area in the centre is free. Starting from values of about 20% on duality violation can explain the difference between experiment and HQE. To see more precisely where the smallest possible value of δ lies, we have zoomed into the overlap region

of δ cannot give an agreement between HQE and experiment, however surprisingly values as low as $\delta^{ss} \approx 0.18$ can explain the current difference. So a duality violation of the order of 20% in the HQE for the charm system is sufficient to explain the huge discrepancy between a naive application of the HQE and the measured value for $\Delta\Gamma_D$.

3.5 Summary

In this chapter we have explored the possibility of duality violations in heavy meson decays. Since the direct measurement of $\Delta\Gamma_s$ in 2012 by the LHCb collaboration huge duality violating effects are excluded [26], but there is still space for duality violating effects of the order of 20%. Because of the constantly improving experimental precision in flavour physics it is crucial to consider corrections of the order of 20% and to investigate whether, and how, such a bound can be improved.

To do so, we introduced in Eqs. 3.2.10–3.2.12 a simple parameterisation of duality violating effects, that relies solely on phase space arguments: the smaller the remaining phase space is in a heavy hadron decay, the larger duality violations might be.

In such a model, decay rate differences depend moderately on the duality violating parameter δ, whereas semi-leptonic asymmetries have a strong δ dependence, as can be seen from our results in Table 3.3. Currently we get the strongest bound on δ from Eq. 3.2.4

$$\frac{\left(\frac{\Delta\Gamma_s}{\Delta M_s}\right)^{\exp}}{\left(\frac{\Delta\Gamma_s}{\Delta M_s}\right)^{\mathrm{SM}}} = 0.96 \pm 0.22 \Rightarrow |\delta| \lesssim 0.1. \tag{3.5.1}$$

If there were agreement to a similar precision between experiment and theory for the semileptonic asymmetries then the bound on δ would go down to ± 0.01. Unfortunately, the semileptonic asymmetries have not yet been observed—we have only experimental bounds.

The same is true for the decay rate difference $\Delta\Gamma_d$, and so we used our bounds on δ from $\Delta\Gamma_s$ to determine the maximal possible size of a_{sl}^q and $\Delta\Gamma_d$, if duality is violated. These regions are compared with current experimental ranges in Fig. 3.3. Any measurement outside the region allowed by duality violation is a clear signal for new physics. We also show a future scenario in which the duality violation is further constrained by more precise values of $\Delta\Gamma_s$ both in experiment and theory.

We have shown that the duality violating parameter δ will also affect the lifetime ratio $\tau(B_s)/\tau(B_d)$, where at the time of writing there was a deviation of about $2.5\,\sigma$ between experiment and theory. Looking at the historical development of this ratio as depicted in Fig. 3.1, one might be tempted to assume a statistical fluctuation in the data (see also Table 2.1 as well as our new calculation in Chap. 6 and the updated experimental value quoted there). Taking that deviation seriously however, it is either a hint for new physics or for a sizeable duality violations of the order of $\delta \sim 0.5$, which is inconsistent with our bounds on δ derived from $\Delta\Gamma_s$. Here a future reduction of the experimental error of $\tau(B_s)/\tau(B_d)$ will give us valuable insight into the correct answer. We have studied two future scenarios in Fig. 3.5, which would either point towards new physics and duality violations or stronger bounds on duality violation. It is very important to note that the theory prediction has a very strong dependence on almost unknown lattice parameters. In particular, we can see from our error budget for the lifetime ratio in Table D.1 that any new calculation of the bag parameters $\epsilon_{1,2}$ would bring large improvements in the theory prediction for $\tau(B_s)/\tau(B_d)$. We make a new calculation in Chap. 6, but our overall error is worse than in this chapter, due to larger uncertainties than the old lattice results for these colour suppressed operators.

As we have mentioned several times, improvements in both experiment and theory for mixing observables and in particular for $\Delta\Gamma_s$ would be extremely helpful in this area. Therefore we presented an update of the SM predictions for the observables $\Delta\Gamma$, ΔM, and a_{sl} in both the B_s and B_d systems, based on the recent Fermilab-MILC lattice results [39] for non-perturbative matrix elements, CKM parameters from CKMfitter [42], and an aggressive error estimate on the unknown bag parameters of dimension-seven operators. With this input the current theory error in the mixing observables could be reduced by a half for ΔM_s or a third for ΔM_d, $\Delta\Gamma_s$ and $\Delta\Gamma_s/\Delta M_s$. In our aggressive scenario, we get for our fundamental relation to establish the possible size of duality violation

$$\frac{\left(\frac{\Delta\Gamma_s}{\Delta M_s}\right)^{\text{exp}}}{\left(\frac{\Delta\Gamma_s}{\Delta M_s}\right)^{\text{SM (agr.)}}} = 0.95 \pm 0.14 . \tag{3.5.2}$$

As expected, the overall uncertainty drops considerably, with a theory uncertainty that comes close to the experimental one—if this can be realised more precision in the experimental values of $\Delta\Gamma_s$ would be most helpful. Along with this improvement in precision, we found in this new analysis that the central values of the mass differences and decay rate differences are enhanced to values of about 20% above the measurements with a significance of around 2 standard deviations. To find out whether this enhancement is real, we need several ingredients: (1) an independent confirmation of the larger values of the matrix element $\langle\mathcal{O}_1\rangle$ found by [39]. (See Chap. 6 for our sum rule calculation, and Appendix H.3 for how that result combined with a sum rule decay constant result affects the mass difference central value). (2) a first principle calculation of $\langle R_{2,3}\rangle$—triggered by the results of [39] we simply assumed small deviations from vacuum saturation approximation. If the new central values turn out to be correct, there will be profound implications for new physics effects and duality violation in the B system. For a further improvement of the theory uncertainties beyond what we have considered, NNLO QCD corrections for mixing have to be calculated.

We finally focussed on the charm system, where a naive application of the HQE gives results that are several orders of magnitude below the experimental values. We found the unexpected result that duality violating effects as low as 20% could solve this discrepancy. Such a result might have profound consequences on the applicability of the HQE. As a decisive test we suggest a lattice calculation of the matrix elements arising in the ratio of charm lifetimes—this ratio is free of any GIM cancellation, which are very severe in mixing, and so allows us to separate the reliability of the underlying HQE from numerical cancellations.

References

1. Poggio EC, Quinn HR, Weinberg S (1976) Smearing the quark model. Phys Rev D 13:1958. https://doi.org/10.1103/PhysRevD.13.1958
2. Bloom ED, Gilman FJ (1970) Scaling, duality, and the behavior of resonances in inelastic electron-proton scattering. Phys Rev Lett 25:1140. https://doi.org/10.1103/PhysRevLett.25.1140
3. Bloom ED, Gilman FJ (1971) Scaling and the behavior of nucleon resonances in inelastic electron-nucleon scattering. Phys Rev D 4:2901. https://doi.org/10.1103/PhysRevD.4.2901
4. Shifman MA (2001) Quark hadron duality. In: At the frontier of particle physics. Handbook of QCD, vol 1–3 (Singapore), pp 1447–1494, World Scientific. https://doi.org/10.1142/9789812810458_0032, arXiv:hep-ph/0009131
5. Bigi IIY, Uraltsev N (2001) A Vademecum on quark hadron duality. Int J Mod Phys A 16:5201–5248. https://doi.org/10.1142/S0217751X01005535, arXiv:hep-ph/0106346
6. Lenz AJ (2011) A simple relation for B_s mixing. Phys Rev D 84:031501. https://doi.org/10.1103/PhysRevD.84.031501, arXiv:1106.3200

7. Grinstein B (2001) Global duality in heavy flavor decays in the 't Hooft model. Phys Rev D 64:094004. https://doi.org/10.1103/PhysRevD.64.094004, arXiv:hep-ph/0106205
8. Grinstein B (2001) Non-local duality in B decays in the t'Hooft model. PoS HEP 099
9. Lebed RF, Uraltsev NG (2000) Precision studies of duality in the 't Hooft model. Phys Rev D 62:094011. https://doi.org/10.1103/PhysRevD.62.094011, arXiv:hep-ph/0006346
10. Bigi IIY, Shifman MA, Uraltsev N, Vainshtein AI (1999) Heavy flavor decays, OPE and duality in two-dimensional 't Hooft model. Phys Rev D 59:054011. https://doi.org/10.1103/PhysRevD.59.054011. arXiv:hep-ph/9805241
11. Shifman MA, Voloshin MB (1986) Hierarchy of lifetimes of charmed and beautiful hadrons. Sov Phys JETP 64:698
12. Lenz A (2015) Lifetimes and heavy quark expansion. Int J Mod Phys A 30:1543005. https://doi.org/10.1142/S0217751X15430058. arXiv:1405.3601
13. Altarelli G, Martinelli G, Petrarca S, Rapuano F (1996) Failure of local duality in inclusive nonleptonic heavy flavor decays. Phys Lett B 382:409–414. https://doi.org/10.1016/0370-2693(96)00637-5, arXiv:hep-ph/9604202
14. Uraltsev NG (1996) On the problem of boosting nonleptonic b baryon decays. Phys Lett B 376:303–308. https://doi.org/10.1016/0370-2693(96)00305-X, arXiv:hep-ph/9602324
15. Neubert M, Sachrajda CT (1997) Spectator effects in inclusive decays of beauty hadrons. Nucl Phys B 483:339–370. https://doi.org/10.1016/S0550-3213(96)00559-7, arXiv:hep-ph/9603202
16. LHCB collaboration, Aaij R et al (2013) Precision measurement of the Λ_b baryon lifetime. Phys Rev Lett 111:102003. https://doi.org/10.1103/PhysRevLett.111.102003, arXiv:1307.2476
17. LHCB collaboration, Aaij R et al (2014) Precision measurement of the ratio of the Λ_b^0 to $\overline{B}^0$ lifetimes. Phys Lett B 734:122–130. https://doi.org/10.1016/j.physletb.2014.05.021, arXiv:1402.6242
18. LHCB collaboration, Aaij R et al (2014) Measurements of the B^+, B^0, B_s^0 meson and Λ_b^0 baryon lifetimes. JHEP 04:114. https://doi.org/10.1007/JHEP04(2014)114, arXiv:1402.2554
19. Bigi IIY, Blok B, Shifman MA, Vainshtein AI (1994) The baffling semileptonic branching ratio of B mesons. Phys Lett B 323:408–416. https://doi.org/10.1016/0370-2693(94)91239-4, arXiv:hep-ph/9311339
20. Falk AF, Wise MB, Dunietz I (1995) Inconclusive inclusive nonleptonic B decays. Phys Rev D 51:1183–1191. https://doi.org/10.1103/PhysRevD.51.1183, arXiv:hep-ph/9405346
21. Buchalla G, Dunietz I, Yamamoto H (1995) Hadronization of $b \to c\bar{c}s$. Phys Lett B 364:188–194. https://doi.org/10.1016/0370-2693(95)01296-6, arXiv:hep-ph/9507437
22. Lenz A (2000) Some comments on the missing charm puzzle. In: Heavy flavours and CP violation. Proceedings, 8th UK Phenomenology workshop, Durham, UK, September 19–24, 2000. arXiv:hep-ph/0011258
23. Krinner F, Lenz A, Rauh T (2013) The inclusive decay $b \to c\bar{c}s$ revisited. Nucl Phys B 876:31–54. https://doi.org/10.1016/j.nuclphysb.2013.07.028, arXiv:1305.5390
24. Czarnecki A, Slusarczyk M, Tkachov FV (2006) Enhancement of the hadronic b quark decays. Phys Rev Lett 96:171803. https://doi.org/10.1103/PhysRevLett.96.171803, arXiv:hep-ph/0511004
25. HFLAV collaboration. https://hflav.web.cern.ch
26. Lenz A (2012) Theoretical update of B-mixing and lifetimes. In: 2012 electroweak interactions and unified theories: proceedings of the 47th Rencontres de Moriond on electroweak interactions and unified theories, La Thuile, March 3–10, 2012. arXiv:1205.1444, https://inspirehep.net/record/1113760/files/arXiv:1205.1444.pdf
27. Artuso M, Borissov G, Lenz A (2016) CP violation in the B_s^0 system. Rev Mod Phys 88:045002. https://doi.org/10.1103/RevModPhys.88.045002, arXiv:1511.09466
28. Lenz A, Nierste U (2007) Theoretical update of $B_s - \bar{B}_s$ mixing. JHEP 06:072. https://doi.org/10.1088/1126-6708/2007/06/072, arXiv:hep-ph/0612167
29. Lenz A, Nierste U (2011) Numerical updates of lifetimes and mixing parameters of B mesons. In: CKM unitarity triangle. Proceedings, 6th international workshop, CKM 2010, Warwick, UK, September 6–10, 2010. arXiv:1102.4274, http://inspirehep.net/record/890169/files/arXiv:1102.4274.pdf

30. Heavy Flavor Averaging Group (HFAG) collaboration, Amhis Y et al (2014) Averages of b-hadron, c-hadron, and τ-lepton properties as of summer 2014. arXiv:1412.7515

31. HFLAV collaboration, B lifetime and oscillation parameters, summer 2015. http://www.slac.stanford.edu/xorg/hflav/osc/summer_2015/

32. LHCB collaboration, Aaij R et al (2016) Measurement of the CP asymmetry in $B_s^0 - \overline{B}_s^0$ mixing. Phys Rev Lett 117:061803. https://doi.org/10.1103/PhysRevLett.118.129903,10.1103/PhysRevLett.117.061803, arXiv:1605.09768

33. Beneke M, Buchalla G, Lenz A, Nierste U (2003) CP asymmetry in flavor specific B decays beyond leading logarithms. Phys Lett B 576:173–183. https://doi.org/10.1016/j.physletb.2003.09.089, arXiv:hep-ph/0307344

34. Palmer WF, Stech B (1993) Inclusive nonleptonic decays of B and D mesons. Phys Rev D 48:4174–4182. https://doi.org/10.1103/PhysRevD.48.4174

35. Dunietz I, Incandela J, Snider FD, Yamamoto H (1998) Large charmless yield in B decays and inclusive B decay puzzles. Eur Phys J C 1:211–219. https://doi.org/10.1007/BF01245810, arXiv:hep-ph/9612421

36. Lenz A, Nierste U, Ostermaier G (1997) Penguin diagrams, charmless B decays and the missing charm puzzle. Phys Rev D 56:7228–7239. https://doi.org/10.1103/PhysRevD.56.7228, arXiv:hep-ph/9706501

37. Franco E, Lubicz V, Mescia F, Tarantino C (2002) Lifetime ratios of beauty hadrons at the next-to-leading order in QCD. Nucl Phys B 633:212–236. https://doi.org/10.1016/S0550-3213(02)00262-6, arXiv:hep-ph/0203089

38. Becirevic D (2001) Theoretical progress in describing the B meson lifetimes. PoS HEP 098. arXiv:hep-ph/0110124

39. Fermilab Lattice, MILC collaboration, Bazavov A et al (2016) $B_{(s)}^0$-mixing matrix elements from lattice QCD for the standard model and beyond. Phys Rev D 93:113016. https://doi.org/10.1103/PhysRevD.93.113016, arXiv:1602.03560

40. Blanke M, Buras AJ (2016) Universal unitarity triangle 2016 and the tension between $\Delta M_{s,d}$ and ε_K in CMFV models. Eur Phys J C 76:197. https://doi.org/10.1140/epjc/s10052-016-4044-6, arXiv:1602.04020

41. Beneke M, Buchalla G, Dunietz I (1996) Width difference in the $B_s - \bar{B}_s$ system. Phys Rev D 54:4419–4431. https://doi.org/10.1103/PhysRevD.54.4419,10.1103/PhysRevD.83.119902, arXiv:hep-ph/9605259

42. CKMfitter collaboration, EPS-HEP 2015 results. http://ckmfitter.in2p3.fr/www/results/plots_eps15/num/ckmEval_results_eps15.html

43. Aoki S et al (2014) Review of lattice results concerning low-energy particle physics. Eur Phys J C 74:2890. https://doi.org/10.1140/epjc/s10052-014-2890-7, arXiv:1310.8555

44. Bobeth C, Haisch U, Lenz A, Pecjak B, Tetlalmatzi-Xolocotzi G (2014) On new physics in $\Delta\Gamma_d$. JHEP 06:040. https://doi.org/10.1007/JHEP06(2014)040, arXiv:1404.2531

45. Asatrian HM, Hovhannisyan A, Nierste U, Yeghiazaryan A (2017) Towards next-to-next-to-leading-log accuracy for the width difference in the $B_s - \bar{B}_s$ system: fermionic contributions to order $(m_c/m_b)^0$ and $(m_c/m_b)^1$. JHEP 10:191. https://doi.org/10.1007/JHEP10(2017)191, arXiv:1709.02160

46. HFLAV collaboration, Global fit for $D^0 - \bar{D}^0$ mixing, CHARM15. http://www.slac.stanford.edu/xorg/hflav/charm/CHARM15/results_mix_cpv.html

47. Particle Data Group (PDG) collaboration. http://pdg.lbl.gov/

48. Georgi H (1992) $D - \bar{D}$ mixing in heavy quark effective field theory. Phys Lett B 297:353–357. https://doi.org/10.1016/0370-2693(92)91274-D, arXiv:hep-ph/9209291

49. Ohl T, Ricciardi G, Simmons EH (1993) D-anti-D mixing in heavy quark effective field theory: the sequel. Nucl Phys B 403:605–632. https://doi.org/10.1016/0550-3213(93)90364-U, arXiv:hep-ph/9301212

50. Falk AF, Grossman Y, Ligeti Z, Petrov AA (2002) SU(3) breaking and $D^0 - \bar{D}^0$ mixing. Phys Rev D 65:054034. https://doi.org/10.1103/PhysRevD.65.054034, arXiv:hep-ph/0110317

51. Falk AF, Grossman Y, Ligeti Z, Nir Y, Petrov AA (2004) The D^0 - $\bar{D}^0$ mass difference from a dispersion relation. Phys Rev D 69:114021. https://doi.org/10.1103/PhysRevD.69.114021, arXiv:hep-ph/0402204

52. Glashow SL, Iliopoulos J, Maiani L (1970) Weak interactions with lepton-hadron symmetry. Phys Rev D 2:1285–1292. https://doi.org/10.1103/PhysRevD.2.1285

53. Bobrowski M, Lenz A, Riedl J, Rohrwild J (2010) How large can the SM contribution to CP violation in $D^0 - \bar{D}^0$ mixing be? JHEP 03:009. https://doi.org/10.1007/JHEP03(2010)009, arXiv:1002.4794

54. Lenz A, Rauh T (2013) D-meson lifetimes within the heavy quark expansion. Phys Rev D 88:034004. https://doi.org/10.1103/PhysRevD.88.034004, arXiv:1305.3588

55. Bigi IIY, Uraltsev NG (2001) D^0 - $\bar{D}^0$ oscillations as a probe of quark hadron duality. Nucl Phys B 592:92–106. https://doi.org/10.1016/S0550-3213(00)00604-0, arXiv:hep-ph/0005089

56. Bobrowski M, Lenz A, Rauh T (2012) Short distance D^0 - $\bar{D}^0$ mixing. In: Proceedings, 5th international workshop on charm physics (Charm 2012), Honolulu, Hawaii, USA, May 14–17, 2012. arXiv:1208.6438

57. Golowich E, Pakvasa S, Petrov AA (2007) New physics contributions to the lifetime difference in D^0-$\bar{D}^0$ mixing. Phys Rev Lett 98:181801. https://doi.org/10.1103/PhysRevLett.98.181801, arXiv:hep-ph/0610039

Chapter 4
Charming Dark Matter

4.1 Introduction

As discussed in Sect. 1.3.1, dark matter has a long history, but the interactions of DM (outside of its gravitational influence) remain elusive, despite concerted efforts. These range from attempts to measure its scattering in terrestrial targets (known as direct detection), its annihilation or decay products in the galaxy or beyond (indirect detection), or through its direct production in colliders (collider searches). A brief history and overview of direct detection (DD) and indirect detection (ID) can be found in [1, 2].

One property of DM that is known to high precision is its abundance in the Universe today. The evolution of the structure of the Universe is well modelled [3] and so the starting point for building a model of a particle DM is to consider how its interactions influence its relic abundance. This leads to the concept of a thermal WIMP, in which the DM achieves its relic abundance by decoupling from thermal equilibrium due to its annihilations or decay into SM particles.

Under the assumption of a WIMP particle interpretation of DM, we have no concrete indications of its mass, spin or interactions, which leaves tremendous freedom when building models. Although many concrete models, e.g. supersymmetric theories, predict the existence of a DM candidate, so far these theories remain unverified and the phenomenology is often complicated by the large parameter spaces. This represents a top-down approach in which DM arises naturally from a UV complete model.

An alternative approach to DM model building is from the bottom up, where a class of simple low energy models or interactions are considered simultaneously. With no theoretical guiding principle, except Lorentz symmetry, on which to build such models, one must consider all possible models within a framework of a few assumptions. This is most easily done using a set of EFT (see Sect. 2.1) operators. Although an EFT may be perfectly valid for low energy experiments such as direct or indirect detection, they face problems with collider searches where the EFT approximation breaks down when heavy ($\mathcal{O}$ (TeV)) states become energetically accessible.

© Springer Nature Switzerland AG 2019
M. J. Kirk, *Charming New Physics in Beautiful Processes?*,
Springer Theses, https://doi.org/10.1007/978-3-030-19197-9_4

To ensure the model is valid up to high energies (or at least above the reach of colliders) a commonly used tool is *simplified models*, where often the mediator between the dark sector and the SM is included as a propagating mode. Simplified models arose first in the context of collider searches for missing energy [4–10], but have recently been applied more widely to indirect and direct detection [4, 11, 12], they allow for a much more broad study since the models themselves are sufficiently simple to contain only a few parameters which dominate the phenomenology of the DM. This approach is not without criticism, and can at times be too simple, for example neglecting gauge symmetries and perturbative unitarity [13–15].

Given the remarkable agreement between the SM and experimentally measured flavour observables it is natural for NP models to enforce the MFV assumption to suppress large NP effects [16, 17]. This assumption limits any quark flavour breaking terms to be at most proportional to the Yukawa couplings, which are responsible for the small violation of the flavour symmetry in the SM. This suppresses FCNCs and avoids strong constraints from rare decays and neutral meson mixing. Nonetheless, some such observables are not reproduced by SM calculations and hence allow room for violations of MFV, for example D^0 mixing which we discuss in Sect. 4.3.1.

Some recent studies of simplified models have begun to go beyond the MFV assumptions. This has been done in the context of down type couplings [18], leptonic couplings [19], and more recently top-like [20], or top and charm-like couplings [21]. Such models allow a continuous change from the MFV assumption to strong MFV breaking and can quantify the degree of MFV breaking permitted by the flavour constraints. Similar scenarios have been studied in [22], taking an overview of both lepton and quark flavoured DM and as well as a more focused study on top DM [23], both in the MFV limit.

In this chapter, we extend the work of [20], taking a more general approach to these kinds of beyond MFV models—by placing fewer restrictions on the parameters of the model we include models with dominant up and charm type couplings, which give non-trivially different exclusion regions for different flavours of DM. We note that a similar scenario, except with scalar dark matter and a fermionic mediator has been studied in [24]. We aim to present statistically robust bounds from the entire parameter space based on a Markov Chain Monte Carlo (MCMC) approach.

We consider the following constraints in detail:

- Relic Density (Sect. 4.2): We calculate the relic density of all three DM particles, including their widths and important coannihilation effects.
- Flavour Bounds (Sect. 4.3): We provide bounds on the model from neutral charm meson mixing, ensuring that the new physics does not exceed the $1\,\sigma$ upper bound of the experimental measurement of the mass difference between the heavy and light state of the D^0. We assess the possibility for constraints on rare decays like $D^+ \to \pi^+ \ell\ell$ but find that the NP is relatively unconstrained compared to mixing.
- Direct Detection (Sect. 4.4): We calculate the event rate for the most excluding DD experiments (LUX and CDMSlite) over a large range of DM masses, including all relevant contributions up to one loop order (including gluon, photon, Z and Higgs exchange) and matching to a full set of non-relativistic form factors.

- Indirect Detection (Sect. 4.5): We include a large collection of constraints from the literature on the thermally averaged annihilation cross section $\langle \sigma v \rangle$ for annihilation into various search targets such as photons, electrons, and protons. We also include a study of gamma ray line searches, generated at the one-loop level in our model.
- Collider Searches (Sect. 4.6): We perform a robust simulation of the dominant signals for a series of monojet, dijet and stop searches for ATLAS and CMS, including the widths of the particles.

We also compute constraints coming from electroweak precision observables, and perturbative unitarity. We calculate the Peskin-Takeuchi parameters [25, 26] as these characterise the NP effects in much of the parameter space of our model, and replicate the literature result for a charged singlet scalar [27]. We find that the S, T, U parameters provide no additional constraints beyond those previously described, and similarly perturbative unitarity calculations prove to be unconstraining and so we make no further mention of them.

Including the various constraints named above we can carry out an MCMC scan in order to identify the parameter space left open to the model—our results are collected in Sect. 4.7. We find that current data can be used to restrict the parameter space where DM of this kind can exist, and go beyond the results of [20] by showing how renormalisation group mixing and running can dramatically improve the direct detection constraints, disfavouring attempts to avoid these limits by predominantly coupling to top quarks.

4.1.1 The DMFV Model

The SM (without Yukawa couplings) has a flavour symmetry amongst the quarks—there are no flavour violating effects such as FCNCs at tree level. *Minimal Flavour Violation* (MFV) is then the statement that the only flavour symmetry breaking terms in the BSM model are the Yukawa terms [17].

In the model of *Dark Minimal Flavour Violation* (DMFV) originally proposed in [18], the SM quark flavour symmetry is increased by the inclusion of a U(3) symmetry in the dark sector,

$$\mathcal{S}_{\text{flavour}} = \text{U}(3)_{Q_L} \times \text{U}(3)_{u_R} \times \text{U}(3)_{d_R} \times \text{U}(3)_{\chi}, \qquad (4.1.1)$$

and the DMFV hypothesis is that this enlarged flavour symmetry is broken only by terms involving the quark Yukawas *and* a new coupling matrix λ. In the original work [18] λ coupled the DM to right-handed down type quarks, whereas in this work we couple the DM to right-handed up type quarks (the choice of right-handed quarks avoids having to introduce any non-trivial $SU(2)$ structure). In this model, we introduce four new particles—a scalar ϕ that is colour and electrically charged, and a flavour triplet χ_i that is a singlet under the SM gauge groups (which allows it to have a standard Dirac mass term). In Table 4.1 we detail the behaviour of the new

Table 4.1 The representation for the relevant symmetries of the particles introduced in the DMFV model, along with the coupling matrix λ and the SM right-handed quarks

	$U(3)_{u_R}$	$U(3)_\chi$	$U(3)_c$	$U(1)_{\mathrm{EM}}$
u_R	3	1	3	$2/3$
χ	1	3	1	0
ϕ	1	1	3	$2/3$
λ	3	$\bar{3}$	1	0

particles and the coupling matrix under various symmetry groups (and their electric charge)—the idea of λ "transforming" under the U(3) flavour symmetries is to be understood by considering λ as a spurion field [17].

The new physics Lagrangian reads

$$\mathcal{L}_{\mathrm{NP}} = \bar{\chi}(i\slashed{\partial} - m_\chi)\chi + D_\mu\phi(D^\mu\phi)^\dagger - m_\phi\phi^\dagger\phi - (\lambda_{ij}\bar{u}_{R,i}\chi_j\phi + \mathrm{h.c.}) , \quad (4.1.2)$$

giving the vertices shown in Fig. 4.1. Note that a coupling between the mediator and the Higgs as well as a mediator self-coupling are allowed by the symmetries of the model, but we neglect them in this work. It was shown in [18] that coupling matrix can be written in the form

$$\lambda = U_\lambda D_\lambda \tag{4.1.3}$$

with the matrices D_λ and U_λ parameterised as (defining $c_{ij} \equiv \cos\theta_{ij}$, $s_{ij} \equiv \sin\theta_{ij}$)

$$U_\lambda = \begin{pmatrix} c_{12}c_{13} & s_{12}c_{13}e^{-i\delta_{12}} & s_{13}e^{-i\delta_{13}} \\ -s_{12}c_{23}e^{i\delta_{12}} - c_{12}s_{23}s_{13}e^{i(\delta_{13}-\delta_{23})} & c_{12}c_{23} - s_{12}s_{23}s_{13}e^{i(\delta_{13}-\delta_{12}-\delta_{23})} & s_{23}c_{13}e^{-i\delta_{23}} \\ s_{12}s_{23}e^{i(\delta_{12}+\delta_{23})} - c_{12}c_{23}s_{13}e^{i\delta_{13}} & -c_{12}s_{23}e^{i\delta_{23}} - s_{12}c_{23}s_{13}e^{i(\delta_{13}-\delta_{12})} & c_{23}c_{13} \end{pmatrix},$$

$$(4.1.4)$$

$$D_\lambda = \begin{pmatrix} D_{11} & 0 & 0 \\ 0 & D_{22} & 0 \\ 0 & 0 & D_{33} \end{pmatrix},$$

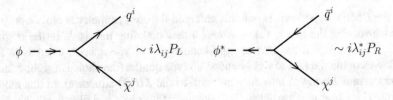

Fig. 4.1 Feynman rules for the interaction in Eq. 4.1.2

where $\theta_{ij} \in [0, \pi/4]$ to avoid double counting the parameter space, and we require $D_{ii} < 4\pi$ for a perturbative theory.

The presence of complex couplings ($\delta_{ij} \neq 0$) creates a violation of CP symmetry (note this is also permissible in the MFV assumption, so long as the complex phases are flavour-blind [28]). Due to the stringent constraints from electric dipole moments (EDM) in the presence of CP violation [17] we will set $\delta_{ij} = 0$ throughout. In total we then have a 10 dimensional parameter space:

$$\{m_{\chi,1}, m_{\chi,2}, m_{\chi,3}, m_\phi, \theta_{12}, \theta_{13}, \theta_{23}, D_{11}, D_{22}, D_{33}\}. \tag{4.1.5}$$

Other than those mentioned above, the only other limit we place on our parameters is $m_\chi, m_\phi \gtrsim 1$ GeV, so that the DM is a conventional WIMP candidate and the mediator is sufficiently heavy to decay to at least the up and charm quarks.

Although the masses of the DM fields and mediator field are in principle arbitrary free parameters, one must impose $m_{\chi,\min} < m_\phi + m_q$ (where m_q is the lightest quark to which $m_{\chi,\min}$ couples) to ensure χ cannot decay. Similarly we must have $m_\phi > m_{\chi,\min} + m_q$, which ensures the mediator has at least one decay channel and prevents it obtaining a relic abundance itself.

It can be shown additionally that a residual $\mathbb{Z}_3$ symmetry exists in the model [18, 29], which prevents either χ or ϕ decaying into purely SM particles. This useful symmetry argument ensures the relic DM (the lightest of the three) is completely stable even once non-renormalisable effects are considered. It is possible for the heavier χ fields to decay to the lightest χ (DM)—in fact the rate of such decays are always large enough to totally erase the relic density of the heaviest two DM.

Finally we briefly mention some interesting behaviour of the widths of our new particles. First, the mediator width Γ_ϕ can be shown to be very narrow, with $\Gamma_\phi/m_\phi \leq \frac{9}{128\pi} \lesssim 1\%$ even in the limit of non-perturbative couplings. Secondly, for small mass splittings ($m_{\chi_i} = m_{\chi_j}(1 + \epsilon)$) the decay rate $\chi_i \to \chi_j + q\bar{q}$ scales as ϵ^5, which is important when we consider the relic abundance of the different DM species.

4.2 Relic Density

4.2.1 Relic Density with Coannihilations

As mentioned in the introduction, the relic density (RD) of DM is currently measured to a very high accuracy by the Planck collaboration [30], and reproducing this is a must for any DM model.

In our model with three possible DM candidates, with potentially almost degenerate masses, we follow the results of [31]—Section III in particular deals with the effects of coannihilations (processes with $\chi_i\chi_j \to$ SM, $i \neq j$). In that work, the authors describe how coannihilations can be very important, and can be included in the "standard" computation [32–34] of relic density through the use of an effective annihilation cross-section $\langle \sigma v \rangle_{\text{eff}}$, defined in Eq. (12) of [31]. We will not reproduce all the detail from that paper here, but summarise the key results.

To compute the relic density, one first finds the freeze-out temperature $x_f \equiv m/T_f$ by solving the equation

$$e^{x_f} = \sqrt{\frac{45}{8} \frac{g_{\text{eff}} m_\chi M_{\text{pl}} \langle \sigma v \rangle_{\text{eff}}}{2\pi^3 g_*^{1/2} x_f^{1/2}}}, \tag{4.2.1}$$

with g_{eff} an effective number of degrees of freedom of the near-degenerate DM candidates, M_{pl} the Planck mass, g_* the total number of relativistic degrees of freedom at freeze-out. The relic density itself can then be written

$$\Omega h^2 = 2 \times 1.04 \times 10^9 \frac{x_f}{\sqrt{g^*} M_{\text{pl}} \left(a_{ii} I_a + 3 b_{ii} I_b / x_f \right)}, \tag{4.2.2}$$

where a_{ii} and b_{ii} are the s-wave and p-wave terms of $\langle \sigma v \rangle_{ii}$ (the cross section for the relic, plus any particles with degenerate mass), and $I_{a,b}$ are temperature integrals.

These results are found through certain simplifications of the full Boltzmann equations, and follow the full numerical calculation very closely.

4.2.2 The Generation of Mass Splitting

Almost degenerate DM masses mean the mass splittings ($\Delta m = m_{\chi_i} - m_{\chi_j}$) between the different χ_i are important to determining the true value of the DM relic density.

We can follow two regimes which distinguish the various possibilities by the dominant effect on the signals they generate:

1. The mass splitting is non-zero, the lightest of the χ_i survives as the relic. This holds as long as the splitting is large enough to accommodate any kind of decay.
2. The masses are truly degenerate, equivalent to a degeneracy which is sufficiently small to prevent decay, i.e. $\Delta m \leq 4 \, \text{MeV}$. In this case, the three DM particles obtain equal relic abundances, with the total affected primarily by their coannihilations.

The difference between the effective cross-section method mentioned above and a full solution of the coupled Boltzmann equations, and the effect of degenerate masses is shown on the left of Fig. 4.2. We see that the effective cross section approach correctly reproduces the relic density of the lightest candidate at late times, and that relic density constraints are not hugely sensitive to the mass splitting if it is non-zero.

As the final relic density depends sensitively on whether a mass splitting in the candidates exists or not, we briefly talk about how such a splitting can arise. Splittings can arise from two sources—a tree-level contribution where m_{χ_i} and m_{χ_j} are split by mass terms of the form $\mathcal{O}(1) \times (\lambda^\dagger \lambda)_{ii}$, or a loop-level contribution from

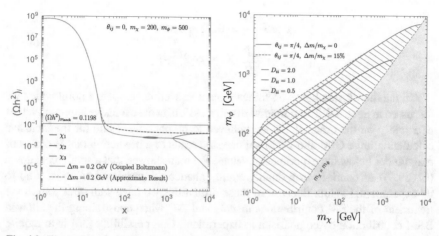

Fig. 4.2 Illustration of relic density over time ($x = m_\chi/T$) as freeze out occurs (left), and the RD bounds with mass splitting calculated with the effective method mentioned in the main text (hatched regions for which the DMFV models allows the correct relic abundance) (right)

renormalisation where the coefficient is instead of the order $N_c/(16\pi^2)\log(\mu^2/\Lambda^2)$ multiplied by the tree-level couplings $(\lambda^\dagger\lambda)_{ii}$ with Λ some high scale at which the masses are universal, and μ a low scale at which we wish to use the mass (e.g. for direct detection this could well be the nuclear scale of around 1 GeV). Explicitly, the resulting shift in the DM mass will be given by

$$m_{\chi_i}(\mu) = m_\chi(\Lambda)\left(1 + \frac{N_c}{16\pi^2}(\lambda^\dagger\lambda)_{ii}\log\left(\frac{\Lambda}{\mu}\right) + \mathcal{O}\left((\lambda^\dagger\lambda)_{ii}^2\right)\right). \qquad (4.2.3)$$

Note that because of our parameterisation of the coupling matrix, $\lambda^\dagger\lambda$ is diagonal, with elements D_{ii}^2.

Relatively large splittings can be generated this way—with a high scale of 100 TeV, then the coefficient of $(\lambda\lambda^\dagger)_{ii}$ can be as large as ~ 0.35. We explore the effect of mass splitting in our work by manually setting the mass splitting $(\Delta m/m_\chi)$ to a large (15%) and small (2%) value.

4.3 Flavour Constraints

4.3.1 Mixing Observables

Since our model introduces couplings to the up type quarks, we would expect new physics effects in the charm meson sector—in particular in neutral D^0 mesons. For the case of D mesons mixing, the current experimental averages from HFLAV are [35]

$$x \equiv \frac{\Delta M}{\Gamma} = 0.32 \pm 0.14\%\,,$$

$$y \equiv \frac{\Delta \Gamma}{2\Gamma} = (0.69^{+0.06}_{-0.07})\%\,. \tag{4.3.1}$$

On the theory side however, things are not so well developed, a point which we discussed in Sect. 3.4. To summarise that discussion, in the exclusive approach values of x and y on the order of 1% are believed to be possible, while for the inclusive calculation huge GIM and CKM suppression leads to a prediction that is orders of magnitudes below the experimental values. We then showed that a small breakdown ($\mathcal{O}\,(20\%)$) of quark-hadron duality could enhance the predicted value of y up to its experimental value. Because of these difficulties we have some freedom in the treatment of the SM contributions to ΔM and $\Delta \Gamma$ when constraining the allowed BSM contribution by comparison to experiment. One possibility [36] is to require that

$$x^{\text{NP}} = \frac{2|M_{12}^{\text{NP}}|}{\Gamma_D} \leq x^{\text{exp, upper limit}}\,, \tag{4.3.2}$$

taking the $1\,\sigma$ upper limit reported by HFLAV (Eq. 4.3.1). This is the limit that would be derived if the NP and SM contributions have roughly the same phase, so that

$$|M_{12}^{\text{NP}} + M_{12}^{\text{SM}}| = |M_{12}^{\text{NP}}| + |M_{12}^{\text{SM}}|\,, \tag{4.3.3}$$

since $\Delta M \leq 2|M_{12}|$. Our NP contribution to M_{12} is given by

$$M_{12}^{\text{NP}} = -\frac{f_D^2 B_D M_D}{384 m_\phi^2 \pi^2} \sum_{i,j=1}^{3} F\left(\frac{m_{\chi_i}^2}{m_\phi^2}, \frac{m_{\chi_j}^2}{m_\phi^2}\right) \lambda_{1i} \lambda_{1j} \lambda_{2i}^* \lambda_{2j}^* \tag{4.3.4}$$

where we take the decay constant f_D from FLAG [37–39], the D mixing bag param-eter B_D from [40], and the loop function F is given by

$$F(x_i, x_j) = \frac{1}{(1 - x_i)(1 - x_j)} + \frac{x_i^2 \log x_i}{(x_i - x_j)(1 - x_i)^2} - \frac{x_j^2 \log x_j}{(x_i - x_j)(1 - x_j)^2}\,.$$

The important result is that $M_{12} \propto ((\lambda \lambda^\dagger)_{12})^2$ for degenerate DM masses. The matrix $(\lambda \lambda^\dagger)$ is diagonal if D_{ii} are all equal, or if $\theta_{ij} = 0$ (no mixing between quark flavours) and then the flavour constraints disappear.

Using the upper $1\,\sigma$ value of the experimentally measured x_D leads to bounds as shown on the left of Fig. 4.3, these bounds can be very strong and significantly exclude almost all masses $m \lesssim 1$ TeV for large couplings $\lambda \gtrsim 0.1$ unless one fine-tunes the model to remove $(\lambda \lambda^\dagger)_{12}$.

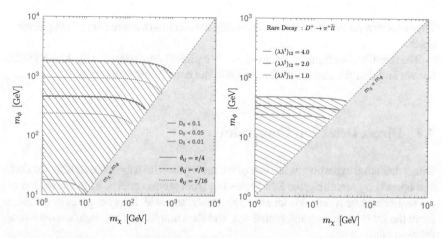

Fig. 4.3 Excluded regions (hatched) for which the value of ΔM from DMFV diagrams exceeds the $+1\,\sigma$ contour of the experimental result (left). The bounds are the most constraining possible given the limits on D_{ii}, and can be made arbitrarily small by adjusting the values (for example with equal values $D_{ii} = D$). The exclusions from $|C'_9| < 1.3$ varying $(\lambda\lambda^\dagger)_{12}$ (right)

4.3.2 Rare Decays

We consider the semileptonic decay $D^+ \to \pi^+ \mu^+ \mu^-$, whose short distance contribution comes from the quark level decay $c \to u\mu^+\mu^-$. This decay is loop and GIM suppressed in the SM, and so should have good sensitivity to new physics. In our model contributions are no longer GIM suppressed, coming from electroweak penguin diagrams with our new particles in the loop.

In [41] rare charm decays are examined to provide limits on the Wilson coefficients of an effective theory—they look at $D \to \mu^+\mu^-$ as well as $D^+ \to \pi^+\mu^+\mu^-$ and find the latter to place the strongest bounds for the coefficients relevant in our model. Matching onto their EFT, and neglecting the Z penguin since the momentum transfer is small, we find only the C'_7, C'_9 coefficients are non-zero, corresponding to the operators.

$$Q'_7 = \frac{em_c}{16\pi^2}(\bar{u}\sigma_{\mu\nu}P_L c)F^{\mu\nu}, \quad Q'_9 = \frac{e^2}{16\pi^2}(\bar{u}\gamma^\mu P_R c)(\bar{\ell}\gamma_\mu \ell), \quad (4.3.5)$$

(our full expressions for the Wilson coefficients can be found in Appendix E.1).

Since the SM branching ratios for the D^0 decay suffer from a strong GIM cancellation, we would expect strong constraints on the flavour breaking terms of the DMFV model. As with the mixing observables, the rare decay process is primarily sensitive to $(\lambda\lambda^\dagger)_{12}$ in the limit of degenerate DM mass. On the right of Fig. 4.3 we show the bounds coming from limits on the Wilson coefficients for $(\lambda\lambda^\dagger)_{12} = 1, 2$ and 4. The bounds on the individual Wilson coefficients are $|C_i| \sim 1$ (see Table II of [41]). Mediators up to $m_\phi \sim 50\,\text{GeV}$ can be ruled out for couplings $D_{ii} \sim (\lambda\lambda^\dagger)_{12} \sim \mathcal{O}(1)$.

These constraints are therefore substantially weaker than from meson mixing observables.

The rare flavour-changing decays $t \to u/c\gamma$ have been measured by ATLAS [42], but we find that the current limits are again not constraining on our model.

4.4 Direct Detection Constraints

Direct detection experiments are one of the most powerful ways of searching for DM, and operate by searching for DM scattering from atomic nuclei. The calculation of the scattering rate is done via an effective theory, where all heavy degrees of freedom (save the DM) have been integrated out, and then amplitudes are matched onto four fermion operators.

We choose to examine data from LUX [43, 44] and CDMSlite [45], which together (at the time of this work) provided the best constraints over the range of DM masses we are looking at. LUX uses liquid xenon as a target, which detects DM with masses above 5 GeV while scattering from DM masses below this is kinematically impossible; CDMSlite is a germanium detector, and best constrains particles with masses between 1.6 GeV and 5.5 GeV. Details of our exact method can be found in Appendix E.2—here we merely state that we use a Poisson probability distribution for both, comparing the number of observed events in each bin to our predicted signal plus background.

At tree level, the only EFT operator which arises from our model is given by a diagram with t-channel ϕ exchange. We only consider the scattering amplitudes in which the incoming and outgoing DM (and quark) are the same flavour, as this avoids the computation of (possibly unknown) hadronic matrix elements of quark currents $\bar{q}_i \Gamma q_j$ for $i \neq j$. The operator in question is

$$\mathcal{L}_{\text{EFT}} = C_{ij}(\bar{\chi}_L^i \gamma^\mu \chi_L^i)(\bar{q}_R^j \gamma_\mu q_R^j) \,, \quad C_{ij}(\mu \sim m_\phi) = \frac{\lambda_{ji} \lambda_{ji}^*}{2((m_\chi - m_q)^2 - m_\phi^2)} \quad (4.4.1)$$

where the Mandelstam variable t has been replaced by its low velocity expansion and we have performed a Fierz transform (see Appendix A).

Vector and axial-vector currents probe the valence quark content and spin distribution respectively of the scattered nucleon, and so would naively be small for non-valence quarks (i.e. c and t). However, there are 1-loop diagrams (see Fig. 4.4) that mix operators with heavy quarks into those with up and down quarks, and in the case of heavy mediators RG running down to the direct detection scale ($\mu \sim 1$ GeV) also alters the relative coupling to nuclei. This calculation has been done in [46, 47], and we find (see Fig. 4.5) that DM that couples to heavy quarks at the mediator scale will mix into up quark coupling at the low scale with up to 10% of its high scale coupling strength; tree-level scattering is therefore substantial (as can be seen in Fig. 4.6), even in the case of only coupling to heavy quarks. The spin-averaged cross

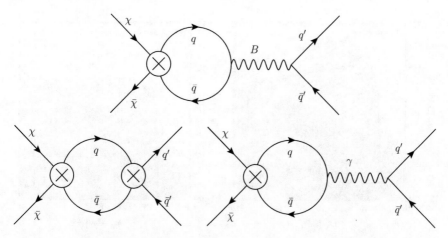

Fig. 4.4 The divergent loop diagrams responsible for mixing between the quark vector and axial vector currents ($\bar{\chi}\Gamma\chi\bar{q}\Gamma q$) above the EW scale (top) and below (bottom). The most important aspect is the mixing of high-scale heavy quark currents $q = c, t$ onto light quark vector currents $q' = u, d$, thus enabling a strong scattering cross section with nuclei

Fig. 4.5 The effect of the RG running from a high scale $\Lambda = m_\phi$ down to the nuclear scattering scale $\mu_N = 1$ GeV

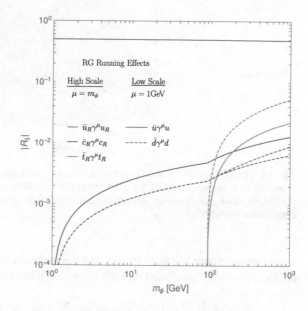

section is parameterised by a series of nuclear form factors $F_{ij}^{(N,N')}$ [48], which are functions of the incident DM velocity squared v^2 and the momentum transfer q^2,

$$\langle |\mathcal{M}|^2 \rangle \sim \sum_{i,j,N,N'} C_i^{(N)} C_j^{(N')} F_{ij}^{(N,N')}(v^2, q^2) \qquad (4.4.2)$$

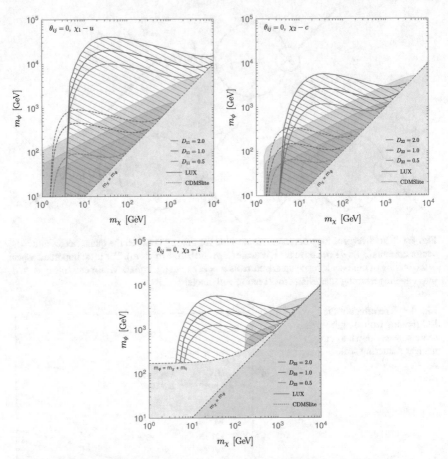

Fig. 4.6 The DD bounds for three coupling choices—χ_1 exclusively coupling to u quarks, χ_2 to c, and χ_3 to t. Bounds for LUX (CDMSlite) are solid (dashed), and the filled region shows the parameters which give the correct relic abundance. Constraints are based on the dominant tree-level contribution to scattering

where we sum over the form factors and the nucleons N, $N' = p, n$. The nucleon coefficients above are related to our Wilson coefficients by

$$C_1^{(p),i}(\mu \sim 1\,\text{GeV}) = 4m_i m_N \sum_j (2R_{ju} + R_{jd})C_{ij}(m_\phi) \qquad (4.4.3)$$

$$C_1^{(n),i}(\mu \sim 1\,\text{GeV}) = 4m_i m_N \sum_j (2R_{jd} + R_{ju})C_{ij}(m_\phi) \qquad (4.4.4)$$

where R_{ju} (R_{jd}) gives the magnitude of the running of operator $\bar{q}_R^j \gamma^\mu q_R^j$ onto $\bar{u}\gamma^\mu u(\bar{d}\gamma^\mu d)$, and we have quoted the $i = j = 1$ relation since the corresponding

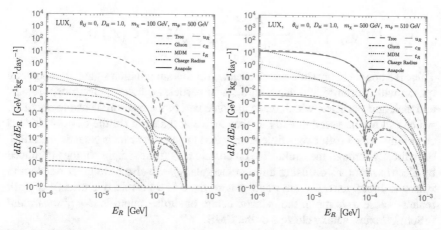

Fig. 4.7 The differential scattering rate in recoil energy for DM-nuclear scattering at LUX. Each of the quark contribution are plotted separately, the rates are also separated according to the way in which they scatter. The right plot represents a model with almost complete degeneracy between the DM and mediator mass, where the loop level interactions become important

form factor has the dominant scaling behaviour. i and j run over the DM and quark flavours respectively. The dependence of the R_{jq} parameters on the high scale (which we take to be the mediator mass) is shown in Fig. 4.5.

At loop-level, there are various new operators that arise—in general these are highly suppressed, but we include them both because they can become dominant in particular regions of parameter space (see Fig. 4.7) and for completeness. The operators we consider are photon operators [49, 50] (which in the non-relativistic limit correspond to the charge-radius, magnetic dipole moment, and anapole moment); Z penguins [49]; and those for DM-gluon [51–53] scattering. We reproduced the quoted literature results as a check.

The more recent null results from XENON1T [54] and PandaX-II [55] push the constraining potential of direct detection even further—nearly an order of magnitude stronger in cross-section, which translates into a factor of ~2 in mediator mass.

4.5 Indirect Detection Constraints

4.5.1 Basics of Indirect Detection

Indirect detection experiments looks for signs of annihilating and/or decaying DM coming from astrophysical sources, such as the centre of galaxies where DM density is largest. The constraints are based around limits on the annihilation cross-section of DM to SM particles—in our model the main limits come from annihilation to quark pairs.

$$\langle \sigma v \rangle_{\bar{\chi}_i \chi_j \to \bar{q}_l q_m} \approx \frac{N_c m_\chi^2}{32\pi (m_\chi^2 + m_\phi^2)^2} \left(\lambda_{mj} \lambda_{li}^* \right)^2 + \mathcal{O}\left(v^2 \right). \tag{4.5.1}$$

There is a bounty of possible search avenues for this annihilation signal; the energetic quarks will hadronise and decay into stable particles (photons, electrons, protons, and their anti-particles, which make up some part of the measured cosmic ray flux), which can be measured directly as they arrive at the earth (in the case of photons especially, which suffer very little energy loss to galactic or inter-galactic material), or indirectly through their influence on cosmic rays (for example photons produced by electrons/protons diffusing through the galaxy). We also have great freedom in where to look; generally anywhere where there is a cosmic overdensity of dark matter—close to home in the galactic centre or further afield in *dwarf spheroidal* (dSph) galaxies, galaxy clusters or the CMB.

Underlying all these is Eq. 4.5.1 and so ID constraints are frequently quoted as confidence limits on the thermally averaged annihilation cross section $\langle \sigma v \rangle_{\bar{f}f}$ into fermions of the same flavour, covering a mass range in m_χ from 1 GeV to 100 TeV. The ID signals from heavy quarks ($q = c, b, t$) are very similar (see Figs. 3 and 4 in [56]), and it is uncommon to find constraints on c, t final states (more common is the b). The primary spectra of electrons, positrons, anti-protons, deuteron and neutrinos are extremely similar between c, b, t quarks, and thus any constraints which look for these particles from DM annihilations will be approximately heavy-flavour independent. The situation is depicted in Fig. 4.8.

It should be noted that the relative strength of these constraints is not robust; different authors use different halo profiles, different astrophysical parameters and are subject to varying degrees of uncertainty, some significantly larger than others, it is beyond the scope of this work to accommodate all these effects and compare

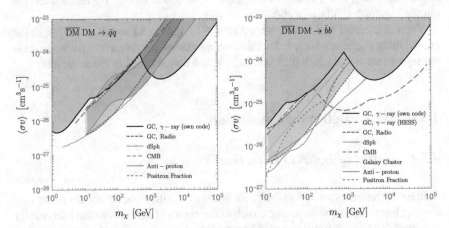

Fig. 4.8 The constraints on $\langle \sigma v \rangle_{\bar{q}q}$ for $q = u, d, s$ (left) and $q = b$ (right) which is representative of $q = c, t$ for $m_\chi > m_{c,t}$. The constraints are taken from many different sources (DSph, galactic centre, clusters) and targets (gamma rays, radio waves, positron, anti-protons)

constraints on a like-for-like basis and so what we present should be taken as representative but not precise. We will use the $\bar{b}b$ final state as representative for constraints based on dSph [57] and anti-proton measurements of AMS-02 [58] which dominate other constraints such as those based on other particle targets, such as the positron fraction [59] or neutrinos [60] and also those based on the galactic centre [61], or galaxy clusters [62].

4.5.2 Gamma Rays (and Other Mono-chromatic Lines)

At the one-loop level, the pair production of quarks from annihilating DM can pair produce photons at a fixed energy $E_\gamma = m_\chi/2$ via a box diagram. We calculate this cross-section using an EFT where the mediator has been integrated out, in which limit only the axial vector operator $(\bar{\chi}\gamma^\mu\gamma^5\chi)(\bar{q}\gamma_\mu\gamma^5 q)$ contributes to the s-wave annihilation, with cross section

$$\langle\sigma v\rangle_{\gamma\gamma} = \frac{16\alpha^2 s}{9468(m_\chi^2 - m_\phi^2)^2\pi^4}\left(1 + 2m_f^2 C_0\right)^2, \qquad (4.5.2)$$

where $s \approx 2m_\chi^2$ is the centre of mass energy of the annihilating DM, and C_0 is the scalar integral $C_0(0, 0, s; m_f^2, m_f^2, m_f^2)$ in LoopTools notation [63].

As well as $\gamma\gamma$ final states, there will be γX final states where $X = Z, h$ for example and these also provide constraints. The presence of a massive particle recoiling against the photon shifts the energy to $E_\gamma = m_\chi(1 - m_X^2/4m_\chi^2)$, but still creates a mono-energetic line signature. We show some results from the indirect searches in Fig. 4.9—we see that indirect searches can be quite powerful, especially in the case of large coupling to top quarks.

4.6 Collider Constraints

Our DMFV model contains a new particle with colour charge, and so we expect there to be significant limits coming from collider experiments. In addition we also have DM which can be searched for in final states with missing energy, and current LHC data can also place limits on the mass of invisible particles. In the past, DM model builders have used effective field theories (EFTs) to analyse NP at colliders, but in recent years it has become clear that the regions of validity of these EFTs at high energy machines such as the LHC are so small as to be almost useless [4, 11, 12]. We briefly detail in the next section this point for our particular model, before moving on to a more complete analysis.

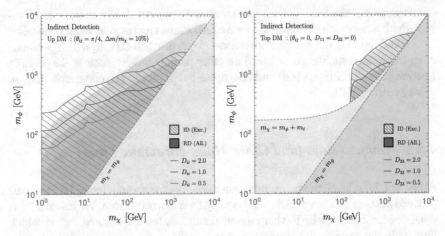

Fig. 4.9 The ID constraints on DMFV model, with 'maximal' mixing $\theta_{ij} = \pi/4$ (left), or for couplings to top quarks only (right), assuming degenerate DM masses. Bounds are produced on individual final states, and therefore scale with the dominant annihilation channel, somewhat surprisingly the top quark channel gives stronger constraints due to the extremely sensitive γ-ray search by H.E.S.S [61]

4.6.1   EFT Limit

In [64] the validity of the EFT approximation for t-channel mediators is quantified by R_Λ, which they define as the ratio of the cross section with the constraint $t < \Lambda^2$ applied to the total cross section (i.e. the total proportion of the cross section which is valid under the EFT assumption). The lines of $R_\Lambda = 0.50$ are plotted alongside the EFT limits taken from ATLAS [65] (the R_Λ contour assumes $|\eta| < 2$ and $p_T < 2$ TeV, the ATLAS results assumed the same range of η, but allow $p_T \lesssim 1.2$ TeV). It is worth noting that the authors of [64] produce results with the limit $g \lesssim 1$, the bounds become significantly weaker by using $g \lesssim 4\pi$ which then permit a small region of validity as shown in Fig. 4.10. The EFT breaks down entirely for $g \lesssim 1$. Thus the EFT approximation cannot be justified in our analysis and we turn to the simulation of the full cross section.

4.6.2   LHC Bounds

To try and cover a large range of constraints, we look at three different LHC processes that could place limits on our model—monojet with missing energy searches, where a single jet recoils off DM pair production; dijet searches with missing energy; and stop searches. The latter are relevant to our model as we have a coloured scalar coupling to top quarks and DM, in analogy with the e.g. stop-top-neutralino vertex

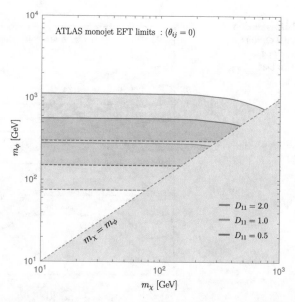

ATLAS monojet EFT limits : $(\theta_{ij} = 0)$

Fig. 4.10 The EFT approximation breaks down beneath the dashed lines (which are the $R_\Lambda = 0.5$ contours with $g \lesssim 4\pi$), while ATLAS excludes below the solid lines, and so only the shaded regions can robustly be excluded using the EFT

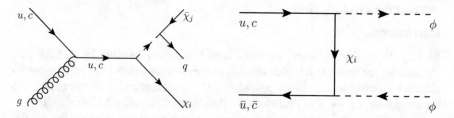

Fig. 4.11 Example Feynman diagram for the monojet (left) and dijet (right) processes

in many supersymmetric theories, and so provide sensitivity to the $\phi - t$ coupling D_{33}.

In Fig. 4.11 one example Feynman diagram that generates monojet and dijet signals is shown—in the dijet case the decay of the mediator into quark plus DM is not shown. Other diagrams that contribute can be seen in Appendix E.3.

We produce our collider constraints using MadGraph [66], replicating, except where noted below, the experimental cuts used by the experiments.

Monojet Searches

In our analysis, we use the most recent monojet search by ATLAS [67] (which uses the Run 2 data ($\sqrt{s} = 13\,\text{TeV}$ and $\mathcal{L} = 3.2\,\text{fb}^{-1}$)), along with a similar analysis performed by CMS [68] with the Run 1 data ($\sqrt{s} = 8\,\text{TeV}$ and $\mathcal{L} = 19.7\,\text{fb}^{-1}$). The total cross section as a function of m_ϕ for a benchmark scenario is shown in

Fig. 4.12 Total cross section for the seven signal regions of the ATLAS monojet search [67] for two DM masses

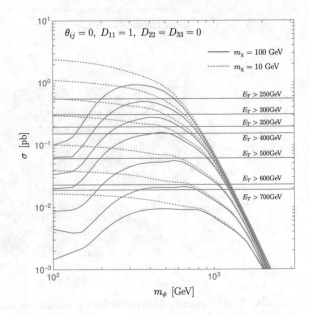

Fig. 4.12 with the ATLAS limits overlaid, and the constraints on our model are shown in the top of Fig. 4.13.

Dijet Searches

Moving on to dijet searches, we use a Run 1 and Run 2 search by ATLAS [69, 70] looking for multiple jets plus missing energy—we restricted our comparison to the 2-jet searches which should provide the strongest constraint. In our model, the process $pp \to \phi\phi \to \bar{\chi}\chi jj$ provides the dominant contribution to this signal.

We replicate all the main selection cuts for both analyses, in particular for the Run 1 comparison: $E_t^{\text{miss}} > 160\,\text{GeV}$, $p_{T,(1,2)} > 130, 60\,\text{GeV}$, $\Delta\phi > 0.4$ (between the jets and missing momentum), and for Run 2 similar cuts are applied (full detail in Table 2 of [70]). The different signals regions (tj1, tjm, tjt) also include a

Table 4.2 Lower limits (at 95% CL) on the visible cross section for three signal regions (SR) in the Run 1 ATLAS dijet plus missing E_T search [69] (top), and ATLAS dijet search from Run 2 [70] (bottom)

SR	N_{obs}	N_{SM}	$N_{\text{n.p.}}$	σ_{obs} / fb
tj1	12 315	13000 ± 1000	15–704	60
tjm	715	760 ± 50	15–59	4.3
tjt	133	125 ± 10	22–50	1.9
tj1	263	283 ± 24	12–37	16
tjm	191	191 ± 21	15–58	15
tjt	26	23 ± 4	10–22	5.2

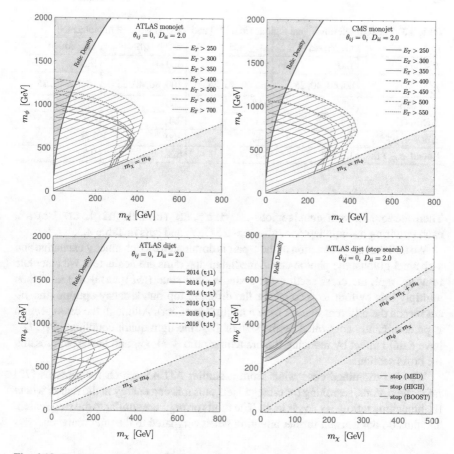

Fig. 4.13 Exclusion regions for different signal regions in the ATLAS (top left) and CMS (top right) monojet analyses, ATLAS dijet searches (bottom left), and ATLAS stop searches (bottom right)

minimum requirement for m_{eff} and $E_T/\sqrt{H_T}$, which are defined as

$$H_T = |p_{T,1}| + |p_{T,2}|,$$
$$m_{\text{eff}} = H_T + E_T,$$

which we implement in MadGraph manually (again, see the respective papers for the cuts in each case). The constraints this places on our model parameters are shown in the bottom left of Fig. 4.13 for the case of no mixing and strong couplings for all DM particles.

ATLAS 2014 Stop Search

Lastly, a study by ATLAS [71] considers a set of cuts optimized for the detection of stops—the signal consists of a lepton in the final state along with four or more jets.

Table 4.3 The four relevant signal regions from [71] and the cuts we have implemented

Cut	tN_diag	tN_med	tN_high	tN_boost
E_T^{miss} / GeV	100	200	320	315
$p_{T,i}^j$ / GeV	60, 60, 40, 25	80, 60, 40, 25	100, 80, 40, 25	75, 65, 40, 25
m_T / GeV	60	140	200	175
$\Delta R(b,l)$	0.4	0.4	0.4	0.4
$\Delta\phi(j_{1,2}, p_T^{\text{miss}})$	60	140	200	175
Bound σ_{vis} / fb	1.8–2.9	0.4	0.3	0.3

There are four relevant signal regions tN_diag, tN_med, tN_high, tN_boost, each requiring a single lepton with $p_T^l > 25$ GeV, and cuts in Table 4.3.[1]

We find that the production of the ϕ pair is dominated by t-channel χ exchange and s-channel gluons; the photon and Z mediated diagrams are neglected. We calculate in MadGraph the cross-section for a single final state $((\bar{b}b)(\bar{d}u) + e^-)$, and then multiply this by four to account for the different top quark decay options (the p_T cut means the different masses have a negligible effect). Although the cross section is predominantly controlled by the size of D_{33}, the light quark couplings D_{11}, D_{22} have a mild affect by reducing the branching ratio $\phi \to t\bar{\chi}_i$ and hence suppressing the cross section.

We also examined constraints from a similar ATLAS search for scharms [72] rather than stops, searching for c-tagged jets plus missing energy in the region where the branching ratio $\phi \to c\chi_i$ is large. The limits on $m_{\phi,\chi}$ are similar to the stop search, and thus do not warrant further attention when compared to the dijet searches.

4.6.3 Collider Constraints Within DMFV

We have now looked at three classes of analysis: monojet searches, dijet searches, and searches optimised for a stop. Within our model we have couplings to u, c, t (which we denote here by $\lambda_{u,c,t}$) and the relative strengths of these dictate which signals will be dominant.

Compared to λ_u, the monojet and dijet processes are suppressed by pure λ_c (due to the charm parton distribution function (PDF)), but generally are enhanced by mixtures of $\lambda_{u,c}$. The coupling λ_t reduces the signals since they dominantly come from s-channel ϕ resonances and thus the branching ratio to u and c jets is $\propto (D_{33})^{-2}$ if $\lambda_t \gg \lambda_{u,c}$. The stop search only becomes relevant for large λ_t with $\lambda_t/\lambda_{u,c} > 1$,

---

[1]We do not include the cuts on the parameters am_{T2} and m_{T2}^τ. From the published cut flows it can be seen that the effect of these cuts is of the order 10% and 2% respectively (although the former cut can have a more pronounced effect $\sim$30% on the tN_med cut choice).

and increasing $\lambda_{u,c}$ suppresses the signal as the branching fraction to top quarks is reduced.

- **Mostly up type**: The dominant signal will come from the monojet processes which have the least QCD suppression and which require an up quark in the initial state. Dijet searches are also sensitive but it tends to be the monojet which sets the better constraint.
- **Mostly charm type**: The monojet processes are enhanced by the presence of charm couplings, however as the up coupling is reduced the monojet processes become suppressed because of the charm PDF by around a factor 10–100. The dijet processes are very similar as for u quarks but the largest contributing diagram is again suppressed by the charm PDF. Both searches provide constraints.
- **Mostly top type**: The monojet signal depends primarily on $\lambda_{u,c}$, only indirectly on λ_t though the widths. λ_t can be probed through stop searches with jet multiplicities of ≥ 4.

Colliders provide very powerful exclusions (up to the TeV scale in mediator mass), and cover the full model parameter space in coupling, although these can be significantly weakened by, for example, strong top couplings. The DM is produced *on shell*, and so the constraints are comparatively weak at high DM mass when compared with searches which depend on the cosmic abundance of DM; on the other hand the fact that the DM is produced in the collider releases any dependence on its abundance in the universe, thus allowing more powerful constraints on DM which has only a fraction of the full relic abundance (or none at all). Similarly, low mass DM is strongly excluded, whereas the most powerful astrophysical probe (direct detection) cannot detect much below the GeV scale due to kinematics.

When compared with the strongest direct detection limits, the collider limits are not as constraining, and this is not likely to change even with more luminosity and higher energy beams.

It is very difficult for a given parameter choice to determine the strongest bound from colliders, except in the extreme cases above, and one should therefore check all available searches as we have done. Due to the interplay between 1 and 2 jet processes, there is no obvious scaling behaviour of the cross section with the coupling parameters, these factors make implementing collider searches in an MCMC scan difficult and slow as each cross section must be numerically computed at each point in phase space.

4.7 Results

We have aimed to produce a robust statistical analysis of the eight dimensional parameter space of the DMFV model, using the Bayesian inference tool Multi-Nest [73–75] and its Python interface PyMultiNest [76] with 5000 live points. The motivation for carrying out this analysis is twofold, firstly from a practical standpoint it enables very quick and efficient algorithms for scanning a large dimensional parameter space, allowing us to include all parameters in one analysis. Secondly, a rudimentary "hit-or-miss" analysis leaves a large region of parameter space allowed, which is not surprising given the flexibility of 8 free parameters, with a statistical result we can quantify the regions of parameter space which are allowed but very improbable given the errors of the experimental data. For clarity, we represent the allowed parameters as contours containing credible regions, using the method of [77]; using the posterior probability density function. The 1 and 2 sigma contours give an indication of the allowed parameter range, with containment probabilities of 68 % and 95 % respectively.

Regarding the use of priors: We make one note of caution regarding the results; the credible regions depend sensitively on the choice of priors for the parameters. This is not surprising since our constraints allow large regions of parameter space to be equally well allowed, and so the use of priors which bias the parameters to lower values (i.e. log-uniform compared with linearly uniform) is reflected in the final result. Nonetheless, we are careful to limit the statements made in the text to those which are independent of the choice of priors. In all figures the log-uniform priors have been used for the masses and for D_{ii}, as this represents the more conservative choice. The ranges and priors for the parameters of the scan are summarized in Table 4.4.

Our results are summarized in Figs. 4.14, 4.15 and 4.16 as $2\,\sigma$ contours, and in Table 4.5 as one-dimensional $1\,\sigma$ intervals. We consider three separate samples in which the DM (the lightest χ) is the first, second and third member of the triplet (denoted 'up', 'charm' and 'top' DM). Within each sample we present a low and high mass splitting (2% and 15%), which primarily distinguish the effects caused by coannihilation in the calculation of relic density, but affect all other bounds to some extent as we have explicitly included the masses in each.

As we see from Fig. 4.14, the masses of the DM and mediator are both required to be in the TeV range, with upper limits in the tens of TeV. The DM and mediator masses are strongly correlated with the D_{ii}, as in Fig. 4.14, due to the relic density and mixing bounds which both scale approximately as $(D/m)^4$ in the high mass limit. Masses in the TeV range favour the D_{ii} to be $\gtrsim \mathcal{O}\,(1)$. The mixing angles are not well constrained in general; $\theta_{ij} = 0$ is favoured, but the full range of angles are usually allowed with $2\,\sigma$ credibility.

Table 4.4 Allowed ranges for the parameters used in the MCMC scan, along with the assumed prior likelihood, which is uniform on either a linear or logarithmic scale

Parameter	Range	Prior
m_χ / GeV	$1\text{--}10^5$	Log-uniform
m_ϕ / GeV	$1\text{--}10^5$	Log-uniform
θ_{ij}	$0\text{--}\frac{\pi}{4}$	Uniform
D_{ii}	$10^{-2}\text{--}4\pi$	Log-uniform

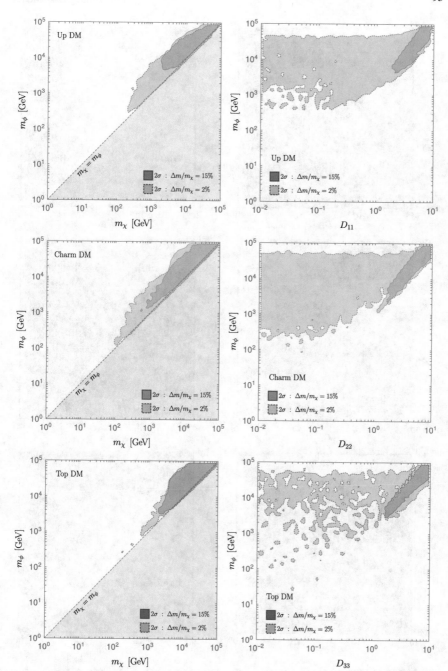

Fig. 4.14 Credible regions ($2\,\sigma$ contours) in the $m_\chi - m_\phi$ plane (left) and $D_{ii} - m_\phi$ (right) where the DM is χ_1 (top), χ_2 (middle) or χ_3 (bottom). Two values of a mass splitting are chosen, shown with solid and dashed contours respectively

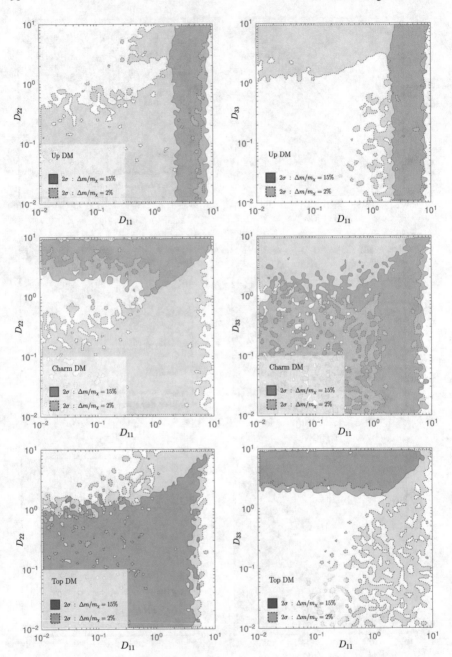

Fig. 4.15 As for Fig. 4.14 but for the $D_{11} - D_{22}$ plane (left) and $D_{11} - D_{33}$ (right), for two values of mass splitting (dashed shaded, and solid darker shaded respectively)

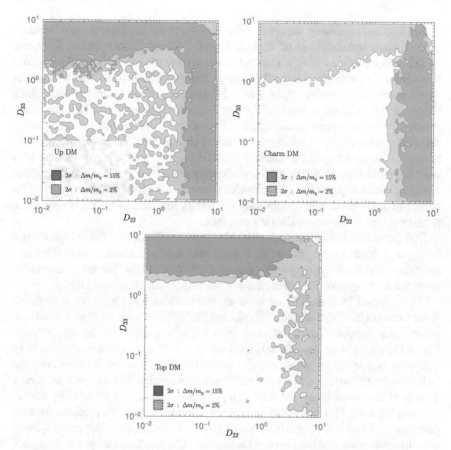

Fig. 4.16 As for Fig. 4.14 but for the $D_{22} - D_{33}$ plane

The D_{ii} themselves are highly correlated from the mixing constraints (see Figs. 4.15 and 4.16) which depend on $(\lambda\lambda^\dagger)_{12}$ which is approximately

$$(\lambda\lambda^\dagger)_{12} \approx \left(s_{13}s_{23}(D_{22}^2 - D_{11}^2) + s_{12}(D_{33}^2 - D_{11}^2)\right), \qquad (4.7.1)$$

and so we see $D_{11} \sim D_{33}$ (and less strongly $D_{11} \sim D_{22}$). Because the correlation between D_{22} and D_{33} is less pronounced, the RD bound controls the behaviour and produces an anti-correlation, since the annihilation cross section scales like

$$\langle\sigma v\rangle_{\text{eff}} \propto (D_{11}^2 + D_{22}^2 + D_{33}^2)^2 \sim 3 \times 10^{-26}\,\text{cm}^3\,\text{s}^{-1} \qquad (4.7.2)$$

due to coannihilations, as such the trend is most pronounced for small mass splitting. This is seen in the range of D_{22} for the small splitting data in Fig. 4.16.

In all cases, increasing the mass splitting reduces the available parameter space of the masses and couplings of the DM since the coannihilations and annihilations of the heavy particles have a reduced effect on the relic density (scaling with a Boltzmann factor $\exp(-\Delta m)$). This allows less flexibility in the DM parameters whilst potentially opening up the allowed parameters of the heavy particles, since their couplings are out of reach of the astronomical constraints (indirect and direct searches) which are proportional to the relic density of the lightest χ (scaling as Ω^2 and Ω respectively). This effect can be clearly seen in the right panels of Fig. 4.14, where the 2% splitting allows much smaller DM couplings compared with the 15% splitting, contrastingly in Fig. 4.15 (middle right panel) the non DM coupling space opens up with a larger splitting. Of course, since we have fixed the mass splitting by hand, the heavy particle parameters are not totally free, and so the parameter space is still reduced by the constraints we consider.

Top quark threshold effects are absent in the MCMC scan, due to the high masses ($m_\chi \gtrsim m_t$). Since $m_\chi, m_\phi \gg m_t$ the three quarks are kinematically equivalent, and so the bounds are not strongly dependent on the flavour of DM. The main differences arise due to the quarks SM interactions which impact the DD and ID limits.

As described in Sect. 4.6, we have studied collider bounds on our model, but these were not directly incorporated into our MultiNest routine as these bounds are much more computationally intensive than the others. However, as we see from Fig. 4.13, the collider bounds only rule out sub-TeV scale masses, even at large couplings and so we do not expect that a full likelihood function incorporating the LHC constraints would give significantly different results. As a test, we checked a sample of the points inside the 68% ($1\,\sigma$) credible regions and found only a small minority (of order 1%) that would be excluded by collider data. We produce, for each parameter, a marginalized posterior integrated over the remaining 7 parameters. From this distribution we find the $1\,\sigma$ credible interval. The results are shown in Table 4.5. This contains results for both uniform and log-uniform priors on D_{ii}, m_χ and m_ϕ; when the two cases are discrepant by $>1\,\sigma$ this is due to a flat posterior, and using the $2\,\sigma$ band instead the two agree.

4.7.1 Constrained Scenarios

We consider two extensions to the previous results:

1. In Sect. 4.2.2 we found that the mass splitting which is generated through RG running of the DM self-energy is approximately proportional to D_{ii}^2, this motivates us to consider a scenario in which the couplings D_{ii} are correlated with the masses (thus introducing a coupling splitting $\Delta D_{ii}/D_\chi \propto \Delta m_{ij}/m_\chi$). The reduced parameter space enforces almost degenerate couplings which leads to two important effects; firstly, it subjects all three χ to the astrophysical constraints of indirect and direct detection, despite the heavier particles having no relic density. By this we mean that, upon fixing the mass splitting, any limits on

Table 4.5 The 1D 1σ credible intervals for all 8 parameters of the MCMC scan. Colours denote the DM flavour (red, green and blue for up, charm and top DM respectively). The two prior choices on D_{ii} and the masses (log-uniform vs uniform) are shown as lines and shaded regions respectively, the modal average of the log-uniform prior choice is shown as a dot. The two mass splitting cases are contained in different panels

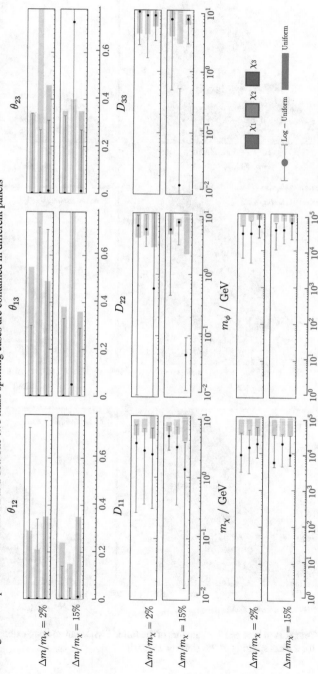

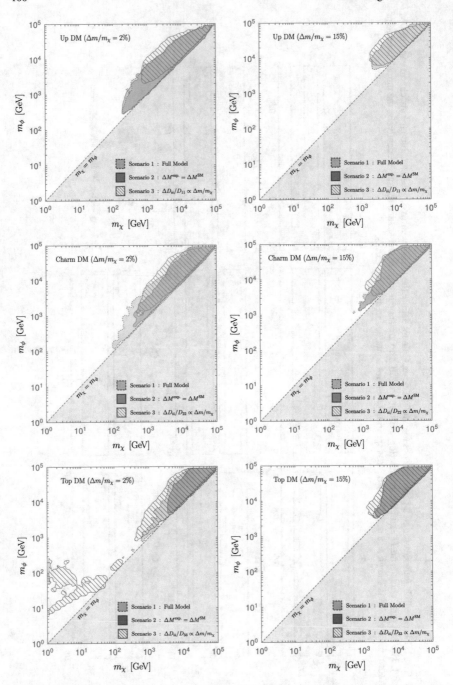

Fig. 4.17 Comparison between $2\,\sigma$ contours of the full MCMC scan and two extensions discussed in the text, for a mass splitting of 2% (left) or 15% (right)

the coupling strength of the relic particle are translated to restrict the non-relic particles. Secondly, because the D_{ii} are equal the mixing effects are naturally small and as a result the mixing angles are much less constrained as they do not need to be small to counteract flavour effects.

This scenario is representative of a model in which MFV is broken only slightly, since the couplings to quark flavours are roughly equal, differing due to the mixing angles and the small differences in the D_{ii}. It is actually only slightly less constrained in both mass and couplings than models in which flavour violation is allowed, which counteracts the naive assumption that without MFV, flavour observables restrict NP very high scales ($\mathcal{O}(100\,\text{TeV})$).

2. When compared with the down type quark sector, flavour bounds are weaker due to D^0 being less well measured and our conservative treatment in which we assume the SM contribution to D^0 mixing is zero and the experimental value comes entirely from the new physics. This is not entirely unreasonable, since current short distance calculations of the observable are known to not reproduce the experimental result, nor is it completely reasonable, since long distance calculations are able to bring the SM into agreement.

To cover this caveat we consider a future scenario in which the SM calculation reproduces the experimental number (but the precision of the measurement stays at its current value). This is also conservative, since any interference terms between the SM and DMFV amplitude are likely to be large. The constraints on the mixing angles are more pronounced

Results for these two further scenarios are shown in Fig. 4.17, and the 1 σ intervals in Tables 4.6 and 4.7.

4.8 Summary

In this chapter we have analysed a model of dark matter, based on [18] but coupling to up type quarks, that goes beyond MFV in order to allow potentially large new effects in the flavour sector, and have seen how the combination of a wide range of constraints can be used to place limits on models of this type. We approached this task of combining many different constraints using the MCMC tool Multinest, which allowed us to place limits on the high dimensional parameter space of our particular model.

As we can see from Fig. 4.14, the MCMC places lower bounds on the new particle masses of at least 1 TeV for Top DM, and a few hundred GeV for Up and Charm DM in certain cases. Our collider bounds (Fig. 4.13) cannot further exclude Top DM, even in the case of strong couplings, but could remove a small area of allowed parameter space from the bottom end of the mass range in the case of Up/Charm DM.

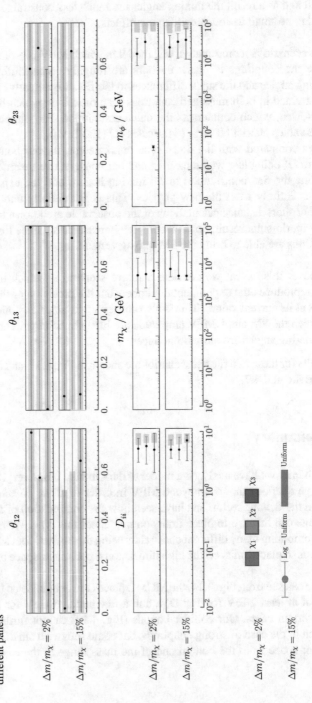

Table 4.6 The 1D 1σ credible intervals for the 6 parameters of constrained scenario in which the D_{ii} splitting are proportional to the $\Delta m / m_\chi$. Colours denote the DM flavour (red, green and blue for up, charm and top DM respectively). The two prior choices on D_{ii} and the masses (log-uniform vs uniform) are shown as lines and shaded regions respectively, the modal average of the log-uniform prior choice is shown as a dot. The two mass splitting cases are contained in different panels

Table 4.7 The 1D 1 σ credible intervals for a future scenario in which the SM prediction for the mixing matches the experimental value. Colours denote the DM flavour (red, green and blue for up, charm and top DM respectively). The two prior choices on D_{ii} and the masses (log-uniform vs uniform) are shown as lines and shaded regions respectively, the modal average of the log-uniform prior choice is shown as a dot. The two mass splitting cases are contained in different panels

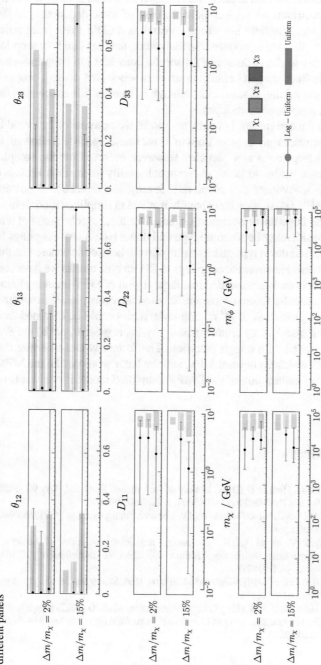

In [20] the authors consider this model, but examined the region of parameter space with dominant top quark couplings. Our results in general agree with their conclusions if we look at their more focused parameter space. For example, they find strong constraints on θ_{12} except in the case of some degeneracy in the D_{ii}, which we replicate. Similarly the strong constraints on DM mass from relic density and direct detection are reproduced. In their work, they explain how loop-level diagrams contributing to direct detection favour the dominant top coupling—however as we explain in Sect. 4.4, RG effects mean even when DM doesn't couple to up quarks directly, the mixing is substantial enough to weaken this conclusion (as long as the mediator mass is large enough).

Given the current level of data, the model we examine of flavoured DM coupling to up type quarks has large sections of its parameter space still allowed, so long as one considers large mass new particles. However, even without the complimentary collider results, the lower mass, phenomenologically interesting, regions of parameter space are disfavoured by flavour, relic density, and direct detection considerations.

The MFV assumption is frequently invoked in simplified models in order to evade potentially large flavour-violating effects. The level of robustness of this assumption varies considerably between up type and down type quark couplings in the DMFV model; for RH down type quarks strong flavour bounds do ensure that the assumption is a good one. However for couplings to RH up type quarks we have seen that in fact the flavour bounds are avoided in a large region of MFV-breaking parameter space.

One particular future development could alter this picture however—if a precise theoretical prediction of D^0 mixing observables could be obtained then either (a) a significant discrepancy requiring new physics is present, or (b) the SM predictions are reproduced with a high precision. The former would motivate the exploration of models which go beyond MFV, and the latter would make the MFV assumption a necessary assumption of the DMFV simplified model if one wants to avoid some fine-tuning.

References

1. Bertone G, Hooper D (2018) History of dark matter. Rev Mod Phys 90: 045002. https://doi. org/10.1103/RevModPhys.90.045002, arXiv:1605.04909
2. Particle Data Group collaboration, Dark Matter. http://pdg.lbl.gov/2017/reviews/rpp2017-rev-dark-matter.pdf
3. Springel V et al (2005) Simulating the joint evolution of quasars, galaxies and their large-scale distribution. Nature 435:629–636. https://doi.org/10.1038/nature03597. arXiv:astro-ph/0504097
4. Abdallah J et al (2014) Simplified Models for Dark Matter and Missing Energy Searches at the LHC. arXiv:1409.2893
5. LHC New Physics Working Group collaboration, Alves D (2012) Simplified models for LHC new physics searches. J Phys G 39:105005. https://doi.org/10.1088/0954-3899/39/10/105005. arXiv:1105.2838

6. Alwall J, Schuster P, Toro N (2009) Simplified models for a first characterization of new physics at the LHC. Phys Rev D 79:075020. https://doi.org/10.1103/PhysRevD.79.075020. arXiv:0810.3921

7. ATLAS collaboration, Aaboud M et al. (2016) Search for new phenomena in events with a photon and missing transverse momentum in pp collisions at $\sqrt{s} = 13$ TeV with the ATLAS detector. JHEP 06:059. https://doi.org/10.1007/JHEP06(2016)059, arXiv:1604.01306

8. De Simone A, Jacques T (2016) Simplified models vs. effective field theory approaches in dark matter searches. Eur Phys J C 76:367. https://doi.org/10.1140/epjc/s10052-016-4208-4, arXiv:1603.08002

9. Goodman J, Shepherd W (2011) LHC bounds on UV-complete models of dark matter. arXiv:1111.2359

10. Dreiner H, Schmeier D, Tattersall J (2013) Contact interactions probe effective dark matter models at the LHC. EPL 102:51001. https://doi.org/10.1209/0295-5075/102/51001. arXiv:1303.3348

11. Abercrombie D et al (2015) Dark matter benchmark models for early LHC Run-2 searches: report of the ATLAS/CMS dark matter forum. arXiv:1507.00966

12. Busoni G et al (2016) Recommendations on presenting LHC searches for missing transverse energy signals using simplified s-channel models of dark matter. arXiv:1603.04156

13. Goncalves D, Machado PAN, No JM (2017) Simplified models for dark matter face their consistent completions. Phys Rev D 95:055027. https://doi.org/10.1103/PhysRevD.95.055027, arXiv:1611.04593

14. Kahlhoefer F, Schmidt-Hoberg K, Schwetz T, Vogl S (2016) Implications of unitarity and gauge invariance for simplified dark matter models. JHEP 02:016. https://doi.org/10.1007/JHEP02(2016)016, arXiv:1510.02110

15. Englert C, McCullough M, Spannowsky M (2016) S-channel dark matter simplified models and unitarity. Phys Dark Univ 14:48–56. https://doi.org/10.1016/j.dark.2016.09.002, arXiv:1604.07975

16. Buras AJ, Gambino P, Gorbahn M, Jäger S, Silvestrini L (2001) Universal unitarity triangle and physics beyond the standard model. Phys Lett B 500:161–167. https://doi.org/10.1016/S0370-2693(01)00061-2. arXiv:hep-ph/0007085

17. D'Ambrosio G, Giudice GF, Isidori G, Strumia A (2002) Minimal flavor violation: an effective field theory approach. Nucl Phys B 645:155–187. https://doi.org/10.1016/S0550-3213(02)00836-2. arXiv:hep-ph/0207036

18. Agrawal P, Blanke M, Gemmler K (2014) Flavored dark matter beyond minimal flavor violation. JHEP 10:72. https://doi.org/10.1007/JHEP10(2014)072. arXiv:1405.6709

19. Chen M-C, Huang J, Takhistov V (2016) Beyond minimal lepton flavored dark matter. JHEP 02:060. https://doi.org/10.1007/JHEP02(2016)060, arXiv:1510.04694

20. Blanke M, Kast S (2017) Top-flavoured dark matter in dark minimal flavour violation. JHEP 05:162. https://doi.org/10.1007/JHEP05(2017)162, arXiv:1702.08457

21. Baek S, Ko P, Wu P (2018) Heavy quark-philic scalar dark matter with a vector-like fermion portal. JCAP 1807:008. https://doi.org/10.1088/1475-7516/2018/07/008, arXiv:1709.00697

22. Agrawal P, Blanchet S, Chacko Z, Kilic C (2012) Flavored dark matter, and its implications for direct detection and colliders. Phys Rev D 86:055002. https://doi.org/10.1103/PhysRevD.86.055002. arXiv:1109.3516

23. Kilic C, Klimek MD, Yu J-H (2015) Signatures of top flavored dark matter. Phys Rev D 91:054036. https://doi.org/10.1103/PhysRevD.91.054036, arXiv:1501.02202

24. Bhattacharya B, London D, Cline JM, Datta A, Dupuis G (2015) Quark-flavored scalar dark matter. Phys Rev D 92:115012. https://doi.org/10.1103/PhysRevD.92.115012, arXiv:1509.04271

25. Peskin ME, Takeuchi T (1990) A new constraint on a strongly interacting Higgs sector. Phys Rev Lett 65:964–967. https://doi.org/10.1103/PhysRevLett.65.964

26. Peskin ME, Takeuchi T (1992) Estimation of oblique electroweak corrections. Phys Rev D 46:381–409. https://doi.org/10.1103/PhysRevD.46.381

27. Grimus W, Lavoura L, Ogreid OM, Osland P (2008) The oblique parameters in multi-Higgs-doublet models. Nucl Phys B 801:81–96. https://doi.org/10.1016/j.nuclphysb.2008.04.019. arXiv:0802.4353

28. Isidori G, Straub DM (2012) Minimal flavour violation and beyond. Eur Phys J C 72:2103. https://doi.org/10.1140/epjc/s10052-012-2103-1. arXiv:1202.0464

29. Batell B, Pradler J, Spannowsky M (2011) Dark matter from minimal flavor violation. JHEP 08:038. https://doi.org/10.1007/JHEP08(2011)038. arXiv:1105.1781

30. Shape Planck collaboration, Ade PAR, Planck, et al (2015) results (2016) Cosmological parameters, XIII. Astron Astrophys 594:A13. https://doi.org/10.1051/0004-6361/201525830, arXiv:1502.01589

31. Griest K, Seckel D (1991) Three exceptions in the calculation of relic abundances. Phys Rev D 43:3191–3203. https://doi.org/10.1103/PhysRevD.43.3191

32. Busoni G, De Simone A, Jacques T, Morgante E, Riotto A (2015) Making the most of the relic density for dark matter searches at the LHC 14 TeV run. JCAP 1503:022. https://doi.org/10.1088/1475-7516/2015/03/022. arXiv:1410.7409

33. Bertone G, Hooper D, Silk J (2005) Particle dark matter: evidence, candidates and constraints. Phys Rept 405:279–390. https://doi.org/10.1016/j.physrep.2004.08.031. arXiv:hep-ph/0404175

34. Gondolo P, Gelmini G (1991) Cosmic abundances of stable particles: improved analysis. Nucl Phys B 360:145–179. https://doi.org/10.1016/0550-3213(91)90438-4

35. HFLAV collaboration, Global Fit for $D^0 - \bar{D}^0$ Mixing, CKM16. http://www.slac.stanford.edu/xorg/hflav/charm/CKM16/results_mix_cpv.html

36. Golowich E, Hewett J, Pakvasa S, Petrov AA (2007) Implications of D^0 - $\bar{D}^0$ mixing for new physics. Phys Rev D 76:095009. https://doi.org/10.1103/PhysRevD.76.095009. arXiv:0705.3650

37. Aoki S, et al. (2017) Review of lattice results concerning low-energy particle physics. Eur Phys J C 77:112. https://doi.org/10.1140/epjc/s10052-016-4509-7, arXiv:1607.00299

38. Na H, Davies CTH, Follana E, Lepage GP, Shigemitsu J (2012) $|V_{cd}|$ from D Meson Leptonic Decays. Phys Rev D 86:054510. https://doi.org/10.1103/PhysRevD.86.054510. arXiv:1206.4936

39. Fermilab Lattice MILC, collaboration, Bazavov A, et al (2012) B- and D-meson decay constants from three-flavor lattice QCD. Phys Rev D 85:114506. https://doi.org/10.1103/PhysRevD.85.114506. arXiv:1112.3051

40. Carrasco N et al (2014) D^0 - $\bar{D}^0$ mixing in the standard model and beyond from N_f =2 twisted mass QCD. Phys Rev D 90:014502. https://doi.org/10.1103/PhysRevD.90.014502. arXiv:1403.7302

41. Fajfer S, Košnik N (2015) Prospects of discovering new physics in rare charm decays. Eur Phys J C 75:567https://doi.org/10.1140/epjc/s10052-015-3801-2, arXiv:1510.00965

42. CMS collaboration, Khachatryan V, et al. (2016) Search for anomalous single top quark production in association with a photon in pp collisions at $\sqrt{s} = 8$ TeV. JHEP 04:035. https://doi.org/10.1007/JHEP04(2016)035, arXiv:1511.03951

43. LUX collaboration, Akerib DS, et al (2014) First results from the LUX dark matter experiment at the Sanford underground research facility. Phys Rev Lett 112:091303. https://doi.org/10.1103/PhysRevLett.112.091303. arXiv:1310.8214

44. LUX collaboration, Akerib DS, et al. (2016) Improved limits on scattering of weakly interacting massive particles from reanalysis of 2013 LUX data. Phys Rev Lett 116:161301. https://doi.org/10.1103/PhysRevLett.116.161301, arXiv:1512.03506

45. SuperCDMS collaboration, Agnese R, et al. (2016) New results from the search for low-mass weakly interacting massive particles with the CDMS low ionization threshold experiment. Phys Rev Lett 116:071301. https://doi.org/10.1103/PhysRevLett.116.071301, arXiv:1509.02448

46. Crivellin A, D'Eramo F, Procura M (2014) New constraints on dark matter effective theories from standard model loops. Phys Rev Lett 112:191304. https://doi.org/10.1103/PhysRevLett.112.191304. arXiv:1402.1173

47. D'Eramo F, Procura M (2015) Connecting dark matter UV complete models to direct detection rates via effective field theory. JHEP 04:054. https://doi.org/10.1007/JHEP04(2015)054. arXiv:1411.3342

48. Fitzpatrick AL, Haxton W, Katz E, Lubbers N, Xu Y (2013) The effective field theory of dark matter direct detection. JCAP 1302:004. https://doi.org/10.1088/1475-7516/2013/02/004. arXiv:1203.3542

49. Ibarra A, Wild S (2015) Dirac dark matter with a charged mediator: a comprehensive one-loop analysis of the direct detection phenomenology. JCAP 1505:047. https://doi.org/10.1088/1475-7516/2015/05/047, arXiv:1503.03382

50. Kahlhoefer F, Wild S (2016) Studying generalised dark matter interactions with extended halo-independent methods. JCAP 1610:032. https://doi.org/10.1088/1475-7516/2016/10/032, arXiv:1607.04418

51. Drees M, Nojiri M (1993) Neutralino-nucleon scattering revisited. Phys Rev D 48:3483–3501. https://doi.org/10.1103/PhysRevD.48.3483. arXiv:hep-ph/9307208

52. Hisano J, Nagai R, Nagata N (2015) Effective Theories for Dark Matter Nucleon Scattering. JHEP 05: 037.https://doi.org/10.1007/JHEP05(2015)037, arXiv:1502.02244

53. Gondolo P, Scopel S (2013) On the sbottom resonance in dark matter scattering. JCAP 1310:032. https://doi.org/10.1088/1475-7516/2013/10/032. arXiv:1307.4481

54. XENON collaboration, Aprile E, et al. (2017) First dark matter search results from the XENON1T experiment. Phys Rev Lett 119:181301. https://doi.org/10.1103/PhysRevLett.119.181301, arXiv:1705.06655

55. PandaX-II collaboration, Cui X, et al. (2017) Dark matter results from 54-Ton-Day exposure of PandaX-II experiment. Phys Rev Lett 119:181302. https://doi.org/10.1103/PhysRevLett.119.181302, arXiv:1708.06917

56. Cirelli M, Corcella G, Hektor A, Hutsi G, Kadastik M, Panci P et al (2011) PPPC 4 DM ID: a poor particle physicist cookbook for dark matter indirect detection. JCAP 1103:051. https://doi.org/10.1088/1475-7516/2012/10/E01, https://doi.org/10.1088/1475-7516/2011/03/051. arXiv:1012.4515

57. MAGIC, Fermi-LAT collaboration, Rico J, Wood M, Drlica-Wagner A, Aleksić J (2016) Limits to dark matter properties from a combined analysis of MAGIC and *Fermi*-LAT observations of dwarf satellite galaxies. PoS ICRC 2015 1206. https://doi.org/10.22323/1.236.1206, arXiv:1508.05827

58. Boudaud M (2015) A fussy revisitation of antiprotons as a tool for Dark Matter searches. arXiv:1510.07500

59. Di Mauro M, Vittino A (2016) AMS-02 electrons and positrons: astrophysical interpretation and Dark Matter constraints. PoS ICRC 2015 1177. https://doi.org/10.22323/1.236.1177, arXiv:1507.08680

60. IceCube collaboration, Aartsen MG, et al. (2016) All-flavour search for neutrinos from Dark Matter annihilations in the milky way with icecube/deepcore. Eur Phys J C 76:531. https://doi.org/10.1140/epjc/s10052-016-4375-3, arXiv:1606.00209

61. H.E.S.S. collaboration, Lefranc V, Moulin E (2016) Dark matter search in the inner Galactic halo with H.E.S.S. I and H.E.S.S. II. PoS ICRC 2015 1208. https://doi.org/10.22323/1.236.1208, arXiv:1509.04123

62. Fermi-LAT collaboration, Ackermann M, et al. (2015) Search for extended gamma-ray emission from the Virgo galaxy cluster with Fermi-LAT. Astrophys J 812:159. https://doi.org/10.1088/0004-637X/812/2/159, arXiv:1510.00004

63. Hahn T, Perez-Victoria M (1999) Automatized one loop calculations in four-dimensions and D-dimensions. Comput Phys Commun 118:153–165. https://doi.org/10.1016/S0010-4655(98)00173-8. arXiv:hep-ph/9807565

64. Busoni G, De Simone A, Jacques T, Morgante E, Riotto A (2014) On the validity of the effective field theory for dark matter searches at the LHC part III: analysis for the *t*-channel. JCAP 1409:022. https://doi.org/10.1088/1475-7516/2014/09/022. arXiv:1405.3101

65. ATLAS collaboration, Aad G, et al. (2015) Search for new phenomena in final states with an energetic jet and large missing transverse momentum in pp collisions at $\sqrt{s} = 8$ TeV with

the ATLAS detector. Eur Phys J C 75:299. https://doi.org/10.1140/epjc/s10052-015-3517-3, https://doi.org/10.1140/epjc/s10052-015-3639-7, arXiv:1502.01518

66. Alwall J, Frederix R, Frixione S, Hirschi V, Maltoni F, Mattelaer O et al (2014) The automated computation of tree-level and next-to-leading order differential cross sections, and their matching to parton shower simulations. JHEP 07:079. https://doi.org/10.1007/JHEP07(2014)079. arXiv:1405.0301

67. ATLAS collaboration, Aaboud M, et al. (2016) Search for new phenomena in final states with an energetic jet and large missing transverse momentum in pp collisions at $\sqrt{s} = 13$ TeV using the ATLAS detector. Phys Rev D 94:032005. https://doi.org/10.1103/PhysRevD.94.032005, arXiv:1604.07773

68. CMS collaboration, Khachatryan V, et al (2015) Search for dark matter, extra dimensions, and unparticles in monojet events in proton-proton collisions at $\sqrt{s} = 8$ TeV. Eur Phys J C 75:235. https://doi.org/10.1140/epjc/s10052-015-3451-4. arXiv:1408.3583

69. ATLAS collaboration, Aad G, et al (2014) Search for squarks and gluinos with the ATLAS detector in final states with jets and missing transverse momentum using $\sqrt{s} = 8$ TeV proton-proton collision data. JHEP 09:176. https://doi.org/10.1007/JHEP09(2014)176. arXiv:1405.7875

70. ATLAS collaboration, Aaboud M, et al. (2016) Search for squarks and gluinos in final states with jets and missing transverse momentum at $\sqrt{s} = 13$ TeV with the ATLAS detector. Eur Phys J C 76:392. https://doi.org/10.1140/epjc/s10052-016-4184-8, arXiv:1605.03814

71. ATLAS collaboration, Aad G, et al (2014) Search for top squark pair production in final states with one isolated lepton, jets, and missing transverse momentum in $\sqrt{s} = 8$ TeV pp collisions with the ATLAS detector. JHEP 11:118. https://doi.org/10.1007/JHEP11(2014)118. arXiv:1407.0583

72. ATLAS collaboration, Aad G, et al. (2015) Search for scalar charm quark pair production in pp collisions at $\sqrt{s} = 8$ TeV with the ATLAS detector. Phys Rev Lett 114:161801. https://doi.org/10.1103/PhysRevLett.114.161801, arXiv:1501.01325

73. Feroz F, Hobson MP (2008) Multimodal nested sampling: an efficient and robust alternative to MCMC methods for astronomical data analysis. Mon Not R Astron Soc 384:449. https://doi.org/10.1111/j.1365-2966.2007.12353.x. arXiv:0704.3704

74. Feroz F, Hobson MP, Bridges M (2009) MultiNest: an efficient and robust Bayesian inference tool for cosmology and particle physics. Mon Not R Astron Soc 398:1601–1614. https://doi.org/10.1111/j.1365-2966.2009.14548.x. arXiv:0809.3437

75. Feroz F, Hobson MP, Cameron E, Pettitt AN, Importance nested sampling and the multinest algorithm. arXiv:1306.2144

76. Buchner J, Georgakakis A, Nandra K, Hsu L, Rangel C, Brightman M et al (2014) X-ray spectral modelling of the AGN obscuring region in the CDFS: Bayesian model selection and catalogue. Astron Astrophys 564:A125. https://doi.org/10.1051/0004-6361/201322971. arXiv:1402.0004

77. Fowlie A, Bardsley MH (2016) Superplot: a graphical interface for plotting and analysing MultiNest output. Eur Phys J Plus 131:391. https://doi.org/10.1140/epjp/i2016-16391-0, arXiv:1603.00555

Chapter 5
Charming New Physics in Rare B_s Decays and Mixing?

5.1 Introduction

As we have discussed in Sect. 1.2, flavour processes such as rare B decays are excellent probes of new physics at the electroweak scale and beyond, due to their strong suppression in the SM. We also mentioned (in Sect. 1.3.4) how there are interesting signs of a BSM effect in $B \to K^{(*)}\mu^+\mu^-$, caused by a contact interaction of the form $(\bar{s}\gamma^\mu P_L b)(\bar{\mu}\gamma_\mu \mu)$. Explaining the effect requires destructive interference from BSM to reduce the Wilson coefficient C_9 by $\mathcal{O}(20\%)$. As was mentioned, a wide variety of models with new particles have been proposed (see e.g. [1–33]), which might in turn be part of a more comprehensive sector of new dynamics. Since the early signs of the anomalies (i.e. before $R_{K^{(*)}}$) suggested a flavour universal interaction, and noting that in the SM about half of C_9 comes from (short-distance) virtual-charm contributions, in this chapter we ask whether new physics affecting the quark-level $b \to c\bar{c}s$ transitions could cause the anomalies by affecting rare B decays through a loop The bulk of these effects would also be captured through an effective shift $\Delta C_9(q^2)$, with a possible dependence on the dilepton mass q^2. At the same time, such a scenario offers the exciting prospect of confirming the rare B decay anomalies through correlated effects in hadronic B decays into charm, with mixing observables such as the B_s meson width difference standing out as precisely measured [34–37] and under reasonable theoretical control. This is in contrast with the Z' and leptoquark models usually considered, where correlated effects are typically restricted to other rare processes and are highly model dependent. Specific scenarios of hadronic new physics in the B widths have been considered previously [38–41], while the possibility of virtual charm BSM physics in rare semileptonic decay has been raised in [42] (see also [43]). As we will show, viable scenarios exist which can mimic a shift $\Delta C_9 = -\mathcal{O}(1)$ while being consistent with all other observables. In particular, very strong renormalisation-group effects can generate large shifts in the (low-energy) effective C_9 coupling from small $b \to c\bar{c}s$ couplings at a high scale without conflicting with the measured $B \to X_s\gamma$ branching ratio [44–51].

© Springer Nature Switzerland AG 2019
M. J. Kirk, *Charming New Physics in Beautiful Processes?*,
Springer Theses, https://doi.org/10.1007/978-3-030-19197-9_5

5.2   Charming New Physics Scenario

We consider a scenario where new physics affects the $b \to c\bar{c}s$ transitions. This could be the case in models containing new scalars or new gauge bosons, or strongly coupled new physics. Such models will typically affect other observables, but in a model-dependent manner. For this paper, we restrict ourselves to studying the new effects induced by modified $b \to c\bar{c}s$ couplings, leaving construction and phenomenology of concrete models for future work. We refer to this as the "charming BSM" (CBSM) scenario. As long as the mass scale M of new physics satisfies $M \gg m_B$, the modifications to the $b \to c\bar{c}s$ transitions can be accounted for through a local effective Hamiltonian,

$$\mathcal{H}_{\text{eff}}^{c\bar{c}} = \frac{4G_F}{\sqrt{2}} V_{cs}^* V_{cb} \sum_{i=1}^{10} (C_i^c Q_i^c + C_i^{c'} Q_i^{c'}). \tag{5.2.1}$$

We choose our operator basis and renormalisation scheme to agree with [52] upon the substitution $d \to b, \bar{s} \to \bar{c}, \bar{u} \to \bar{s}$:

$$
\begin{aligned}
Q_1^c &= (\bar{c}^i \gamma_\mu P_L b^j)(\bar{s}^j \gamma^\mu P_L c^i), & Q_2^c &= (\bar{c}^i \gamma_\mu P_L b^i)(\bar{s}^j \gamma^\mu P_L c^j), \\
Q_3^c &= (\bar{c}^i P_L b^j)(\bar{s}^j P_R c^i), & Q_4^c &= (\bar{c}^i P_L b^i)(\bar{s}^j P_R c^j), \\
Q_5^c &= (\bar{c}^i \gamma_\mu P_R b^j)(\bar{s}^j \gamma^\mu P_L c^i), & Q_6^c &= (\bar{c}^i \gamma_\mu P_R b^i)(\bar{s}^j \gamma^\mu P_L c^j), \\
Q_7^c &= (\bar{c}^i P_R b^j)(\bar{s}^j P_R c^i), & Q_8^c &= (\bar{c}^i P_R b^i)(\bar{s}^j P_R c^j), \\
Q_9^c &= (\bar{c}^i \sigma_{\mu\nu} P_R b^j)(\bar{s}^j \sigma^{\mu\nu} P_R c^i), & Q_{10}^c &= (\bar{c}^i \sigma_{\mu\nu} P_R b^i)(\bar{s}^j \sigma^{\mu\nu} P_R c^j).
\end{aligned}
\tag{5.2.2}
$$

The $Q_i^{c'}$ are obtained by changing all the quark chiralities. These chirality flipped operators do not give any effect in C_9, and so we discard the $Q_i^{c'}$ below. We split the Wilson coefficients into SM and BSM parts,

$$C_i^c(\mu) = C_i^{c,\text{SM}}(\mu) + \Delta C_i(\mu), \tag{5.2.3}$$

where $C_i^{c,\text{SM}} = 0$ except for $i = 1, 2$ and μ is the renormalisation scale.

5.3   Rare B Decays

The leading-order (LO), one-loop CBSM effects in radiative and rare semileptonic decays may be expressed through "effective" Wilson coefficient contributions $\Delta C_9^{\text{eff}}(q^2)$ and $\Delta C_7^{\text{eff}}(q^2)$ in an effective local Hamiltonian

$$\mathcal{H}_{\text{eff}}^{rsl} = -\frac{4G_F}{\sqrt{2}} V_{ts} V_{tb} \left(C_7^{\text{eff}}(q^2) Q_{7\gamma} + C_9^{\text{eff}}(q^2) Q_{9V} \right), \tag{5.3.1}$$

Fig. 5.1 Leading CBSM contributions to rare decays (left), and to width difference $\Delta\Gamma_s$ and lifetime ratio $\tau(B_s)/\tau(B_d)$ (right)

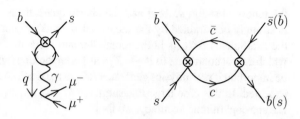

where q^2 is the dilepton mass and

$$Q_{7\gamma} = \frac{em_b}{16\pi^2}(\bar{s}\sigma_{\mu\nu}P_R b)F^{\mu\nu}, \quad Q_{9V} = \frac{\alpha}{4\pi}(\bar{s}\gamma_\mu P_L b)(\bar{\ell}\gamma^\mu\ell).$$

For small q^2 (in particular, well below the charm resonances), $\Delta C_9^{\text{eff}}(q^2)$ and $\Delta C_7^{\text{eff}}(q^2)$ govern the theoretical predictions for both exclusive ($B \to K^{(*)}\ell^+\ell^-$, $B_s \to \phi\ell^+\ell^-$, etc.) and inclusive $B \to X_s\ell^+\ell^-$ decay, up to $\mathcal{O}(\alpha_s)$ QCD corrections and power corrections to the heavy quark limit that we neglect in our leading-order analysis. Similarly, $\Delta C_7^{\text{eff}}(0)$ determines radiative B decay rates. We will neglect the small CKM combination $V_{*us}V_{ub}$, implying $V_{*cs}V_{cb} = -V_{*ts}V_{tb}$, and focus on real (CP-conserving) values for the C_i^c. From the diagram shown in Fig. 5.1 (left) we then obtain

$$\Delta C_9^{\text{eff}}(q^2) = \left(C_{1,2}^c - \frac{C_{3,4}^c}{2}\right)h - \frac{2}{9}C_{3,4}^c, \tag{5.3.2}$$

$$\Delta C_7^{\text{eff}}(q^2) = \frac{m_c}{m_b}\left[(4C_{9,10}^c - C_{7,8}^c)\,y + \frac{4C_{5,6}^c - C_{7,8}^c}{6}\right], \tag{5.3.3}$$

with $C_{x,y}^c = 3\Delta C_x + \Delta C_y$ and the loop functions

$$h(q^2, m_c, \mu) = -\frac{4}{9}\left[\ln\frac{m_c^2}{\mu^2} - \frac{2}{3} + (2+z)a(z) - z\right], \tag{5.3.4}$$

$$y(q^2, m_c, \mu) = -\frac{1}{3}\left[\ln\frac{m_c^2}{\mu^2} - \frac{3}{2} + 2a(z)\right], \tag{5.3.5}$$

where $a(z) = \sqrt{|z-1|}\arctan\frac{1}{\sqrt{z-1}}$ and $z = 4m_c^2/q^2$. Our numerical evaluation employs the charm pole mass.

We note that only the four Wilson coefficients ΔC_{1-4} enter $\Delta C_9^{\text{eff}}(q^2)$. Conversely, $\Delta C_7^{\text{eff}}(q^2)$ is given in terms of the other six Wilson coefficients ΔC_{5-10}. The appearance of a one-loop, q^2-dependent contribution to C_7^{eff} is a novel feature in the CBSM scenario. Numerically, the loop function $a(z)$ equals one at $q^2 = 0$ and vanishes at $q^2 = (2m_c)^2$. The constant terms and the logarithm accompanying $y(q^2, m_c)$ partially cancel the contribution from $a(z)$ and they introduce a sizeable dependence on the renormalisation scale μ and the charm quark mass. Since a shift of $\Delta C_7^{\text{eff}}(q^2)$ is strongly constrained by the measured $B \to X_s\gamma$ decay rate, we do not consider the coefficients ΔC_{5-10} in the remainder and focus on the four coefficients ΔC_{1-4}, which do not contribute to $B \to X_s\gamma$ at 1-loop order. Higher-order contributions can be important if new physics generates ΔC_i at the weak scale or beyond, as is typically expected. In this case large logarithms $\ln M/m_B$ occur, requiring resummation. To leading-logarithmic accuracy, we find

$$\Delta C_7^{\text{eff}} = 0.02\Delta C_1 - 0.19\Delta C_2 - 0.01\Delta C_3 - 0.13\Delta C_4, \qquad (5.3.6)$$

$$\Delta C_9^{\text{eff}} = 8.48\Delta C_1 + 1.96\Delta C_2 - 4.24\Delta C_3 - 1.91\Delta C_4, \qquad (5.3.7)$$

if ΔC_i are understood to be renormalized at $\mu = M_W$ and $\Delta C_{7,9}^{\text{eff}}$ at $\mu = 4.2$ GeV. It is clear that ΔC_1 and ΔC_3 contribute (strongly) to rare semileptonic decay but only weakly to $B \to X_s\gamma$.

5.4 Mixing and Lifetime Observables

A distinctive feature of the CBSM scenario is that non-zero ΔC_i affect not only radiative and rare semileptonic decays, but also tree-level hadronic $b \to c\bar{c}s$ transitions. While the theoretical control over exclusive $b \to c\bar{c}s$ modes is very limited at present, the decay width difference $\Delta\Gamma_s$ and the lifetime ratio $\tau(B_s)/\tau(B_d)$ stand out as being calculable in the HQE, via the diagrams shown in Fig. 5.1 (right). For both observables, the heavy quark expansion gives rise to an operator product expansion in terms of local $\Delta B = 2$ (for the width difference) or $\Delta B = 0$ (for the lifetime ratio) operators. For the B_s width difference, we have $\Delta\Gamma_s = 2|\Gamma_{12}^{s,\text{SM}} + \Gamma_{12}^{c\bar{c}}|\cos\phi_{12}^s$, where the phase ϕ_{12}^s is small. Neglecting the strange quark mass, we find

$$
\begin{aligned}
\Gamma_{12}^{cc} = &-G_F^2 (V_{cs}^* V_{cb})^2 m_b^2 M_{B_s} f_{B_s}^2 \frac{\sqrt{1 - 4x_c^2}}{576\pi} \times \\
&\left\{ \left[\; 16(1 - x_c^2)(4C_2^{c,2} + C_4^{c,2}) \right.\right. \\
&\quad +8(1 - 4x_c^2)(12C_1^{c,2} + 8C_1^c C_2^c + 2C_3^c C_4^c + 3C_3^{c,2}) \\
&\quad \left. -192x_c^2(3C_1^c C_3^c + C_1^c C_4^c + C_2^c C_3^c + C_2^c C_4^c)\right] B_1 \\
&\quad +2(1 + 2x_c^2)\left(\frac{M_{B_s}}{m_b + m_s}\right)^2 \times \\
&\quad (4C_2^{c,2} - 8C_1^c C_2^c - 12C_1^{c,2} - 3C_3^{c,2} - 2C_3^c C_4^c + C_4^{c,2})B_3 \right\},
\end{aligned}
\qquad (5.4.1)
$$

with $x_c = m_c/m_b$. B_1 and B_3 are defined as in Eq. C.0.2 with values taken from [53]. For our numerical evaluation of Γ_{12}^{cc}, we split the Wilson coefficients according to

Eq. 5.2.3, subtract from the LO expression (Eq. 5.4.1) the pure SM contribution and add the NLO SM expressions from [54–59]. In general, a modification of Γ_{12}^{cc} also affects the semi-leptonic CP asymmetries. However, since we consider CP-conserving new physics in this paper and since the corresponding experimental uncertainties are still large, the semi-leptonic asymmetries will not lead to an additional constraint.

In a similar manner, for the lifetime ratio we find

$$\frac{\tau(B_s)}{\tau(B_d)} = \left(\frac{\tau(B_s)}{\tau(B_d)}\right)_{SM} + \left(\frac{\tau(B_s)}{\tau(B_d)}\right)_{NP}, \tag{5.4.2}$$

where the SM contribution is taken from [60] and

$$\left(\frac{\tau(B_s)}{\tau(B_d)}\right)_{NP} = G_F^2 |V_{cb} V_{cs}|^2 m_b^2 M_{B_s} f_{B_s}^2 \tau_{B_s} \frac{\sqrt{1 - 4x_c^2}}{144\pi} \times$$
$$\left\{(1 - x_c^2)[(4C_{1,2}^{c,2} + C_{3,4}^{c,2})B_1 + 6(4C_2^{c,2} + C_4^{c,2})\epsilon_1]\right.$$
$$- 12x_c^2[C_{1,2}^c C_{3,4}^c B_1 + 6C_2^c C_4^c \epsilon_1] \tag{5.4.3}$$
$$- (1 + 2x_c^2)\left(\frac{M_{B_s}}{m_b + m_s}\right)^2 \times$$
$$\left. [(4C_{1,2}^{c,2} + C_{3,4}^{c,2})B_2 + 6(4C_2^{c,2} + C_4^{c,2})\epsilon_2]\right\},$$

subtracting the SM part and defining $B_1, B_2, \epsilon_1, \epsilon_2$ as in Eqs. C.0.3 and C.0.4 with values taken from our work in Chap. 6. We interpret the quark masses as $\overline{MS}$ parameters at $\mu = 4.2$ GeV.

5.5 Rare Decays Versus Lifetimes—Low-Scale Scenario

We are now in a position to confront the CBSM scenario with rare decay and mixing observables, as long as we consider renormalisation scales $\mu \sim m_B$. Then the logarithms inside the h function entering Eq. 5.3.2 are small and our leading-order calculation should be accurate. Such a scenario is directly applicable if the mass scale M of the physics generating the ΔC_i is not too far above m_B, such that $\ln(M/m_B)$ is small. Figure 5.2 (left) shows the experimental 1σ allowed regions for the width difference [61] and lifetime ratio [62] in the $\Delta C_1 - \Delta C_2$ plane. The central values are attained on the brown (solid) and green (dashed) curves, respectively. The measured lifetime ratio and the width difference measurement can be simultaneously accommodated for different values of the Wilson coefficients: in the $\Delta C_1 - \Delta C_2$ plane, we find the SM solution, as well as a solution around $\Delta C_1 = -0.5$ and $\Delta C_2 \approx 0$. In the $\Delta C_3 - \Delta C_4$ plane, we have a relatively broad allowed range, roughly covering the

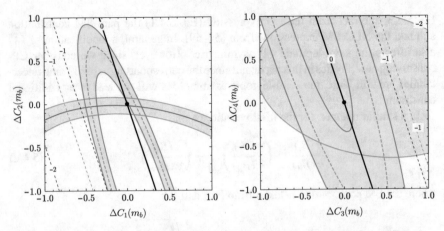

Fig. 5.2 Mixing observables versus rare decays in the CBSM scenario, in the $\Delta C_1 - \Delta C_2$ plane (left) and $\Delta C_3 - \Delta C_4$ plane (right). All Wilson coefficients are renormalized at $\mu = 4.2$ GeV and those not corresponding to either axis set to zero. The black dot corresponds to the SM, i.e. $\Delta C_i = 0$. The measured central value for the width difference is shown as brown (solid) line together with the $1\,\sigma$ allowed region. The lifetime ratio measurement is depicted as green (dashed) line and band. Overlaid are contours of $\Delta C_9^{\mathrm{eff}}(5\ \mathrm{GeV}^2) = -1, -2$ (black, dashed) and $\Delta C_9^{\mathrm{eff}}(2\ \mathrm{GeV}^2) = -1, -2$ (red, dotted), as computed from Eq. 5.3.2, and of $\Delta C_9^{\mathrm{eff}} = 0$ (black, solid)

interval $[-0.9 + 0.7]$ for ΔC_3 and $[-0.6 + 1.1]$ for ΔC_4. For further conclusions, a considerably higher precision in experiment and theory is required for $\Delta\Gamma_s$ and $\tau(B_s)/\tau(B_d)$. Also shown in the plot are contour lines for the contribution to the effective semileptonic coefficient $\Delta C_9^{\mathrm{eff}}(q^2)$, both for $q^2 = 2$ and $q^2 = 5\ \mathrm{GeV}^2$. We see that sizeable negative shifts are possible while respecting the measured width difference and the lifetime ratio. For example, a shift $\Delta C_9^{\mathrm{eff}} \sim -1$ as data may suggest could be achieved through $\Delta C_1 \sim -0.5$ alone. Such a value for ΔC_1 may well be consistent with CP-conserving exclusive $b \to c\bar{c}s$ decay data, where no accurate theoretical predictions exist. On the other hand, $\Delta C_9^{\mathrm{eff}}$ only exhibits a mild q^2-dependence. Distinguishing this from possible long-distance contributions would require substantial progress on the theoretical understanding of the latter.

We can also consider other Wilson coefficients, such as the pair $(\Delta C_3, \Delta C_4)$ (right panel in Fig. 5.2). A shift $\Delta C_9^{\mathrm{eff}} \sim -1$ is equally possible and consistent with the width difference, requiring only $\Delta C_3 \sim 0.5$.

5.6 High-Scale Scenario and RGE

5.6.1 RG Enhancement of ΔC_9^{eff}

If the CBSM operators are generated at a high scale then large logarithms $\ln M/m_B$ appear. Their resummation is achieved by evolving the initial (matching) conditions $C_i(\mu_0 \sim M)$ to a scale $\mu \sim M_B$ according to the coupled renormalisation-group equations (RGE),

$$\mu \frac{dC_j}{d\mu}(\mu) = \gamma_{ij}(\mu)C_i(\mu), \qquad (5.6.1)$$

where γ_{ij} is the anomalous dimension matrix. As is well known, the operators Q_i^c mix not only with Q_7 and Q_9, but also with the 4 QCD penguin operators P_{3-6} and the chromodipole operator Q_{8g} (defined as in [63]), which in turn mix into Q_7. Hence the index j runs over 11 operators with $\Delta B = -\Delta S = 1$ flavour quantum numbers in order to account for all contributions to $C_7(\mu)$ that are proportional to $\Delta C_i(\mu_0)$. Most entries of γ_{ij} are known at LO [52, 63–77]; our novel results are ($i = 3, 4$)

$$\gamma_{Q_i^c \tilde{Q}_9}^{(0)} = \left(\frac{4}{3}, \frac{4}{9}\right)_i, \quad \gamma_{Q_i^c P_4}^{(0)} = \left(0, -\frac{2}{3}\right)_i, \quad \gamma_{Q_i^c Q_7}^{eff(0)} = \left(0, \frac{224}{81}\right)_i, \qquad (5.6.2)$$

where $\tilde{Q}_9 = (4\pi/\alpha_s)Q_{9V}(\mu)$ and $\gamma_{Q_i^c Q_7}^{eff(0)}$ requires a two-loop calculation. (See Appendix F for further technical information.) Solving the RGE for $\mu_0 = M_W$, $\mu = 4.2$ GeV, and $\alpha_s(M_Z) = 0.1181$ results in the CBSM contributions

$$
\begin{pmatrix} \Delta C_1(\mu) \\ \Delta C_2(\mu) \\ \Delta C_3(\mu) \\ \Delta C_4(\mu) \\ \Delta C_7^{eff}(\mu) \\ \Delta C_9^{eff}(\mu) \end{pmatrix} =
\begin{pmatrix}
1.12 & -0.27 & 0 & 0 \\
-0.27 & 1.12 & 0 & 0 \\
0 & 0 & 0.92 & 0 \\
0 & 0 & 0.33 & 1.91 \\
0.02 & -0.19 & -0.01 & -0.13 \\
8.48 & 1.96 & -4.24 & -1.19
\end{pmatrix}
\begin{pmatrix} \Delta C_1(\mu_0) \\ \Delta C_2(\mu_0) \\ \Delta C_3(\mu_0) \\ \Delta C_4(\mu_0) \end{pmatrix}. \qquad (5.6.3)
$$

A striking feature are the large coefficients in the ΔC_9^{eff} case, which are $\mathcal{O}(1/\alpha_s)$ in the logarithmic counting. The largest coefficients appear for ΔC_1 and ΔC_3, which at the same time practically do not mix into C_7^{eff}. This means that small values $\Delta C_1 \sim -0.1$ or $\Delta C_3 \sim 0.2$ can generate $\Delta C_9^{eff}(\mu) \sim -1$ while having essentially no impact on the $B \to X_s\gamma$ decay rate. Conversely, values for ΔC_2 or ΔC_4 that lead to $\Delta C_9^{eff} \sim -1$ lead to large effects in C_7^{eff} and $B \to X_s\gamma$.

5.6.2 Phenomenology for High NP Scale

The situation in various two-parameter planes is depicted in Fig. 5.3, where the 1σ constraint from $B \to X_s\gamma$ is shown as blue, straight bands. (We implement it by splitting $\mathcal{B}(B \to X_s\gamma)$ into SM and BSM parts and employ the numerical result and theory error from [78] for the former. The experimental result is taken from [79].) The top row corresponds to Fig. 5.2, but contours of given ΔC_9 lie much closer to the origin. All six panels testify to the fact that the SM is consistent with all data when leaving aside the question of rare semileptonic B decays – the largest pull stems from the fact that the experimental value for $\tau(B_s)/\tau(B_d)$ is just under 1.5 standard deviations below the SM expectation, such that the black (SM) point is less than 0.5σ outside the green area. Our main question now is: can we have a new contribution $\Delta C_9^{\text{eff}} \sim -1$ to rare semileptonic decays, while being consistent with the bounds stemming from $b \to s\gamma$, $\Delta\Gamma_s$ and $\tau(B_s)/\tau(B_d)$? This is clearly possible (indicated by the yellow star in the plots) if we have a new contribution $\Delta C_3 \approx 0.2$, see the three plots of the $\Delta C_i - \Delta C_3$ planes in Fig. 5.3 (right on the top row, left on the middle row and left on the lower row). In these cases, the $\Delta C_9^{\text{eff}} \sim -1$ solution is even favoured compared to the SM solution. A joint effect in $\Delta C_2 \approx -0.1$ and $\Delta C_4 \approx 0.3$ can also accommodate our desired scenario (see the right plot on the lower row), while new BSM effects in the pairs ΔC_1, ΔC_2 and ΔC_1, ΔC_4 alone are less favoured. One could also consider three or all four ΔC_i simultaneously.

5.6.3 Implications for UV Physics

Our model-independent results are well suited to study the rare B decay and lifetime phenomenology of UV completions of the Standard Model. Any such completion may include extra UV contributions to $C_7(M)$ and $C_9(M)$, correlations with other flavour observables, collider phenomenology, etc.; the details are highly model-dependent and beyond the scope of our model-independent analysis. Here we restrict ourselves to some basic sanity checks.

Taking the case of $\Delta C_1(M) \sim -0.1$ corresponds to a naive ultraviolet scale

$$\Lambda \sim \left(\frac{4G_F}{\sqrt{2}}|V_{cs}^*V_{cb}| \times 0.1\right)^{-1/2} \sim 3\,\text{TeV}.$$

This effective scale could arise in a weakly-coupled scenario from tree-level exchange of new scalar or vector mediators, or at loop level in addition from fermions; or the effective operator could arise from strongly-coupled new physics. For a tree-level exchange, $\Lambda \sim M/g_*$ where $g_* = \sqrt{g_1 g_2}$ is the geometric mean of the relevant couplings. For weak coupling $g_* \sim 1$, this then gives $M \sim 3$ TeV. Particles of such mass are certainly allowed by collider searches if they do not couple (or only sufficiently weakly) to leptons and first-generation quarks. Multi-TeV weakly coupled particles

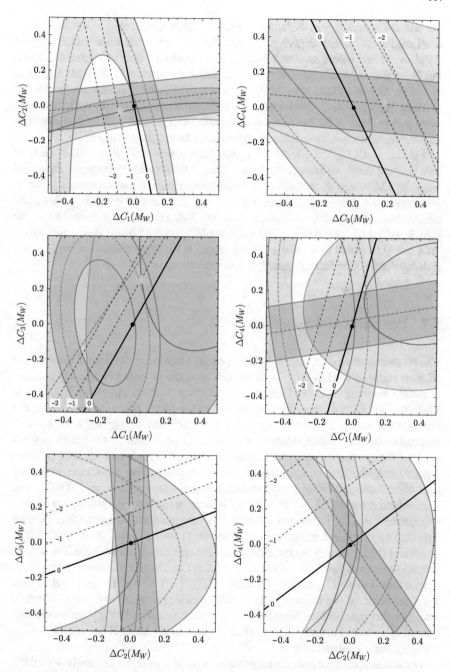

Fig. 5.3 Mixing observables versus rare decays, for ΔC_i renormalized at $\mu_0 = M_W$. Colour coding as in Fig. 5.2, $B \to X_s \gamma$ constraint shown in addition (straight blue bands)

are also not generically in violation of electroweak precision tests of the SM. Loop-level mediation would require mediators close to the weak scale which may be problematic and would require a specific investigation; this is of course unsurprising given that $b \to c\bar{c}s$ transitions are mediated at tree level in the SM. The same would be true in a BSM scenario that mimics the flavour suppressions in the SM (such as MFV models). Conversely, in a strongly-coupled scenario we would have $M \sim g_* \Lambda \sim 4\pi \Lambda \sim 30$ TeV. This is again safe from generic collider and precision constraints, and a model-specific analysis would be required to say more.

Finally, as all CBSM effects are lepton-flavour-universal, they cannot on their own account for departures of the lepton flavour universality parameters $R_{K^{(*)}}$ [80] from the SM values as suggested by current experimental measurements [81–83]. However, even if those departures are real, they may still be caused by direct UV contributions to ΔC_9. For example, as shown in [84], a scenario with a muon-specific contribution $\Delta C_9^\mu = -\Delta C_{10}^\mu \sim -0.6$ and in addition a lepton-universal contribution $\Delta C_9 \sim -0.6$, which may have a CBSM origin, is perfectly consistent with all rare B decay data, and in fact marginally preferred.

5.7   Prospects and Summary

The preceding discussion suggests that a precise knowledge of width difference and lifetime ratio, as well as $\mathcal{B}(B \to X_s \gamma)$, can have the potential to identify and discriminate between different CBSM scenarios, or rule them out altogether. This is illustrated in Fig. 5.4, showing contour values for future precision both in mixing and lifetime observables. In each panel, the solid (brown and green) contours correspond to the SM central values of the width difference and lifetime ratio (respectively). The spacing of the accompanying contours is such that the area between any two neighbouring contours corresponds to a prospective $1\,\sigma$-region, assuming a combined (theoretical and experimental) error on the lifetime ratio of 0.001 and a combined error on $\Delta\Gamma_s$ of 5% (theory is already there for the lifetime ratio, and experiment for $\Delta\Gamma$). The assumed future errors are ambitious but seem feasible with expected experimental and theoretical progress. Overlaid is the (current) $B \to X_s\gamma$ constraint (blue). The figure indicates that a discrimination between the SM and the scenario where $\Delta C_9 \approx -1$, while $\mathcal{B}(B \to X_s\gamma)$ is SM-like is clearly possible. A crucial role is played by the lifetime ratio $\tau(B_s)/\tau(B_d)$: in e.g. the $\Delta C_3 - \Delta C_4$ case a $1\,\sigma$ deviation of the lifetime ratio almost coincides with the $\Delta C_9 = -1$ contour line; a further precise determination of $\Delta\Gamma_s$ could then identify the point on this line chosen by nature. Further progress on $B \to X_s\gamma$ in the Belle II era would provide complementary information.

In summary, we have given a comprehensive, model-independent analysis of BSM effects in partonic $b \to c\bar{c}s$ transitions (CBSM scenario) in the CP-conserving case, focusing on those observables that can be computed in a heavy quark expansion. An effect in rare semileptonic B decays compatible with hints from current LHCb and B-factory data can be generated, while satisfying the $B \to X_s\gamma$ constraint. It can

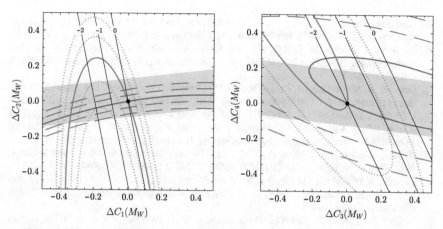

Fig. 5.4 Future prospects for mixing observables. Dashed: contours of constant width difference, dotted: contours of constant lifetime ratio. See text for discussion

originate from different combinations of $b \to c\bar{c}s$ operators. The required Wilson coefficients are so small that constraints from B decays into charm are not effective, particularly if new physics enters at a high scale; then large renormalisation-group enhancements are present. Likewise, there are no obvious model-independent conflicts with collider searches or electroweak precision observables. A more precise measurement of mixing observables and lifetime ratios, at a level achievable at LHCb, may be able to confirm (or rule out) the CBSM scenario, and to discriminate between different BSM couplings. Finally, all CBSM effects are lepton-flavour-universal; the current R_K and R_{K^*} anomalies would either have to be mismeasurements or require additional lepton-flavour-specific UV contribution to C_9; such a combined scenario has been shown elsewhere [84] to be consistent with all rare B decay data and also presents the most generic way for UV physics to affect rare decays. With the stated caveats, our conclusions are rather model independent. It would be interesting to construct concrete UV realizations of the CBSM scenario, which almost certainly will affect other observables in a correlated, but model-dependent manner.

References

1. Buras AJ, Girrbach J (2013) Left-handed Z' and Z FCNC quark couplings facing new $b \to s\mu^+\mu^-$ data. JHEP 12:009. https://doi.org/10.1007/JHEP12(2013)009, arXiv:1309.2466
2. Gauld R, Goertz F, Haisch U (2014) An explicit Z'-boson explanation of the $B \to K^*\mu^+\mu^-$ anomaly. JHEP 01:069. https://doi.org/10.1007/JHEP01(2014)069, arXiv:1310.1082
3. Buras AJ, De Fazio F, Girrbach J (2014) 331 models facing new $b \to s\mu^+\mu^-$ data. JHEP 02:112. https://doi.org/10.1007/JHEP02(2014)112, arXiv:1311.6729
4. Datta A, Duraisamy M, Ghosh D (2014) Explaining the $B \to K^*\mu^+\mu^-$ data with scalar interactions. Phys Rev D 89:071501. https://doi.org/10.1103/PhysRevD.89.071501, arXiv:1310.1937

5. Altmannshofer W, Gori S, Pospelov M, Yavin I (2014) Quark flavor transitions in $L_\mu - L_\tau$ models. Phys Rev D 89:095033. https://doi.org/10.1103/PhysRevD.89.095033, arXiv:1403.1269

6. Hiller G, Schmaltz M (2014) R_K and future $b \to s\ell\ell$ physics beyond the standard model opportunities. Phys Rev D 90:054014. https://doi.org/10.1103/PhysRevD.90.054014, arXiv:1408.1627

7. Gripaios B, Nardecchia M, Renner SA (2015) Composite leptoquarks and anomalies in B-meson decays. JHEP 05:006. https://doi.org/10.1007/JHEP05(2015)006, arXiv:1412.1791

8. Crivellin A, D'Ambrosio G, Heeck J (2015) Explaining $h \to \mu^\pm \tau^\mp$, $B \to K^* \mu^+ \mu^-$ and $B \to K \mu^+ \mu^- / B \to K e^+ e^-$ in a two-Higgs-doublet model with gauged $L_\mu - L_\tau$. Phys Rev Lett 114:151801. https://doi.org/10.1103/PhysRevLett.114.151801, arXiv:1501.00993

9. de Medeiros Varzielas I, Hiller G (2015) Clues for flavor from rare lepton and quark decays. JHEP 06:072. https://doi.org/10.1007/JHEP06(2015)072, arXiv:1503.01084

10. Crivellin A, D'Ambrosio G, Heeck J (2015) Addressing the LHC flavor anomalies with horizontal gauge symmetries. Phys Rev D 91:075006. https://doi.org/10.1103/PhysRevD.91.075006, arXiv:1503.03477

11. Bečirević D, Fajfer S, Košnik N (2015) Lepton flavor nonuniversality in $b \to s\mu^+\mu^-$ processes. Phys Rev D 92:014016. https://doi.org/10.1103/PhysRevD.92.014016, arXiv:1503.09024

12. Celis A, Fuentes-Martin J, Jung M, Serôdio H (2015) Family nonuniversal Z' models with protected flavor-changing interactions. Phys Rev D 92:015007. https://doi.org/10.1103/PhysRevD.92.015007, arXiv:1505.03079

13. Alonso R, Grinstein B, Martin Camalich J (2015) Lepton universality violation and lepton flavor conservation in B-meson decays. JHEP 10:184. https://doi.org/10.1007/JHEP10(2015)184, arXiv:1505.05164

14. Belanger G, Delaunay C, Westhoff S (2015) A dark matter relic from muon anomalies. Phys Rev D 92:055021. https://doi.org/10.1103/PhysRevD.92.055021, arXiv:1507.06660

15. Falkowski A, Nardecchia M, Ziegler R (2015) Lepton flavor non-universality in B-meson decays from a U(2) flavor model. JHEP 11:173. https://doi.org/10.1007/JHEP11(2015)173, arXiv:1509.01249

16. Gripaios B, Nardecchia M, Renner SA (2016) Linear flavour violation and anomalies in B physics. JHEP 06:083. https://doi.org/10.1007/JHEP06(2016)083, arXiv:1509.05020

17. Bauer M, Neubert M (2016) Minimal Leptoquark explanation for the $R_{D^{(*)}}$, R_K, and $(g - 2)_g$ anomalies. Phys Rev Lett 116:141802. https://doi.org/10.1103/PhysRevLett.116.141802, arXiv:1511.01900

18. Fajfer S, Košnik N (2016) Vector leptoquark resolution of R_K and $R_{D^{(*)}}$ puzzles. Phys Lett B 755:270–274. https://doi.org/10.1016/j.physletb.2016.02.018, arXiv:1511.06024

19. Boucenna SM, Celis A, Fuentes-Martin J, Vicente A, Virto J (2016) Phenomenology of an $SU(2) \times SU(2) \times U(1)$ model with lepton-flavour non-universality. JHEP 12:059. https://doi.org/10.1007/JHEP12(2016)059, arXiv:1608.01349

20. Arnan P, Hofer L, Mescia F, Crivellin A (2017) Loop effects of heavy new scalars and fermions in $b \to s\mu^+\mu^-$. JHEP 04:043. https://doi.org/10.1007/JHEP04(2017)043, arXiv:1608.07832

21. Bečirević D, Fajfer S, Košnik N, Sumensari O (2016) Leptoquark model to explain the B-physics anomalies, R_K and R_D. Phys Rev D 94:115021. https://doi.org/10.1103/PhysRevD.94.115021, arXiv:1608.08501

22. Crivellin A, Fuentes-Martin J, Greljo A, Isidori G (2017) Lepton flavor non-universality in B decays from dynamical Yukawas. Phys Lett B 766:77–85. https://doi.org/10.1016/j.physletb.2016.12.057, arXiv:1611.02703

23. Garcia IG (2017) LHCb anomalies from a natural perspective. JHEP 03:040. https://doi.org/10.1007/JHEP03(2017)040, arXiv:1611.03507

24. Cline JM (2018) B decay anomalies and dark matter from vectorlike confinement. Phys Rev D 97:015013. https://doi.org/10.1103/PhysRevD.97.015013, arXiv:1710.02140

25. Baek S (2018) Dark matter contribution to $b \to s\mu^+\mu^-$ anomaly in local $U(1)_{L_\mu - L_\tau}$ model. Phys Lett B 781:376–382. https://doi.org/10.1016/j.physletb.2018.04.012, arXiv:1707.04573

26. Cline JM, Camalich JM (2017) B decay anomalies from nonabelian local horizontal symmetry. Phys Rev D 96:055036. https://doi.org/10.1103/PhysRevD.96.055036, arXiv:1706.08510

27. Kawamura J, Okawa S, Omura Y (2017) Interplay between the $b \to s\ell\ell$ anomalies and dark matter physics. Phys Rev D 96:075041. https://doi.org/10.1103/PhysRevD.96.075041, arXiv:1706.04344

28. Di Chiara S, Fowlie A, Fraser S, Marzo C, Marzola L, Raidal M et al (2017) Minimal flavor-changing Z' models and muon $g - 2$ after the R_{K^*} measurement. Nucl Phys B 923:245–257. https://doi.org/10.1016/j.nuclphysb.2017.08.003, arXiv:1704.06200

29. Kamenik JF, Soreq Y, Zupan J (2018) Lepton flavor universality violation without new sources of quark flavor violation. Phys Rev D 97:035002. https://doi.org/10.1103/PhysRevD.97.035002, arXiv:1704.06005

30. Crivellin A, Mller D, Ota T (2017) Simultaneous explanation of R(D$^{(*)}$) and $b \to s\mu^+\mu^-$: the last scalar leptoquarks standing. JHEP 09:040. https://doi.org/10.1007/JHEP09(2017)040, arXiv:1703.09226

31. Ko P, Omura Y, Shigekami Y, Yu C (2017) LHCb anomaly and B physics in flavored Z' models with flavored Higgs doublets. Phys Rev D 95:115040. https://doi.org/10.1103/PhysRevD.95.115040, arXiv:1702.08666

32. Ko P, Nomura T, Okada H (2017) Explaining $B \to K^{(*)}\ell^+\ell^-$ anomaly by radiatively induced coupling in $U(1)_{\mu-\tau}$ gauge symmetry. Phys Rev D 95:111701. https://doi.org/10.1103/PhysRevD.95.111701, arXiv:1702.02699

33. Di Luzio L, Greljo A, Nardecchia M (2017) Gauge leptoquark as the origin of B-physics anomalies. Phys Rev D 96:115011. https://doi.org/10.1103/PhysRevD.96.115011, arXiv:1708.08450

34. ATLAS collaboration, Aad G et al (2016) Measurement of the CP-violating phase ϕ_s and the B_s^0 meson decay width difference with $B_s^0 \to J/\psi\phi$ decays in ATLAS. JHEP 08:147. https://doi.org/10.1007/JHEP08(2016)147, arXiv:1601.03297

35. CMS collaboration, Khachatryan V et al (2016) Measurement of the CP-violating weak phase ϕ_s and the decay width difference $\Delta\Gamma_s$ using the $B_s^0 \to J/\psi\phi(1020)$ decay channel in pp collisions at $\sqrt{s} = 8$ TeV. Phys Lett B 757:97–120. https://doi.org/10.1016/j.physletb.2016.03.046, arXiv:1507.07527

36. LHCb collaboration, Aaij R et al (2015) Precision measurement of CP violation in $B_s^0 \to J/\psi K^+ K^-$ decays. Phys Rev Lett 114:041801. https://doi.org/10.1103/PhysRevLett.114.041801, arXiv:1411.3104

37. LHCb collaboration, Aaij R et al (2014) Measurements of the B^+, B^0, B_s^0 meson and Λ_b^0 baryon lifetimes. JHEP 04:114. https://doi.org/10.1007/JHEP04(2014)114, arXiv:1402.2554

38. Bobeth C, Haisch U, Lenz A, Pecjak B, Tetlalmatzi-Xolocotzi G (2014) On new physics in $\Delta\Gamma_d$. JHEP 06:040. https://doi.org/10.1007/JHEP06(2014)040, arXiv:1404.2531

39. Bobeth C, Gorbahn M, Vickers S (2015) Weak annihilation and new physics in charmless $B \to MM$ decays. Eur Phys J C 75:340. https://doi.org/10.1140/epjc/s10052-015-3535-1, arXiv:1409.3252

40. Brod J, Lenz A, Tetlalmatzi-Xolocotzi G, Wiebusch M (2015) New physics effects in tree-level decays and the precision in the determination of the quark mixing angle γ, Phys Rev D 92:033002. https://doi.org/10.1103/PhysRevD.92.033002, arXiv:1412.1446

41. Bauer CW, Dunn ND (2011) Comment on new physics contributions to Γ_{12}^s. Phys Lett B 696:362–366. https://doi.org/10.1016/j.physletb.2010.12.039, arXiv:1006.1629

42. Lyon J, Zwicky R (2014) Resonances gone topsy turvy-the charm of QCD or new physics in $b \to s\ell^+\ell^-$?. arXiv:1406.0566

43. He XG, Tandean J, Valencia G (2009) Probing new physics in charm couplings with FCNC. Phys Rev D 80:035021. https://doi.org/10.1103/PhysRevD.80.035021, arXiv:0904.2301

44. CLEO collaboration, Chen S et al (2001) Branching fraction and photon energy spectrum for $b \to s\gamma$. Phys Rev Lett 87:251807. https://doi.org/10.1103/PhysRevLett.87.251807, arXiv:hep-ex/0108032

45. Belle collaboration, Abe K et al (2001) A measurement of the branching fraction for the inclusive $B \to X(s)\gamma$ decays with BELLE. Phys Lett B 511:151–158.https://doi.org/10.1016/S0370-2693(01)00626-8, arXiv:hep-ex/0103042

46. BaBar collaboration, Aubert B et al (2008) Measurement of the $B \to X_s\gamma$ branching fraction and photon energy spectrum using the recoil method. Phys Rev D 77:051103. https://doi.org/10.1103/PhysRevD.77.051103, arXiv:0711.4889
47. Belle collaboration, Limosani A et al (2009) Measurement of inclusive radiative B-meson decays with a photon energy threshold of 1.7-GeV. Phys Rev Lett 103:241801. https://doi.org/10.1103/PhysRevLett.103.241801, arXiv:0907.1384
48. BaBar collaboration, Lees JP et al (2012) Precision measurement of the $B \to X_s\gamma$ photon energy spectrum, branching fraction, and direct CP asymmetry $A_{CP}(B \to X_{s+d}\gamma)$. Phys Rev Lett 109:191801.https://doi.org/10.1103/PhysRevLett.109.191801, arXiv:1207.2690
49. BaBar collaboration, Lees JP et al (2012) Measurement of B $(B \to X_s\gamma)$, the $B \to X_s\gamma$ photon energy spectrum, and the direct CP asymmetry in $B \to X_{s+d}\gamma$ decays. Phys Rev D 86:112008. https://doi.org/10.1103/PhysRevD.86.112008, arXiv:1207.5772
50. BaBar collaboration, Lees JP et al (2012) Exclusive measurements of $b \to s\gamma$ transition rate and photon energy spectrum. Phys Rev D 86:052012. https://doi.org/10.1103/PhysRevD.86.052012, arXiv:1207.2520
51. Belle collaboration, Saito T et al (2015) Measurement of the $\bar{B} \to X_s\gamma$ branching fraction with a sum of exclusive decays. Phys Rev D 91:052004. https://doi.org/10.1103/PhysRevD.91.052004, arXiv:1411.7198
52. Buras AJ, Misiak M, Urban J (2000) Two loop QCD anomalous dimensions of flavor changing four quark operators within and beyond the standard model. Nucl Phys B 586:397–426. https://doi.org/10.1016/S0550-3213(00)00437-5, arXiv:hep-ph/0005183
53. Aoki S et al (2014) Review of lattice results concerning low-energy particle physics. Eur Phys J C 74:2890. https://doi.org/10.1140/epjc/s10052-014-2890-7, arXiv:1310.8555
54. Beneke M, Buchalla G, Dunietz I (1996) Width difference in the $B_s - \bar{B}_s$ system. Phys Rev D 54:4419–4431. https://doi.org/10.1103/PhysRevD.54.4419, https://doi.org/10.1103/PhysRevD.83.119902, arXiv:hep-ph/9605259
55. Dighe AS, Hurth T, Kim CS, Yoshikawa T (2002) Measurement of the lifetime difference of B_d mesons: possible and worthwhile? Nucl Phys B 624:377–404. https://doi.org/10.1016/S0550-3213(01)00655-1, arXiv:hep-ph/0109088
56. Beneke M, Buchalla G, Greub C, Lenz A, Nierste U (1999) Next-to-leading order QCD corrections to the lifetime difference of B_s mesons. Phys Lett B 459:631–640. https://doi.org/10.1016/S0370-2693(99)00684-X, arXiv:hep-ph/9808385
57. Beneke M, Buchalla G, Lenz A, Nierste U (2003) CP asymmetry in flavor specific B decays beyond leading logarithms. Phys Lett B 576:173–183. https://doi.org/10.1016/j.physletb.2003.09.089, arXiv:hep-ph/0307344
58. Ciuchini M, Franco E, Lubicz V, Mescia F, Tarantino C (2003) Lifetime differences and CP violation parameters of neutral B mesons at the next-to-leading order in QCD. JHEP 08:031. https://doi.org/10.1088/1126-6708/2003/08/031, arXiv:hep-ph/0308029
59. Lenz A, Nierste U (2007) Theoretical update of $B_s - \bar{B}_s$ mixing. JHEP 06:072. https://doi.org/10.1088/1126-6708/2007/06/072, arXiv:hep-ph/0612167
60. Franco E, Lubicz V, Mescia F, Tarantino C (2002) Lifetime ratios of beauty hadrons at the next-to-leading order in QCD. Nucl Phys B 633:212–236. https://doi.org/10.1016/S0550-3213(02)00262-6, arXiv:hep-ph/0203089
61. HFLAV collaboration (2016) B lifetime and oscillation parameters. Spring 2016. http://www.slac.stanford.edu/xorg/hflav/osc/spring_2016/
62. HFLAV collaboration (2017) B lifetime and oscillation parameters. Summer 2017. http://www.slac.stanford.edu/xorg/hflav/osc/summer_2017/
63. Chetyrkin KG, Misiak M, Munz M (1997) Weak radiative B meson decay beyond leading logarithms. Phys Lett B 400:206–219. https://doi.org/10.1016/S0370-2693(97)00324-9, arXiv:hep-ph/9612313
64. Gaillard MK, Lee BW (1974) $\Delta I = 1/2$ rule for nonleptonic decays in asymptotically free field theories. Phys Rev Lett 33:108. https://doi.org/10.1103/PhysRevLett.33.108
65. Altarelli G, Maiani L (1974) Octet enhancement of nonleptonic weak interactions in asymptotically free gauge theories. Phys Lett B 52:351–354. https://doi.org/10.1016/0370-2693(74)90060-4

66. Gilman FJ, Wise MB (1979) Effective Hamiltonian for $\Delta S = 1$ weak nonleptonic decays in the six quark model. Phys Rev D 20:2392. https://doi.org/10.1103/PhysRevD.20.2392

67. Shifman MA, Vainshtein AI, Zakharov VI (1977) Nonleptonic decays of K mesons and hyperons. Sov Phys JETP 45:670

68. Gilman FJ, Wise MB (1980) $K \rightarrow \pi e^+ e^-$ in the six quark model. Phys Rev D 21:3150. https://doi.org/10.1103/PhysRevD.21.3150

69. Guberina B, Peccei RD (1980) Quantum chromodynamic effects and CP violation in the Kobayashi-Maskawa model. Nucl Phys B 163:289–311. https://doi.org/10.1016/0550-3213(80)90404-6

70. Ciuchini M, Franco E, Martinelli G, Reina L, Silvestrini L (1993) Scheme independence of the effective Hamiltonian for $b \rightarrow s\gamma$ and $b \rightarrow sg$ decays. Phys Lett B 316:127–136. https://doi.org/10.1016/0370-2693(93)90668-8, arXiv:hep-ph/9307364

71. Ciuchini M, Franco E, Martinelli G, Reina L, Silvestrini L (1994) $b \rightarrow s\gamma$ and $b \rightarrow sg$: a theoretical reappraisal. Phys Lett B 334:137–144. https://doi.org/10.1016/0370-2693(94)90602-5, arXiv:hep-ph/9406239

72. Ciuchini M, Franco E, Reina L, Silvestrini L (1994) Leading order QCD corrections to $b \rightarrow s\gamma$ and $b \rightarrow sg$ decays in three regularization schemes. Nucl Phys B 421:41–64. https://doi.org/10.1016/0550-3213(94)90223-2, arXiv:hep-ph/9311357

73. Bertolini S, Borzumati F, Masiero A (1987) QCD enhancement of radiative b decays. Phys Rev Lett 59:180. https://doi.org/10.1103/PhysRevLett.59.180

74. Grinstein B, Springer RP, Wise MB (1988) Effective Hamiltonian for weak radiative B meson decay. Phys Lett B 202:138–144. https://doi.org/10.1016/0370-2693(88)90868-4

75. Grinstein B, Springer RP, Wise MB (1990) Strong interaction effects in weak radiative $\bar{B}$ meson decay. Nucl Phys B 339:269–309. https://doi.org/10.1016/0550-3213(90)90350-M

76. Misiak M (1991) QCD corrected effective Hamiltonian for the $b \rightarrow s\gamma$ decay. Phys Lett B 269:161–168. https://doi.org/10.1016/0370-2693(91)91469-C

77. Chetyrkin KG, Misiak M, Munz M (1998) Beta functions and anomalous dimensions up to three loops. Nucl Phys B 518:473–494. https://doi.org/10.1016/S0550-3213(98)00122-9, arXiv:hep-ph/9711266

78. Misiak M et al (2015) Updated NNLO QCD predictions for the weak radiative B-meson decays. Phys Rev Lett 114:221801. https://doi.org/10.1103/PhysRevLett.114.221801, arXiv:1503.01789

79. HFLAV collaboration (2014) Rare B decay parameters. http://www.slac.stanford.edu/xorg/hflav/rare/2014/radll/OUTPUT/HTML/radll_table3.html

80. Hiller G, Kruger F (2004) More model-independent analysis of $b \rightarrow s$ processes. Phys Rev D 69:074020. https://doi.org/10.1103/PhysRevD.69.074020, arXiv:hep-ph/0310219

81. LHCb collaboration, Aaij R et al (2014) Test of lepton universality using $B^+ \rightarrow K^+\ell^+\ell^-$ decays. Phys Rev Lett 113:151601. https://doi.org/10.1103/PhysRevLett.113.151601, arXiv:1406.6482

82. LHCb collaboration, Aaij R et al (2017) Test of lepton universality with $B^0 \rightarrow K^{*0}\ell^+\ell^-$ decays. JHEP 08:055. https://doi.org/10.1007/JHEP08(2017)055, arXiv:1705.05802

83. LHCb collaboration, Aaij R et al (2018) Measurement of the ratio of branching fractions $\mathcal{B}(B_c^+ \rightarrow J/\psi\tau^+\nu_\tau)/\mathcal{B}(B_c^+ \rightarrow J/\psi\mu^+\nu_\mu)$. Phys Rev Lett 120:121801. https://doi.org/10.1103/PhysRevLett.120.121801, arXiv:1711.05623

84. Geng LS, Grinstein B, Jäger S, Camalich JM, Ren X-L, Shi R-X (2017) Towards the discovery of new physics with lepton-universality ratios of $b \rightarrow s\ell\ell$ decays. Phys Rev D 96:093006. https://doi.org/10.1103/PhysRevD.96.093006, arXiv:1704.05446

Chapter 6
Dimension-Six Matrix Elements from Sum Rules

6.1 Introduction

As we have seen in Sect. 2.4.2, the SM contribution to meson mixing arises at the 1-loop level and is both CKM and GIM suppressed. This makes these observables highly sensitive to new physics contributions (an issue which we will explore further in Chap. 7), and so a precise knowledge of the theoretical predictions is very important. The SM calculation can be factorised into a perturbative calculation of the Wilson coefficients and a non perturbative part coming from the matrix elements of effective operators that contribute to mixing. While the perturbative part (of which we showed the leading order calculation at the start of this thesis) is known up to NLO in QCD, and the first steps towards the NNLO QCD corrections have recently been done [1], the dominant uncertainty comes from the non perturbative part of the calculation. These hadronic matrix elements are usually determined by lattice simulations and results for the leading dimension-six operators are available from several collaborations [2–4]. Using the most recent result [4], a small tension at the level of $2\,\sigma$ arises for B_s mixing (see [4] and Chaps. 3 and 7). Given the important nature of B_s mixing observables in constraining new physics with a non-trivial flavour structure, it is important to try and settle this issue. An independent determination of the matrix elements is an important step in this direction, which we will take in this chapter.

QCD sum rules [5, 6] provide an alternative way to determine hadronic quantities– this approach employs quark-hadron duality and the analyticity of Green functions, as opposed to a numerical solution of the QCD path integral as done by lattice groups. Since its sources of uncertainty are entirely different from lattice simulations, sum rule analyses can provide truly independent results.

We determine the hadronic matrix elements of the dimension-six $\Delta B = 2$ operators for B mixing from a sum rule for three-point correlators first introduced in [7]. Since the sum rule is valid at scales $\mu_\rho \sim 1.5\,\text{GeV}$ which are much smaller than the bottom quark mass, we formulate the sum rule in HQET. We then run the HQET matrix elements up to a scale μ_m of the order of the bottom quark mass where the matching to QCD can be performed without introducing large logarithms. Earlier

© Springer Nature Switzerland AG 2019
M. J. Kirk, *Charming New Physics in Beautiful Processes?*,
Springer Theses, https://doi.org/10.1007/978-3-030-19197-9_6

sum rule results are available for the SM operator Q_1 [8, 9] and condensate corrections have been computed for dimension-six [9–12] and dimension-seven [10] operators. We can apply the same technique for a variety of other operators, and so we also study dimension-six operators that contribute to mixing in the charm sector and to the lifetimes of B and D mesons, since, as we explain below, these are much needed results.

The theory calculation for the mixing induced width differences and the meson lifetimes are based on the HQE, which was covered in Sect. 2.2. As we saw in Chap. 3, violations of quark-hadron duality in the HQE are currently limited to around 20 % by the current agreement between experiment and theory. As was discussed there as well, lifetimes provide an excellent testing ground for the HQE, but this is hindered by the availability of up to date calculations of the relevant non-perturbative parameters as these are the dominant uncertainties here (for a review of the calculation of meson lifetimes, see e.g. [13]). For B mesons, the most recent calculation comes from a conference proceedings in 2001 [14], while for D mesons the calculation has never been done and so the corresponding SM prediction has a very large uncertainty [15].

In the charm sector the validity of the HQE is rather uncertain due to the smaller charm mass ($m_c \sim m_b/3$). The direct translation of the predictions for B mixing fails by several orders of magnitude [16], but it has been argued that higher-dimensional contributions can lift the severe GIM suppression and potentially explain the size of mixing observables [16–20]. For D meson lifetimes, while the uncertainty is large, the central value agrees well with experiment and so there is no immediate sign of a breakdown of the HQE.

The outline of this chapter is as follows: In Sect. 6.2 we describe the details of the QCD-HQET matching computation focussing on $\Delta B = 2$ operators. The sum rule and the calculation of the three-point correlators are discussed in Sect. 6.3. Our results for the matrix elements are presented in Sect. 6.4 and compared to other recent works. In Sect. 6.5 we study $\Delta B = 0$ operators and ratios of B meson lifetimes. We determine the matrix elements of $\Delta C = 0, 2$ operators in Sect. 6.6 and update the HQE result for the $D^+ - D^0$ lifetime ratio using these results. Finally, we summarise our results in Sect. 6.7.

6.2 QCD-HQET Matching for $\Delta B = 2$ Operators

We perform the matching computation between QCD and HQET operators at the 1-loop level. The details of the computation are described in Sect. 6.2.1 for the $\Delta B = 2$ operators. Our results for the matching of the operators and bag parameters are given in Sect. 6.2.2 and Sect. 6.2.3, respectively.

6.2.1 Setup

The matching calculation for the SM operator Q_1 appearing in ΔM_s has been performed in [21–23]. We compute the matching coefficients of the full dimension-six

$\Delta B = 2$ operator basis needed for ΔM_s in BSM theories and for $\Delta \Gamma_s$ in the SM. We work in dimensional regularization with $d = 4 - 2\epsilon$ and an anti-commuting γ^5 (NDR scheme). We consider the following operators in QCD

$$
\begin{aligned}
Q_1 &= (\bar{b}_i \gamma_\mu (1 - \gamma^5) q_i)(\bar{b}_j \gamma^\mu (1 - \gamma^5) q_j), \\
Q_2 &= (\bar{b}_i (1 - \gamma^5) q_i)(\bar{b}(1 - \gamma^5) b), \quad Q_3 = (\bar{b}_i (1 - \gamma^5) q_j)(\bar{b}_j (1 - \gamma^5) q_i), \\
Q_4 &= (\bar{b}_i (1 - \gamma^5) q_i)(\bar{b}(1 + \gamma^5) b), \quad Q_5 = (\bar{b}_i (1 - \gamma^5) q_j)(\bar{b}_j (1 + \gamma^5) q_i).
\end{aligned}
$$
$$(6.2.1)$$

To fix the renormalisation scheme we also have to specify a basis of evanescent operators [24–26]—we do this following [27]. Evanescent operators are those with Dirac structures that vanish in $d = 4$ dimensions, but whose matrix elements do not necessarily vanish beyond tree-level. The explicit form of our choice of evanescent operators can be found in Appendix G.1. On the HQET side, we have the operators

$$
\begin{aligned}
\tilde{Q}_1 &= (\bar{h}_i^{\{(+)} \gamma_\mu (1 - \gamma^5) q_i)(\bar{h}_j^{(-)\}} \gamma^\mu (1 - \gamma^5) q_j), \\
\tilde{Q}_2 &= (\bar{h}_i^{\{(+)} (1 - \gamma^5) q_i)(\bar{h}_j^{(-)\}} (1 - \gamma^5) q_j), \\
\tilde{Q}_4 &= (\bar{h}_i^{\{(+)} (1 - \gamma^5) q_i)(\bar{h}_j^{(-)\}} (1 + \gamma^5) q_j), \\
\tilde{Q}_5 &= (\bar{h}_i^{\{(+)} (1 - \gamma^5) q_j)(\bar{h}_j^{(-)\}} (1 + \gamma^5) q_i),
\end{aligned}
$$
$$(6.2.2)$$

where the HQET field $h^{(+)}(x)$ annihilates a bottom quark, $h^{(-)}(x)$ creates an anti-bottom and we have introduced the notation

$$
(\bar{h}^{\{(+)} \Gamma_A q)(\bar{h}^{(-)\}} \Gamma_B q) = (\bar{h}^{(+)} \Gamma_A q)(\bar{h}^{(-)} \Gamma_B q) + (\bar{h}^{(-)} \Gamma_A q)(\bar{h}^{(+)} \Gamma_B q). \quad (6.2.3)
$$

Note that no operator $\tilde{Q}_3$ appears on the HQET side because it is not linearly independent, just like its QCD equivalent at leading order in $1/m_b$ [28]. We define the evanescent HQET operators up to three constants a_i with $i = 1, 2, 3$ which allow us to keep track of the scheme dependence. Again the explicit basis of the evanescent operators can be found in Appendix G.1.

The matching condition for the $\Delta B = 2$ operators is given by

$$
\langle Q_i \rangle(\mu) = \sum C_{Q_i \tilde{Q}_j}(\mu) \langle \tilde{Q}_j \rangle(\mu) + \mathcal{O}\left(\frac{1}{m_b}\right), \quad (6.2.4)
$$

where $\langle A \rangle = \langle \bar{B}|A|B \rangle$. The matching coefficients can be expanded in perturbation theory and take the form

$$
C_{Q_i \tilde{Q}_j}(\mu) = C_{Q_i \tilde{Q}_j}^{(0)} + \frac{\alpha_s(\mu)}{4\pi} C_{Q_i \tilde{Q}_j}^{(1)}(\mu) + \dots. \quad (6.2.5)
$$

The partonic QCD matrix elements are

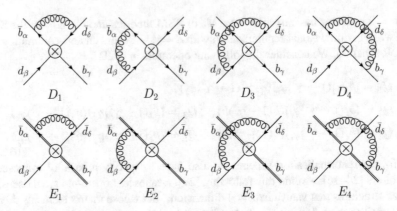

Fig. 6.1 QCD (D_i) and HQET (E_i) diagrams that enter the matching. Symmetric diagrams are not shown

$$\langle Q \rangle = \frac{\delta_{\alpha\beta}\delta_{\gamma\delta}}{N_c} \left[Z_b^{OS} Z_q^{OS} Z_{QO} \left(\begin{array}{c} \bar{b}_\alpha \qquad \bar{d}_\delta \\ \otimes O \\ d_\beta \qquad b_\gamma \end{array} + \begin{array}{c} \bar{b}_\alpha \qquad \bar{d}_\delta \\ \otimes \\ d_\beta \quad O \quad b_\gamma \end{array} \right) + \mathcal{O}\left(\alpha_s\right) \right],$$

(6.2.6)

where we sum over $\mathcal{O}$, including all physical and evanescent operators, and the colour singlet initial and final state have been projected out.

The two tree-level contractions appear with a relative minus sign, as we saw in Eq. 2.4.18. The gluon corrections are shown in Fig. 6.1 and do not contain self-energy insertions on the external legs, since the quark fields are renormalized in the on-shell scheme. The HQET matrix elements follow from the replacements $Q \to \tilde{Q}$, $\mathcal{O} \to \tilde{\mathcal{O}}$, $Z_b^{OS} \to Z_h^{OS}$ and using HQET propagators instead of the full QCD ones for the bottom quark. The heavy quark on-shell renormalisation constants are

$$Z_b^{OS} = 1 - \frac{\alpha_s C_F}{4\pi} \left(\frac{3}{\epsilon} + 4 + 3\ln\frac{\mu^2}{m_b^2} \right) + \mathcal{O}\left(\alpha_s^2\right), \qquad Z_h^{OS} = 1.$$ (6.2.7)

The light quark renormalisation is trivial in the massless case: $Z_q^{OS} = 1$. For the renormalisation of the physical operators the $\overline{MS}$ scheme is used. In accordance with [24–26] the evanescent operators are renormalized by a finite amount such that their physical matrix elements vanish. Consequently the Wilson coefficients $C_{Q_i\tilde{E}_j}$ are not required for the determination of the hadronic matrix elements and are omitted in the results shown below. However, in the matching computation itself the matrix elements are taken between external on-shell quark states and are therefore not IR finite. While the IR divergences cancel in the matching of the QCD to the HQET operators there are non-vanishing contributions to the physical matching coefficients $C_{Q\tilde{Q}}$ from matrix elements of the evanescent operators that are multiplied by IR poles since the evanescent operators are defined differently in QCD and HQET (see Appendix G.1).

We also find that the NLO matching coefficients $C^{(1)}_{Q_3\tilde{Q}_j}$ of the operator Q_3 are affected by the finite renormalisation of the evanescent operator $\tilde{E}_2$ which contains contributions proportional to the physical operators. This usually only happens at NNLO (as is the case for the other operators) but is already present here at NLO because the tree-level matching coefficient $C^{(0)}_{Q_3\tilde{E}_2}$ of this operator is non-vanishing and, therefore, the NLO matrix element of the evanescent HQET operator $\tilde{E}_2$ already appears at NLO in the matching calculation.

In the computation we have used both a manual approach and an automated setup utilizing QGRAF [29] and Mathematica to generate the amplitudes. The Dirac algebra has been performed with a customized version of TRACER [30] as well as with Package-X [31, 32] and the QCD loop integrals have been evaluated using Package-X. We have also checked our results by performing the calculation with a gluon mass as an IR regulator and found full agreement.

6.2.2 Results

We write the LO QCD anomalous dimension matrix (ADM) as

$$\gamma^{(0)} = \begin{pmatrix} \gamma^{(0)}_{QQ} & \gamma^{(0)}_{QE} \\ \gamma^{(0)}_{EQ} & \gamma^{(0)}_{EE} \end{pmatrix}, \tag{6.2.8}$$

where $\gamma^{(0)}_{QQ}$ is the ADM for the physical set of operators (Eq. 6.2.1), $\gamma^{(0)}_{QE}$ describes the mixing of the physical operators into the evanescent ones (Eq. G.1.1), $\gamma^{(0)}_{EQ}$ vanishes (see [26]) and $\gamma^{(0)}_{EE}$ is not required. We decompose the LO HQET ADM $\tilde{\gamma}^{(0)}$ analogously. Our results for the non-vanishing entries are given in Appendix G.1.

The non-vanishing Wilson coefficients at LO are

$$C^{(0)}_{Q_1\tilde{Q}_1} = 1, \quad C^{(0)}_{Q_2\tilde{Q}_2} = 1, \quad C^{(0)}_{Q_3\tilde{Q}_1} = -\frac{1}{2},$$
$$C^{(0)}_{Q_3\tilde{Q}_2} = -1, \quad C^{(0)}_{Q_4\tilde{Q}_4} = 1, \quad C^{(0)}_{Q_5\tilde{Q}_5} = 1. \tag{6.2.9}$$

The NLO corrections to the matching coefficients read

$$C^{(1)}_{Q\tilde{Q}} = \begin{pmatrix} -\frac{41}{3} + \frac{a_2}{12} - 6L_\mu & -8 & 0 & 0 \\ \frac{3}{2} - \frac{a_1}{12} + L_\mu & 8 + 4L_\mu & 0 & 0 \\ 5 + \frac{2a_1 - a_2}{24} + 4L_\mu & 4 + 4L_\mu & 0 & 0 \\ 0 & 0 & 8 - \frac{a_3}{24} + \frac{9L_\mu}{2} & -4 + \frac{a_3}{8} - \frac{3L_\mu}{2} \\ 0 & 0 & 4 + \frac{a_3}{8} + \frac{3L_\mu}{2} & -8 - \frac{a_3}{24} - \frac{9L_\mu}{2} \end{pmatrix}, \tag{6.2.10}$$

where $L_\mu = \ln(\mu^2/m_b^2)$ and we have set $N_c = 3$ to keep the results compact.

6.2.3 Matching of QCD and HQET Bag Parameters

We define the QCD bag parameters B_Q as

$$\langle Q(\mu) \rangle = A_Q f_B^2 M_B^2 B_Q(\mu), \tag{6.2.11}$$

where the coefficients read

$$A_{Q_1} = 2 + \frac{2}{N_c},$$

$$A_{Q_2} = \frac{M_B^2}{(m_b + m_q)^2} \left(-2 + \frac{1}{N_c} \right), \quad A_{Q_3} = \frac{M_B^2}{(m_b + m_q)^2} \left(1 - \frac{2}{N_c} \right), \tag{6.2.12}$$

$$A_{Q_4} = 2 \frac{M_B^2}{(m_b + m_q)^2} + \frac{1}{N_c}, \qquad A_{Q_5} = \frac{2}{N_c} \frac{M_B^2}{(m_b + m_q)^2} + 1,$$

and the B meson decay constant f_B is defined as

$$\langle 0 | \bar{b} \gamma^\mu \gamma^5 q | B(p) \rangle = i f_B p^\mu, \tag{6.2.13}$$

which match our definitions in Appendix C (up to a factor four due to the normalisation of the $P_{L,R}$ projection operators). We note that the quark masses appearing in Eq. 6.2.12 are *not* $\overline{\text{MS}}$ masses which is the usual convention today [4, 33], but pole masses. We prefer the definition in Eq. 6.2.11 for the analysis because the use of $\overline{\text{MS}}$ masses makes the LO ADM of the bag parameters explicitly μ-dependent and prohibits an analytic solution of the RGE. At the end we convert our results to the convention of [4, 33] which we denote as

$$\langle Q(\mu) \rangle = \overline{A}_Q(\mu) f_B^2 M_B^2 \overline{B}_Q(\mu), \tag{6.2.14}$$

where the $\overline{A}_Q(\mu)$ follow from A_Q with the replacements $m_b \to \overline{m}_b(\mu)$ and $m_q \to \overline{m}_q(\mu)$. Similar to Eq. 6.2.11, we use for the HQET operators

$$\langle \tilde{Q}(\mu) \rangle = A_{\tilde{Q}} F^2(\mu) B_{\tilde{Q}}(\mu), \tag{6.2.15}$$

where

$$A_{\tilde{Q}_1} = 2 + \frac{2}{N_c}, \quad A_{\tilde{Q}_2} = -2 + \frac{1}{N_c}, \quad A_{\tilde{Q}_4} = 2 + \frac{1}{N_c}, \quad A_{\tilde{Q}_5} = 1 + \frac{2}{N_c}, \tag{6.2.16}$$

and the matrix elements have been taken between non-relativistically normalized states $\langle \tilde{Q}_i(\mu) \rangle \equiv \langle \overline{\mathbf{B}} | \tilde{Q}_i(\mu) | \mathbf{B} \rangle$ with

$$|B(p)\rangle = \sqrt{2M_B} |\mathbf{B}(v)\rangle + \mathcal{O}(1/m_b) , \qquad (6.2.17)$$

such that

$$\langle \mathbf{B}(v') | \mathbf{B}(v) \rangle = \frac{v^0}{M_B^3} (2\pi)^3 \delta^{(3)}(v' - v) . \qquad (6.2.18)$$

The parameter $F(\mu)$ is defined as

$$\langle 0 | \bar{h}^{(-)} \gamma^\mu \gamma^5 q | \mathbf{B}(v) \rangle = i F(\mu) v^\mu , \qquad (6.2.19)$$

and related to the decay constant by

$$f_B = \sqrt{\frac{2}{M_B}} C(\mu) F(\mu) + \mathcal{O}(1/m_b) , \qquad (6.2.20)$$

with [34]

$$C(\mu) = 1 - 2 C_F \frac{\alpha_s(\mu)}{4\pi} + \mathcal{O}\left(\alpha_s^2\right) . \qquad (6.2.21)$$

From Eqs. 6.2.11–6.2.15, we obtain (by use of Eqs. 6.2.4, 6.2.17 and 6.2.20)

$$B_{Q_i}(\mu) = \sum_j \frac{A_{\tilde{Q}_j}}{A_{Q_i}} \frac{C_{Q_i \tilde{Q}_j}(\mu)}{C^2(\mu)} B_{\tilde{Q}_j}(\mu) + \mathcal{O}(1/m_b) . \qquad (6.2.22)$$

The HQET bag parameters $B_{\tilde{Q}}$ are determined from a sum rule analysis, which we describe in the next section.

6.3 HQET Sum Rule

The HQET sum rule is introduced in Sect. 6.3.1. We give results for the double-discontinuity of the three-point correlators in Sect. 6.3.2 and describe the determination of HQET and QCD bag parameters in Sect. 6.3.3.

6.3.1 The Sum Rule

We define the three-point correlator

$$K_{\tilde{Q}}(\omega_1, \omega_2) = \int d^d x_1 \, d^d x_2 \, e^{ip_1 \cdot x_1 - ip_2 \cdot x_2} \langle 0|T\left[\tilde{j}_+(x_2)\tilde{Q}(0)\tilde{j}_-(x_1)\right]|0\rangle, \quad (6.3.1)$$

where $\omega_{1,2} = p_{1,2} \cdot v$ and

$$\tilde{j}_+ = \bar{q}\gamma^5 h^{(+)}, \qquad \tilde{j}_- = \bar{q}\gamma^5 h^{(-)}, \quad (6.3.2)$$

are interpolating currents for the pseudoscalar $\bar{B}$ and B mesons. The correlator (Eq. 6.3.1) is analytic in $\omega_{1,2}$ apart from discontinuities for positive real ω. This allows us to construct a dispersion relation

$$K_{\tilde{Q}}(\omega_1, \omega_2) = \int_0^\infty d\eta_1 \, d\eta_2 \, \frac{\rho_{\tilde{Q}}(\eta_1, \eta_2)}{(\eta_1 - \omega_1)(\eta_2 - \omega_2)} + \text{subtraction terms}, \quad (6.3.3)$$

where $\rho_{\tilde{Q}}$ is the *double* discontinuity of $K_{\tilde{Q}}$ in ω_1 and ω_2. The second term on the right originates from the integration of $K_{\tilde{Q}}$ along the circle at infinity in the complex η_1 and/or η_2 planes and is therefore polynomial in ω_1 and/or ω_2. The correlator $K_{\tilde{Q}}$ can be computed by means of an OPE

$$K_{\tilde{Q}}^{\text{OPE}}(\omega_1, \omega_2) = K_{\tilde{Q}}^{\text{pert}}(\omega_1, \omega_2) + K_{\tilde{Q}}^{\langle\bar{q}q\rangle}(\omega_1, \omega_2)\langle\bar{q}q\rangle$$
$$+ K_{\tilde{Q}}^{\langle\alpha_s G^2\rangle}(\omega_1, \omega_2)\langle\alpha_s G^2\rangle + \dots \quad (6.3.4)$$

for values of $\omega_{1,2}$ that lie far away from the physical cut. Assuming quark-hadron duality, we can equate the correlator $K_{\tilde{Q}}^{\text{OPE}}$ with its hadronic counterpart

$$K_{\tilde{Q}}^{\text{had}}(\omega_1, \omega_2) = \int_0^\infty d\eta_1 \, d\eta_2 \, \frac{\rho_{\tilde{Q}}^{\text{had}}(\eta_1, \eta_2)}{(\eta_1 - \omega_1)(\eta_2 - \omega_2)} + \text{subtraction terms}, \quad (6.3.5)$$

which is obtained from integration over the hadronic spectral function

$$\rho_{\tilde{Q}}^{\text{had}}(\omega_1, \omega_2) = F^2(\mu)\langle\tilde{Q}(\mu)\rangle\delta(\omega_1 - \overline{\Lambda})\delta(\omega_2 - \overline{\Lambda}) + \rho_{\tilde{Q}}^{\text{cont}}(\omega_1, \omega_2). \quad (6.3.6)$$

We use a double Borel transformation with respect to $\omega_{1,2}$ to remove the contribution from the integration over the circle at infinity and to suppress the sensitivity to the continuum part $\rho_{\tilde{Q}}^{\text{cont}}$ of the spectral function, which yields the sum rule

$$\int_0^\infty d\omega_1 \, d\omega_2 \, e^{-\frac{\omega_1}{t_1} - \frac{\omega_2}{t_2}} \rho_{\tilde{Q}}^{\text{OPE}}(\omega_1, \omega_2) = \int_0^\infty d\omega_1 \, d\omega_2 \, e^{-\frac{\omega_1}{t_1} - \frac{\omega_2}{t_2}} \rho_{\tilde{Q}}^{\text{had}}(\omega_1, \omega_2). \quad (6.3.7)$$

Fig. 6.2 Leading order
diagram for the three-point
HQET correlator (Eq. 6.3.1).
The sum over the two
possible contractions of the
operator $\tilde{Q}$ is implied

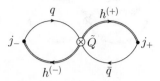

In principle one can proceed by modelling the continuum $\rho_{\tilde{Q}}^{\text{cont}}$. The desired matrix element of the operator $\tilde{Q}$ between the mesonic ground state can then be disentangled by varying the Borel parameters. However, the continuum contribution is exponentially suppressed in the Borel sum rule and it is safe to simply "cut off" the sum rule by assuming that

$$\rho_{\tilde{Q}}^{\text{cont}}(\omega_1, \omega_2) = \rho_{\tilde{Q}}^{\text{OPE}}(\omega_1, \omega_2)\left[1 - \theta(\omega_c - \omega_1)\theta(\omega_c - \omega_2)\right], \qquad (6.3.8)$$

which directly yields a finite-energy sum rule for the matrix elements

$$F^2(\mu)\langle\tilde{Q}(\mu)\rangle e^{-\frac{\bar{\Lambda}}{t_1} - \frac{\bar{\Lambda}}{t_2}} = \int_0^{\omega_c} d\omega_1\, d\omega_2\, e^{-\frac{\omega_1}{t_1} - \frac{\omega_2}{t_2}} \rho_{\tilde{Q}}^{\text{OPE}}(\omega_1, \omega_2). \qquad (6.3.9)$$

From this expression, we see that a determination of the HQET bag parameters requires the computation of the spectral functions $\rho_{\tilde{Q}}^{\text{OPE}}$. The leading condensate corrections have been determined in [10]. We compute the $\mathcal{O}(\alpha_s)$ corrections to the perturbative contribution below.

6.3.2   Spectral Functions at NLO

We determine the spectral functions by first computing the correlator

$$K_{\tilde{Q}}^{\text{pert}}(\omega_1, \omega_2) = K_{\tilde{Q}}^{(0)}(\omega_1, \omega_2) + \frac{\alpha_s}{4\pi} K_{\tilde{Q}}^{(1)}(\omega_1, \omega_2) + \dots \qquad (6.3.10)$$

and then taking its double discontinuity. At LO we have to evaluate the diagram in Fig. 6.2 which factorizes into two two-point functions. We obtain[1]

$$K_{\tilde{Q}_i}^{(0)}(\omega_1, \omega_2) = \left(A_{\tilde{Q}_i} - \delta_{i1}\frac{2\epsilon}{N_c}\right)\Pi^{(0)}(\omega_1)\Pi^{(0)}(\omega_2), \qquad (6.3.11)$$

---

[1]As discussed below the sum rule reproduces the VSA at LO. Therefore the factors $A_{\tilde{Q}_i}$ appear at leading order in the expansion of the results in ϵ. However, the correlator is computed in d dimensions and corrections can appear. We find that this happens only for $\tilde{Q}_1$ where the contraction of the two γ matrices inside the trace yields a d-dimensional factor.

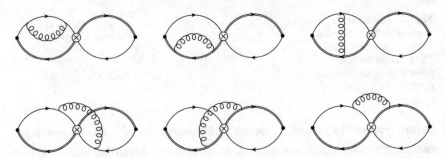

Fig. 6.3 Diagrams contributing to the three-point HQET correlator (Eq. 6.3.1) at NLO. Symmetric diagrams are not shown

where

$$\Pi^{(0)}(\omega) = -\frac{4N_c}{(4\pi)^{2-\epsilon}} \tilde{\mu}^{2\epsilon} (-2\omega)^{2-2\epsilon} \Gamma(2-\epsilon)\Gamma(-2+2\epsilon) \tag{6.3.12}$$

is the LO result for the two-point correlator

$$\Pi(\omega) = i \int d^d x \, e^{ip\cdot x} \langle 0|T\left[\tilde{j}_+^\dagger(0)\tilde{j}_+(x)\right]|0\rangle, \tag{6.3.13}$$

where $\omega = p\cdot v$ and the use of $\tilde{\mu}^2 = \mu^2 \exp(\gamma_E)/(4\pi)$ corresponds to the $\overline{\text{MS}}$ scheme. The bare NLO correction $K_{\tilde{Q}}^{(1),\text{bare}}$ is given by the diagrams shown in Fig. 6.3. At this order we get corrections that do not factorize due to gluon exchange between the left and right-hand side. These genuine three-loop contributions—given by the diagrams in the second row of Fig. 6.3—are the most computationally challenging. The Dirac traces have been evaluated with both TRACER and Package-X. We use the code FIRE [35–37] to find IBP relations [38] between the three-loop integrals and to reduce them to a set of master integrals via the Laporta algorithm [39]. The relevant master integrals have been computed analytically in [9, 40].

The renormalized NLO correlators are given by

$$K_{\tilde{Q}_i}^{(1)} = K_{\tilde{Q}_i}^{(1),\text{bare}} + \frac{1}{2\epsilon}\left[\left(2\tilde{\gamma}_{\tilde{j}}^{(0)}\delta_{ij} + \tilde{\gamma}_{\tilde{Q}_i\tilde{Q}_j}^{(0)}\right)K_{\tilde{Q}_j}^{(0)} + \tilde{\gamma}_{\tilde{Q}_i\tilde{E}_j}^{(0)}K_{\tilde{E}_j}^{(0)}\right]. \tag{6.3.14}$$

where $\tilde{\gamma}_{\tilde{j}}^{(0)} = -3C_F$ is the LO anomalous dimension of the currents $\tilde{j}_\pm$. The contributions from the evanescent operators modify the double discontinuities of the correlators by a finite amount and introduce a dependence in the correlator on the choice of basis of the HQET evanescent operators. This dependence propagates to the HQET bag parameters extracted in the sum rule and cancels with the HQET evanescent scheme dependence of the matching coefficients (as seen in Eq. 6.2.10) in the matching equation (Eq. 6.2.22) for the QCD bag parameters.

Methods to compute the double discontinuities of the correlators have been described in [9, 41]. The results take the form

$$\rho_{\tilde{Q}_i}^{\text{pert}}(\omega_1, \omega_2) = A_{\tilde{Q}_i} \rho_\Pi(\omega_1) \rho_\Pi(\omega_2) + \Delta\rho_{\tilde{Q}_i}, \qquad (6.3.15)$$

where

$$\rho_\Pi(\omega) \equiv \frac{\Pi(\omega + i0) - \Pi(\omega - i0)}{2\pi i}$$
$$= \frac{N_c \omega^2}{2\pi^2} \left[1 + \frac{\alpha_s C_F}{4\pi} \left(17 + \frac{4\pi^2}{3} + 3\ln\frac{\mu^2}{4\omega^2} \right) + \mathcal{O}\left(\alpha_s^2\right) \right] \qquad (6.3.16)$$

is the discontinuity of the two-point correlator (Eq. 6.3.13) up to two-loop order [42–44]. The non-factorizable contributions are

$$\Delta\rho_{\tilde{Q}_i} \equiv \frac{N_c C_F}{4} \frac{\omega_1^2 \omega_2^2}{\pi^4} \frac{\alpha_s}{4\pi} r_{\tilde{Q}_i}(x, L_\omega), \qquad (6.3.17)$$

where $x = \omega_2/\omega_1$, $L_\omega = \ln(\mu^2/(4\omega_1\omega_2))$ and we obtain

$$r_{\tilde{Q}_1}(x, L_\omega) = 8 - \frac{a_2}{2} - \frac{8\pi^2}{3},$$
$$r_{\tilde{Q}_2}(x, L_\omega) = 25 + \frac{a_1}{2} - \frac{4\pi^2}{3} + 6L_\omega + \phi(x),$$
$$r_{\tilde{Q}_4}(x, L_\omega) = 16 - \frac{a_3}{4} - \frac{4\pi^2}{3} + 3L_\omega + \frac{\phi(x)}{2}, \qquad (6.3.18)$$
$$r_{\tilde{Q}_5}(x, L_\omega) = 29 - \frac{a_3}{2} - \frac{8\pi^2}{3} + 6L_\omega + \phi(x),$$

where

$$\phi(x) = \begin{cases} x^2 - 8x + 6\ln(x), & x \leq 1, \\ \frac{1}{x^2} - \frac{8}{x} - 6\ln(x), & x > 1. \end{cases} \qquad (6.3.19)$$

Taking $a_2 = -4$ in accordance with [9] we reproduce their result for $r_{\tilde{Q}_1}$ up to a factor of 2 which is due to the different normalization of the HQET operators.

6.3.3 Sum Rule for the Bag Parameters

Inserting the decomposition given in Eq. 6.3.15 into the sum rule in Eq. 6.3.9 allows us to subtract the factorized contribution, by making use of the sum rule [42–44] for the HQET decay constant

$$F^2(\mu)e^{-\frac{\overline{\Lambda}}{t}} = \int_0^{\omega_c} d\omega \, e^{-\frac{\omega}{t}} \rho_{\Pi}(\omega) + \dots . \tag{6.3.20}$$

The factorizable part of Eq. 6.3.15 exactly reproduces the VSA for the matrix elements. After subtracting it, we obtain a sum rule for the deviation $\Delta B_{\tilde{Q}} = B_{\tilde{Q}} - 1$ from the VSA. In the traditional sum rule approach this gives

$$\Delta B_{\tilde{Q}_i} = \frac{1}{A_{\tilde{Q}_i} F(\mu)^4} \int_0^{\omega_c} d\omega_1 \, d\omega_2 \, e^{\frac{\overline{\Lambda}-\omega_1}{t_1}+\frac{\overline{\Lambda}-\omega_2}{t_2}} \Delta \rho_{\tilde{Q}_i}(\omega_1, \omega_2) \tag{6.3.21}$$

$$= \frac{1}{A_{\tilde{Q}_i}} \frac{\int_0^{\omega_c} d\omega_1 \, d\omega_2 \, e^{-\frac{\omega_1}{t_1}-\frac{\omega_2}{t_2}} \Delta \rho_{\tilde{Q}_i}(\omega_1, \omega_2)}{\left(\int_0^{\omega_c} d\omega_1 \, e^{-\frac{\omega_1}{t_1}} \rho_{\Pi}(\omega_1)\right) \left(\int_0^{\omega_c} d\omega_2 \, e^{-\frac{\omega_2}{t_2}} \rho_{\Pi}(\omega_2)\right)} . \tag{6.3.22}$$

The stability of the sum rule in Eq. 6.3.22 can then be assessed numerically by variation of the cutoff ω_c and the Borel parameters t_i (see e.g. [41, 43]).

In our analysis we instead follow a different approach that allows us to obtain analytic results for the HQET bag parameters. We exploit the fact that the dispersion relation given in Eq. 6.3.3 is not violated by the introduction of an arbitrary weight function $w(\omega_1, \omega_2)$ in the integration, as long as w is chosen such that no additional discontinuities appear in the complex plane.[2] In the presence of such a weight function w, the square of the sum rule for the decay constant (Eq. 6.3.20) takes the form

$$F^4(\mu)e^{-\frac{\overline{\Lambda}}{t_1}-\frac{\overline{\Lambda}}{t_2}} w(\overline{\Lambda}, \overline{\Lambda}) = \int_0^{\omega_c} d\omega_1 \, d\omega_2 \, e^{-\frac{\omega_1}{t_1}-\frac{\omega_2}{t_2}} w(\omega_1, \omega_2) \rho_{\Pi}(\omega_1) \rho_{\Pi}(\omega_2) + \dots .$$
$$\tag{6.3.23}$$

Since the condensate contributions have already been taken into account in [9, 10] and are in the sub-percent range we only focus on the perturbative contribution to the sum rule. By using Eq. 6.3.23 with the choice[3]

$$w_{\tilde{Q}_i}(\omega_1, \omega_2) = \frac{\Delta \rho_{\tilde{Q}_i}^{\text{pert}}(\omega_1, \omega_2)}{\rho_{\Pi}^{\text{pert}}(\omega_1) \rho_{\Pi}^{\text{pert}}(\omega_2)} = \frac{C_F}{N_c} \frac{\alpha_s}{4\pi} r_{\tilde{Q}_i}(x, L_\omega), \tag{6.3.24}$$

we can remove the integration in Eq. 6.3.21 altogether and find the simple result

---

[2] The arbitrariness of the weight function is a mathematical statement which holds for the dispersion relation. The sum rule in Eq. 6.3.7 does however also assume quark-hadron duality and breaks down if pathological weight functions are used, e.g. rapidly oscillating ones. In the following we only use slowly varying weight functions with support on the complete integration domain.

[3] This choice, while technically original, is a relatively straightforward modification of the previous usage of weight functions in sum rule calculations.

$$\Delta B_{\tilde{Q}_i}^{\text{pert}}(\mu_\rho) = \frac{C_F}{N_c A_{\tilde{Q}_i}} \frac{\alpha_s(\mu_\rho)}{4\pi} r_{\tilde{Q}_i}\left(1, \log \frac{\mu_\rho^2}{4\overline{\Lambda}^2}\right). \tag{6.3.25}$$

This sum rule is valid at a low scale $\mu_\rho \sim 2\omega_i \sim 2\overline{\Lambda}$ where the logarithms that appear in the spectral functions are small. From there we have to evolve the results for the bag parameters up to the scale $\mu_m \sim m_b$ where the matching (which we described in Sect. 6.2.3) to the QCD bag parameters can be performed without introducing large logarithms. From Eq. 6.2.15 and the running of the HQET operators and decay constant

$$\frac{d\vec{\tilde{Q}}}{d\ln\mu} = -\hat{\tilde{\gamma}}_{\tilde{Q}\tilde{Q}}\,\vec{\tilde{Q}}\,, \qquad \frac{dF(\mu)}{d\ln\mu} = -\tilde{\gamma}_{\tilde{j}}F(\mu)\,, \tag{6.3.26}$$

we obtain the RG equations for the HQET bag parameters

$$\frac{d\vec{B}_{\tilde{Q}}}{d\ln\mu} = -\left(\hat{A}_{\tilde{Q}}^{-1}\hat{\tilde{\gamma}}_{\tilde{Q}\tilde{Q}}\hat{A}_{\tilde{Q}} - 2\tilde{\gamma}_{\tilde{j}}\right)\vec{B}_{\tilde{Q}} \equiv -\hat{\tilde{\gamma}}_{\tilde{B}}\vec{B}_{\tilde{Q}}\,, \tag{6.3.27}$$

where $\hat{A}_{\tilde{Q}}$ is the diagonal matrix with entries $A_{\tilde{Q}}$ given in Eq. 6.2.16. The LO solution to Eq. 6.3.27 takes the form

$$\vec{B}_{\tilde{Q}}(\mu) = \hat{U}_{\tilde{B}}^{(0)}(\mu, \mu_0)\,\vec{B}_{\tilde{Q}}(\mu_0)\,. \tag{6.3.28}$$

Here $\hat{U}_{\tilde{B}}^{(0)}$ is the LO evolution matrix

$$\hat{U}_{\tilde{B}}^{(0)}(\mu, \mu_0) = \left(\frac{\alpha_s(\mu)}{\alpha_s(\mu_0)}\right)^{\frac{\hat{\tilde{\gamma}}_{\tilde{B}}^{(0)}}{2\beta_0}} = \hat{V}\left(\frac{\alpha_s(\mu)}{\alpha_s(\mu_0)}\right)^{\frac{\vec{\tilde{\gamma}}_{\tilde{B}}^{(0)}}{2\beta_0}}\hat{V}^{-1}\,, \tag{6.3.29}$$

with $\hat{V}$ the transformation that diagonalizes the ADM $\hat{\tilde{\gamma}}_{\tilde{B}}^{(0)}$

$$\hat{\tilde{\gamma}}_{\tilde{B}}^{(0),D} = \hat{V}^{-1}\hat{\tilde{\gamma}}_{\tilde{B}}^{(0)}\hat{V}\,, \tag{6.3.30}$$

and the vector $\vec{\tilde{\gamma}}_{\tilde{B}}^{(0)}$ contains the diagonal entries of $\hat{\tilde{\gamma}}_{\tilde{B}}^{(0),D}$. As part of our error analysis we allow the matching scale μ_m to differ from $\overline{m}_b(\overline{m}_b)$ and then evolve the QCD bag parameters back to $\overline{m}_b(\overline{m}_b)$. The LO evolution matrix has the same form as its HQET counterpart (Eq. 6.3.29) while the anomalous dimension matrix of the QCD bag parameters is given by

$$\hat{\gamma}_B = \hat{A}_Q^{-1}\hat{\gamma}_{QQ}\hat{A}_Q\,. \tag{6.3.31}$$

We only resum the leading logarithms because the NLO anomalous dimensions in HQET are currently not known. This implies that dependence of the QCD matrix elements on the basis of evanescent HQET operators does not fully cancel. As discussed below, we use variation of the parameters a_i to estimate the effects of NLL resummation. We expect this effect to be small since the scales μ_ρ and μ_m are not very widely separated and so $\ln(\mu_m/\mu_\rho) \sim \mathcal{O}(1)$.

6.4 Results for $\Delta B = 2$ Operators

We describe our analysis in Sect. 6.4.1, then give the results for the bag parameters, together with a comparison with other works, in Sect. 6.4.2. In Sect. 6.4.3 the results for the mixing observables with our bag parameters are shown.

6.4.1 Details of the Analysis

We determine the HQET bag parameters from the sum rule in Eq. 6.3.25] with the central values $\mu_\rho = 1.5\,\text{GeV}$ and $\overline{\Lambda} = 0.5\,\text{GeV}$. We use RunDec [45, 46] to evolve $\alpha_s(M_Z) = 0.1181$ [47] down to the bottom quark $\overline{\text{MS}}$ mass $\overline{m}_b(\overline{m}_b) = 4.203\,\text{GeV}$ [48, 49] with five-loop accuracy [50–54]. From there we use two-loop running with four and five flavours in HQET and QCD, respectively. The decoupling of the bottom quark is trivial at this accuracy.

The HQET bag parameters are then evolved from the scale μ_ρ up to the scale $\mu_m = \overline{m}_b(\overline{m}_b)$ using Eq. 6.3.28. There the matching to the QCD bag parameters is performed. The factors $C_{Q_i \tilde{Q}_j}(\mu)/C^2(\mu)$ are expanded in α_s and truncated after the linear term. We also expand the ratios $A_{\tilde{Q}_j}/A_{Q_i}$ strictly in $\overline{\Lambda}/m_b$ and m_q/m_b. Up to higher order perturbative corrections, this is equivalent to the use of the VSA for the power-suppressed HQET operators that arise in the QCD-HQET matching (Eq. 6.2.4).

A small dependence on the choice of basis for the evanescent HQET operators remains in the QCD bag parameters because the RG evolution of the HQET bag parameters is only known at the LL level. We have checked that the a_i-dependence fully cancels when the scales μ_ρ and μ_m are identified and the matching (Eq. 6.2.22) is strictly expanded in the strong coupling, which serves as a strong cross-check of our calculation. For different scales μ_ρ and μ_m the remaining a_i-dependence can be removed by a future computation of the NLO ADMs.

Finally, we convert the QCD bag parameters B_Q to the usual convention $\overline{B}_Q$ defined in Eq. 6.2.14. This is done by expanding the ratios of the prefactors $A_Q/\overline{A}_Q(\overline{m}_b(\overline{m}_b))$ in α_s and truncating them after the linear term.

To estimate the errors of the bag parameters we take the following sources of uncertainties into account:

- The uncertainty in the analytic form of the sum rule given in Eq. 6.3.25 is estimated through variation of the residual mass $\overline{\Lambda}$ in the range 0.4–0.6 GeV. In addition we include an intrinsic sum rule uncertainty of 0.02 in the HQET bag parameters. This numerical value is determined by comparing the analytic values from Eq. 6.3.25 with results obtained from the traditional sum rule approach (Eq. 6.3.22).

- The condensate contributions to $B_{\tilde{Q}_1}$ and $B_{\tilde{Q}_2}$ are taken from [10] and are in the sub-percent range. For $B_{\tilde{Q}_4}$ and $B_{\tilde{Q}_5}$, which have not been determined in that work, we therefore add an additional error of ± 0.01 to the perturbative results.
- To assign an uncertainty from the unknown α_s^2 contributions to the spectral densities we vary the scale μ_ρ in the range 1–2 GeV.
- As discussed above we implicitly include higher-order corrections in $1/m_b$ in the VSA approximation. The non-factorizable corrections of this kind are of the order $(\alpha_s/\pi) \times (\overline{\Lambda}/m_b) \sim 0.01$, which we take as an estimate for the error.
- Higher order perturbative contributions to the QCD-HQET matching relation and the RG evolution of the bag parameters are estimated through variation of μ_m in the range 3–6 GeV and variation of the a_i in the range ± 10. The QCD bag parameters are then evolved to the central scale $\overline{m}_b(\overline{m}_b)$ with LL accuracy as described in Sect. 6.3.3. The variation of μ_m by the usual factors of $1/2$ and 2 would lead to a doubling of the matching uncertainty estimates given below, which would significantly exceed the effect of the NLO matching at the central scale. We therefore use a less conservative range but cannot exclude larger matching effects at NNLO at present, while a full calculation is not available.
- The parametric uncertainty from $\alpha_s(M_Z)$ is in the 0.1% range and is hence neglected.

The individual errors are then summed in quadrature. We also divide the uncertainties into a sum rule uncertainty which contains the first three items in the list above and a matching uncertainty which contains the remaining three.

6.4.2 Results and Comparison

From the sum rule we obtain the HQET bag parameters

$$
\begin{aligned}
B_{\tilde{Q}_1}(1.5\,\text{GeV}) &= 0.910^{+0.023}_{-0.031} \\
&= 0.910^{+0.000}_{-0.000}(\overline{\Lambda}) \,^{+0.020}_{-0.020}(\text{intr.}) \,^{+0.005}_{-0.005}(\text{cond.}) \,^{+0.011}_{-0.024}(\mu_\rho), \\
B_{\tilde{Q}_2}(1.5\,\text{GeV}) &= 0.923^{+0.029}_{-0.035} \\
&= 0.923^{+0.016}_{-0.020}(\overline{\Lambda}) \,^{+0.020}_{-0.020}(\text{intr.}) \,^{+0.004}_{-0.004}(\text{cond.}) \,^{+0.013}_{-0.020}(\mu_\rho), \\
B_{\tilde{Q}_4}(1.5\,\text{GeV}) &= 1.009^{+0.024}_{-0.023} \\
&= 1.009^{+0.007}_{-0.006}(\overline{\Lambda}) \,^{+0.020}_{-0.020}(\text{intr.}) \,^{+0.010}_{-0.010}(\text{cond.}) \,^{+0.003}_{-0.003}(\mu_\rho), \\
B_{\tilde{Q}_5}(1.5\,\text{GeV}) &= 1.004^{+0.030}_{-0.028} \\
&= 1.004^{+0.020}_{-0.016}(\overline{\Lambda}) \,^{+0.020}_{-0.020}(\text{intr.}) \,^{+0.010}_{-0.010}(\text{cond.}) \,^{+0.004}_{-0.006}(\mu_\rho),
\end{aligned}
\tag{6.4.1}
$$

where we have set $a_i = 0$ in order to uniquely specify a basis of evanescent HQET operators. The individual uncertainties were determined as described above and added in quadrature. The corrections to the VSA for scales in the range 1–2 GeV are at the level of 5–11% for $\tilde{Q}_{1,2}$ and 0–4% for $\tilde{Q}_{4,5}$. We find that the total sum rule uncertainties of the bag parameters are quite small. This is because the sum rule (Eq. 6.3.25) is formulated for the deviation from the VSA and the substantial relative uncertainties of the sum rule itself are small in comparison with the VSA contribution to the bag parameters.

Following the steps outlined in Sect. 6.4.1 we obtain the following results for the QCD bag parameters

$$\overline{B}_{Q_1}(\overline{m}_b(\overline{m}_b)) = 0.868^{+0.051}_{-0.050}$$
$$= 0.868^{+0.021}_{-0.029}(\text{sum rule})^{+0.046}_{-0.041}(\text{matching}),$$
$$\overline{B}_{Q_2}(\overline{m}_b(\overline{m}_b)) = 0.842^{+0.078}_{-0.073}$$
$$= 0.842^{+0.028}_{-0.033}(\text{sum rule})^{+0.073}_{-0.065}(\text{matching}),$$
$$\overline{B}_{Q_3}(\overline{m}_b(\overline{m}_b)) = 0.818^{+0.162}_{-0.159}$$
$$= 0.818^{+0.126}_{-0.132}(\text{sum rule})^{+0.102}_{-0.087}(\text{matching}), \quad (6.4.2)$$
$$\overline{B}_{Q_4}(\overline{m}_b(\overline{m}_b)) = 1.049^{+0.092}_{-0.084}$$
$$= 1.049^{+0.025}_{-0.025}(\text{sum rule})^{+0.089}_{-0.080}(\text{matching}),$$
$$\overline{B}_{Q_5}(\overline{m}_b(\overline{m}_b)) = 1.073^{+0.083}_{-0.075}$$
$$= 1.073^{+0.028}_{-0.026}(\text{sum rule})^{+0.078}_{-0.070}(\text{matching}).$$

The evolution to the scale $\overline{m}_b(\overline{m}_b))$ and the matching to QCD increase the deviations from the VSA to up to 18%. With the exception of $\overline{B}_{Q_3}$ the uncertainties of the bag parameters are dominated by the matching. A detailed list of the uncertainties can be found in Appendix G.2.

In Fig. 6.4 we compare our results to other recent determinations from lattice simulations [2–4, 55] and sum rules [9]. We find excellent agreement for the bag parameters of the operators Q_1, Q_2 and Q_3. The uncertainties of our sum rule analysis are similar to those obtained on the lattice. We observe that the uncertainty of the bag parameter $\overline{B}_{Q_3}$ is significantly larger than those of $\overline{B}_{Q_1}$ and $\overline{B}_{Q_2}$. This is related to the small colour factor $A_{Q_3} = 1/3 + \mathcal{O}(1/m_b)$ which implies that the sum rule uncertainties get enhanced by the factors $A_{\tilde{Q}_1}/A_{Q_3} = 8 + \mathcal{O}(1/m_b)$ and $A_{\tilde{Q}_2}/A_{Q_3} = -5 + \mathcal{O}(1/m_b)$ when we match from HQET to QCD (Eq. 6.2.22)—the absolute sum rule uncertainty of the matrix element of Q_3 is of a similar size as that of the other operators.

The tiny difference of the central value of $\overline{B}_{Q_1}$ compared to the sum rule determination [9] is mostly due to different scale choices. Since $\overline{B}_{Q_1}$ does not run at the LL order, [9] sets all scales equal to the bottom quark mass. We on the other hand, evaluate the sum rule at a lower scale $\mu_\rho \sim 1.5$ GeV where the strong coupling is larger and causes a bigger deviation from the VSA.

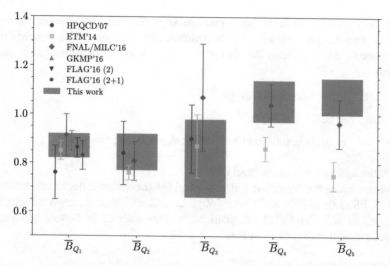

Fig. 6.4 Comparison of our results for the $\Delta B = 2$ bag parameters at the scale $\overline{m}_b(\overline{m}_b)$ to the lattice values of HPQCD'07 [2], ETM'14 [3] and FNAL/MILC'16 [4], the FLAG averages [55] and the sum rule result GKMP'16 [9]

Only two previous lattice results [3, 4] exist for the matrix elements of the operators Q_4 and Q_5, and they differ at the level of more than two sigma. Our results are in very good agreement with those of [4] and show an even higher level of tension with [3] in $\overline{B}_{Q_5}$.

6.4.3 B_s and B_d Mixing Observables

In this section, we consider the observables ΔM_s, $\Delta \Gamma_s$, and a_{sl}^s (see Eqs. 2.4.16 and 2.5.2 for definitions). Using our values for the bag parameters, we give predictions for these observables and compare them to the current experimental averages given by HFLAV [56]. In our sum rule determination we have assumed the light quark q in the B_q meson to be massless – the corrections to Eq. 6.3.25 from a non-zero strange quark mass are of the order $(\alpha_s/\pi)(m_s/(2\overline{\Lambda})) \approx 0.02$, and so we add another uncertainty of ± 0.02 in quadrature to the results (Eq. 6.4.2) to account for these unknown corrections. (This point has recently been discussed in more detail in [57].) The effect on the total uncertainty is small.

We find excellent agreement between experiment and the SM prediction for the mass difference:

$$
\begin{aligned}
\Delta M_s^{\text{exp}} &= (17.757 \pm 0.021)\,\text{ps}^{-1}, \\
\Delta M_s^{\text{SM}} &= \left(18.1^{+1.9}_{-1.8}\right)\text{ps}^{-1} \\
&= \left(18.1^{+1.3}_{-1.2}(\text{had.}) \pm 0.1(\text{scale})^{+1.4}_{-1.3}(\text{param.})\right)\text{ps}^{-1},
\end{aligned}
\tag{6.4.3}
$$

where we have used the input values given in Appendix G.2. The 10% uncertainty of the SM prediction is dominated by the hadronic and parametric CKM uncertainties which are of the same size. We also give results for the mass difference in the B_d system

$$\Delta M_d^{\text{exp}} = (0.5065 \pm 0.0019) \, \text{ps}^{-1},$$
$$\Delta M_d^{\text{SM}} = (0.61 \pm 0.09) \, \text{ps}^{-1} \tag{6.4.4}$$
$$= (0.61 \pm 0.04(\text{had.}) \pm 0.00(\text{scale}) \pm 0.08(\text{param.})) \, \text{ps}^{-1},$$

where the agreement is at the level of $1.1 \, \sigma$.

We determine the decay rate difference and the semileptonic decay asymmetry in the $\overline{\text{MS}}$, PS [58], 1S [59] and kinetic [60] mass schemes with the mass values given in Appendix G.2. The $\overline{\text{MS}}$ charm quark mass at the scale of the bottom quark mass has been used throughout. We obtain

$$\Delta \Gamma_s^{\text{exp}} = (0.090 \pm 0.005) \, \text{ps}^{-1},$$
$$\Delta \Gamma_s^{\overline{\text{MS}}} = \left(0.080_{-0.023}^{+0.018}\right) \text{ps}^{-1}$$
$$= \left(0.080 \pm 0.016(\text{had.})_{-0.015}^{+0.006}(\text{scale}) \pm 0.006(\text{param.})\right) \text{ps}^{-1},$$
$$\Delta \Gamma_s^{\text{PS}} = \left(0.079_{-0.026}^{+0.020}\right) \text{ps}^{-1}$$
$$= \left(0.079 \pm 0.018(\text{had.})_{-0.018}^{+0.007}(\text{scale}) \pm 0.006(\text{param.})\right) \text{ps}^{-1}, \tag{6.4.5}$$
$$\Delta \Gamma_s^{\text{1S}} = \left(0.075_{-0.028}^{+0.021}\right) \text{ps}^{-1}$$
$$= \left(0.075 \pm 0.019(\text{had.})_{-0.020}^{+0.008}(\text{scale}) \pm 0.006(\text{param.})\right) \text{ps}^{-1},$$
$$\Delta \Gamma_s^{\text{kin}} = (0.076_{-0.027}^{+0.020}) \, \text{ps}^{-1}$$
$$= \left(0.076 \pm 0.018(\text{had.})_{-0.019}^{+0.008}(\text{scale}) \pm 0.006(\text{param.})\right) \text{ps}^{-1},$$

and

$$a_{\text{sl}}^{s,\text{exp}} = (-60 \pm 280) \times 10^{-5},$$
$$a_{\text{sl}}^{s,\overline{\text{MS}}} = (2.1 \pm 0.3) \times 10^{-5}$$
$$= \left(2.1 \pm 0.1(\text{had.})_{-0.1}^{+0.0}(\text{scale})_{-0.3}^{+0.2}(\text{param.})\right) \times 10^{-5},$$
$$a_{\text{sl}}^{s,\text{PS}} = (2.0_{-0.3}^{+0.2}) \times 10^{-5}$$
$$= \left(2.0 \pm 0.1(\text{had.})_{-0.1}^{+0.0}(\text{scale}) \pm 0.2(\text{param.})\right) \times 10^{-5},$$
$$a_{\text{sl}}^{s,\text{1S}} = (2.0_{-0.3}^{+0.2}) \times 10^{-5}$$
$$= \left(2.0 \pm 0.0(\text{had.})_{-0.1}^{+0.0}(\text{scale}) \pm 0.2(\text{param.})\right) \times 10^{-5},$$

$$a_{sl}^{s,kin} = (2.0^{+0.2}_{-0.3}) \times 10^{-5}$$
$$= \left(2.0 \pm 0.1(had.)^{+0.0}_{-0.1}(scale) \pm 0.2(param.)\right) \times 10^{-5}. \qquad (6.4.6)$$

The different mass schemes are in good agreement with each other and we adopt the PS mass scheme as our central result. The SM value for the decay rate difference is in good agreement with the experimental average. The theory uncertainty is currently at the level of 30%, and is dominated by the matrix elements of the dimension-seven operators, in particular the VSA estimate $\overline{B}_{R_2} = 1 \pm 0.5$ which contributes $\pm 0.016 \, \text{ps}^{-1}$ to the uncertainty. The second largest contribution is the scale variation. A detailed overview is given in Appendix G.2. To achieve a significant reduction of the combined uncertainties, a determination of the dimension-seven matrix elements and a NNLO calculation of the perturbative matching are needed (see [57, 61] for the first work in this direction).

The experimental uncertainty for the semileptonic decay asymmetry is two orders of magnitude larger than the SM prediction, making it a clear null test for the SM [62], while the decay rate difference and the semileptonic decay asymmetry in the B_d system have not yet been measured. The current experimental averages and our predictions are

$$\Delta\Gamma_d^{exp} = (-1.3 \pm 6.6) \times 10^{-3} \, \text{ps}^{-1},$$
$$\Delta\Gamma_d^{PS} = \left(2.7^{+0.8}_{-0.9}\right) \times 10^{-3} \text{ps}^{-1}$$
$$= \left(2.7^{+0.6}_{-0.6}(had.)^{+0.2}_{-0.6}(scale)^{+0.4}_{-0.4}(param.)\right) \times 10^{-3} \text{ps}^{-1},$$
$$a_{sl}^{d,exp} = (-21 \pm 17) \times 10^{-4}, \qquad (6.4.7)$$
$$a_{sl}^{d,PS} = (-4.0 \pm 0.5) \times 10^{-4}$$
$$= \left(-4.0 \pm 0.1(had.)^{+0.2}_{-0.1}(scale) \pm 0.5(param.)\right) \times 10^{-4}.$$

The results obtained in different mass schemes are compatible and the relative uncertainties of the predictions are of the same magnitude as in the B_s system.

6.5 $\Delta B = 0$ Operators and Ratios of B Meson Lifetimes

The dominant contribution to lifetime differences between the mesons B_q with $q = u, d, s$ is due to spectator effects which first appear as dimension-six contributions in the HQE (see the bullet points in Sect. 2.2). The NLO Wilson coefficients have been computed in [63–65]. The dimension-seven contributions are known at LO [15, 66]. We define the set of operators in Sect. 6.5.1 and present the results for their bag parameters in Sect. 6.5.2. The updated HQE results for the B meson lifetime ratios are given in Sect. 6.5.3.

Fig. 6.5 Leading order eye
contraction

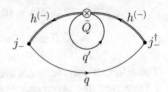

6.5.1 Operators and Matrix Elements

The following QCD operators enter at dimension six:

$$
\begin{aligned}
Q_1^q &= (\bar{b}\gamma_\mu(1 - \gamma^5)q)(\bar{q}\gamma^\mu(1 - \gamma^5)b)\,, \\
Q_2^q &= (\bar{b}(1 - \gamma^5)q)(\bar{q}\gamma^\mu(1 + \gamma^5)b)\,, \\
T_1^q &= (\bar{b}\gamma_\mu(1 - \gamma^5)t^a q)(\bar{q}\gamma^\mu(1 - \gamma^5)t^a b)\,, \\
T_2^q &= (\bar{b}(1 - \gamma^5)t^a q)(\bar{q}(1 + \gamma^5)t^a b)\,.
\end{aligned}
\tag{6.5.1}
$$

On the HQET side they match onto

$$
\begin{aligned}
\tilde{Q}_1^q &= (\bar{h}\gamma_\mu(1 - \gamma^5)q)(\bar{q}\gamma^\mu(1 - \gamma^5)h)\,, \\
\tilde{T}_1^q &= (\bar{h}\gamma_\mu(1 - \gamma^5)t^a q)(\bar{q}\gamma^\mu(1 - \gamma^5)t^a h)\,, \\
\tilde{Q}_2^q &= (\bar{h}(1 - \gamma^5)q)(\bar{q}(1 + \gamma^5)h)\,, \\
\tilde{T}_2^q &= (\bar{h}(1 - \gamma^5)t^a q)(\bar{q}(1 + \gamma^5)t^a h)\,.
\end{aligned}
\tag{6.5.2}
$$

Our basis of evanescent operators and the results of the matching computation can
be found in Appendix G.1.2. We only consider the isospin-breaking combinations
of operators

$$
Q_i = Q_i^u - Q_i^d\,, \qquad T_i = T_i^u - T_i^d\,,
\tag{6.5.3}
$$

and their analogues in HQET. This implies that the eye contractions displayed in Fig.
6.5 cancel in the limit of exact isospin symmetry.

The matrix elements are

$$
\langle Q_i(\mu)\rangle = A_i\, f_B^2 M_B^2\, B_i(\mu)\,, \qquad \langle T_i(\mu)\rangle = A_i\, f_B^2 M_B^2\, \epsilon_i(\mu)\,,
\tag{6.5.4}
$$

where $\langle Q\rangle = \langle B^-|Q|B^-\rangle$, the coefficients read

$$
A_1 = 1\,, \qquad A_2 = \frac{M_B^2}{(m_b + m_q)^2}\,,
\tag{6.5.5}
$$

and $B_i = 1$, $\epsilon_i = 0$ corresponds to the VSA approximation, matching our definitions
in Eqs. C.0.3 and C.0.4. Similarly we obtain for the HQET operators

$$\langle\!\langle \tilde{Q}_i(\mu) \rangle\!\rangle = \tilde{A}_i \, F^2(\mu) \, \tilde{B}_i(\mu) \,, \qquad \langle\!\langle \tilde{T}_i(\mu) \rangle\!\rangle = \tilde{A}_i \, F^2(\mu) \, \tilde{\epsilon}_i(\mu) \,, \tag{6.5.6}$$

where

$$\tilde{A}_1 = 1 \,, \qquad \tilde{A}_2 = 1 \,. \tag{6.5.7}$$

6.5.2 Results for the Spectral Functions and Bag Parameters

For the $\Delta B = 0$ operators we use the same conventions for the decomposition of the three-point correlator and the sum rule as for the $\Delta B = 2$ operators above. We obtain for the double discontinuities of the non-factorizable contributions

$$r_{\tilde{Q}_i}(x, L_\omega) = 0 \,,$$

$$r_{\tilde{T}_1}(x, L_\omega) = -8 + \frac{a_1}{8} + \frac{2\pi^2}{3} - \frac{3}{2}L_\omega - \frac{1}{4}\phi(x) \,, \tag{6.5.8}$$

$$r_{\tilde{T}_2}(x, L_\omega) = -\frac{29}{4} + \frac{a_2}{8} + \frac{2\pi^2}{3} - \frac{3}{2}L_\omega - \frac{1}{4}\phi(x) \,.$$

The leading condensate contributions have been determined in [11]. From their results we deduce that

$$\rho_{\tilde{Q}_i}^{\text{cond}}(\omega_1, \omega_2) = 0 + \dots \,,$$

$$\rho_{\tilde{T}_1}^{\text{cond}}(\omega_1, \omega_2) = \frac{\langle g_s \bar{q} \sigma_{\mu\nu} G^{\mu\nu} q \rangle}{128\pi^2} \left[\delta(\omega_1) + \delta(\omega_2) \right] + \dots \,,$$

$$\rho_{\tilde{T}_2}^{\text{cond}}(\omega_1, \omega_2) = -\frac{1}{64\pi^2} \left[\left\langle \frac{\alpha_s}{\pi} G^2 \right\rangle + \left\langle g_s \bar{q} \sigma_{\mu\nu} G^{\mu\nu} q \right\rangle \left[\delta(\omega_1) + \delta(\omega_2) \right] \right] + \dots \,, \tag{6.5.9}$$

where the dots indicate factorizable contributions, α_s corrections and contributions from condensates of dimension six and higher. To determine the condensate contributions to the HQET parameters we have used the traditional form of the sum rule, because the appearance of the δ-functions prevents the application of a weight function analogous to Eq. 6.3.24. We find

$$\Delta \tilde{B}_i^{\text{cond}}(1.5\,\text{GeV}) = 0.000 \pm 0.002 \,,$$

$$\Delta \tilde{\epsilon}_1^{\text{cond}}(1.5\,\text{GeV}) = -0.005 \pm 0.003 \,, \tag{6.5.10}$$

$$\Delta \tilde{\epsilon}_2^{\text{cond}}(1.5\,\text{GeV}) = 0.006 \pm 0.004 \,.$$

The associated errors were determined from an uncertainty of ± 0.002 for missing higher-dimensional condensates, variations of the Borel parameters and the continuum cutoff and the uncertainty in the condensates

$$\left\langle \frac{\alpha_s}{\pi} G^2 \right\rangle = (0.012 \pm 0.006)\, \text{GeV}^4 \,,$$

$$\left\langle g_s \bar{q} \sigma_{\mu\nu} G^{\mu\nu} q \right\rangle = (-0.011 \pm 0.002)\, \text{GeV}^5 \,. \tag{6.5.11}$$

We note that our results for the contributions of the condensate corrections to the deviation of the bag parameters from the VSA are much smaller than those of [11]. This is mostly due to the choice of the Borel parameter. We use $t \sim 1\,\text{GeV}$ where the sum rule is stable against variations of the Borel parameter, while the Borel region of [11] translates to $t = 0.35$–$0.5\,\text{GeV}$, where the sum rule becomes unstable as can be seen in their plots. Our choice is also preferred by other modern sum rule analyses [10, 67, 68].

Following the same analysis strategy for the perturbative contributions as described for the $\Delta B = 2$ bag parameters in Sect. 6.4.1, we find the HQET bag parameters

$$\tilde{B}_1(1.5\,\text{GeV}) = 1.000^{+0.020}_{-0.020}$$
$$= 1.000^{+0.000}_{-0.000}(\overline{\Lambda})^{+0.020}_{-0.020}(\text{intr.})^{+0.002}_{-0.002}(\text{cond.})^{+0.000}_{-0.001}(\mu_\rho)\,,$$
$$\tilde{B}_2(1.5\,\text{GeV}) = 1.000^{+0.020}_{-0.020}$$
$$= 1.000^{+0.000}_{-0.000}(\overline{\Lambda})^{+0.020}_{-0.020}(\text{intr.})^{+0.002}_{-0.002}(\text{cond.})^{+0.000}_{-0.001}(\mu_\rho)\,,$$
$$\tilde{\epsilon}_1(1.5\,\text{GeV}) = -0.016^{+0.021}_{-0.022}$$
$$= -0.016^{+0.007}_{-0.008}(\overline{\Lambda})^{+0.020}_{-0.020}(\text{intr.})^{+0.003}_{-0.003}(\text{cond.})^{+0.003}_{-0.003}(\mu_\rho)\,,$$
$$\tilde{\epsilon}_2(1.5\,\text{GeV}) = 0.004^{+0.022}_{-0.022}$$
$$= 0.004^{+0.007}_{-0.008}(\overline{\Lambda})^{+0.020}_{-0.020}(\text{intr.})^{+0.004}_{-0.004}(\text{cond.})^{+0.002}_{-0.002}(\mu_\rho)\,. \tag{6.5.12}$$

where we have set $a_1 = a_2 = 0$. At the considered order, there is no deviation from the VSA for the bag parameters of the colour singlet operators, as can be seen from Eqs. 6.5.8 and 6.5.9, because the corresponding colour factors vanish. The deviations for the colour octet operators are in the range 0–2% for scales μ_ρ between 1 GeV and 2 GeV. In QCD we obtain

$$\overline{B}_1(\mu = \overline{m}_b(\overline{m}_b)) = 1.028^{+0.064}_{-0.056}$$
$$= 1.028^{+0.019}_{-0.019}(\text{sum rule})^{+0.061}_{-0.053}(\text{matching})\,,$$
$$\overline{B}_2(\mu = \overline{m}_b(\overline{m}_b)) = 0.988^{+0.087}_{-0.079}$$
$$= 0.988^{+0.020}_{-0.020}(\text{sum rule})^{+0.085}_{-0.077}(\text{matching})\,,$$
$$\overline{\epsilon}_1(\mu = \overline{m}_b(\overline{m}_b)) = -0.107^{+0.028}_{-0.029}$$
$$= -0.107^{+0.023}_{-0.024}(\text{sum rule})^{+0.015}_{-0.017}(\text{matching})\,,$$
$$\overline{\epsilon}_2(\mu = \overline{m}_b(\overline{m}_b)) = -0.033^{+0.021}_{-0.021}$$
$$= -0.033^{+0.018}_{-0.018}(\text{sum rule})^{+0.011}_{-0.011}(\text{matching})\,. \tag{6.5.13}$$

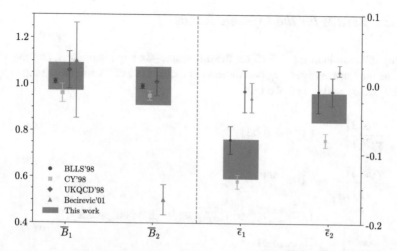

Fig. 6.6 Comparison of our results for the $\Delta B = 0$ bag parameters at the scale $\overline{m}_b(\overline{m}_b)$ to the HQET sum rule results BLLS'98 [11] and CY'98 [12], and the lattice values of UKQCD'98 [69] and Becirevic'01 [14]

The RG evolution and the perturbative matching cause larger deviations from the VSA, but these do not exceed 11%.

In Fig. 6.6 we compare our results to previous ones from sum rules [11, 12] and the lattice [14, 69]. The results of [11, 12, 69] were obtained within HQET. For the comparison we match their results to QCD at tree level while expanding factors of $\tilde{A}_i/\tilde{A}_Q(\overline{m}_b(\overline{m}_b))$ in $1/m_b$. As discussed in Sect. 6.4.1 this effectively includes $1/m_b$ corrections in the VSA approximation.

The $\overline{B}_i$ are in good agreement, with the exception of the value for $\overline{B}_2$ from [14], which differs from the other results and the VSA by a factor of about two. While the other sum rule results for the $\overline{\epsilon}_i$ agree reasonably well with ours, the lattice results for $\overline{\epsilon}_1$ show significantly smaller deviations from the VSA. The similarity between the sum rule results [11, 12] and ours appears to be mostly coincidental, because of the following reasons. We find that the bulk of the deviation from the VSA in the $\overline{\epsilon}_i$ is due to the RG running and matching, while the latter was not considered in [11, 12]. In their analyses, there is instead a sizeable deviation at the hadronic scale, originating from the condensate contributions. In comparison with [11] we find that this is due to the choice of very small values of the Borel parameter which lie outside of the stability region as discussed above. To properly assess the origin of the smallness of the lattice results [14, 69] for the $\overline{\epsilon}_i$ would need comparison with a state-of-the-art lattice simulation which currently does not exist, as many of the approximations made in [14, 69], like quenching, have since been reappraised.

6.5.3 Results for the Lifetime Ratios

Using our results in Eq. 6.5.13 for the dimension-six bag parameters, and the VSA for the dimension-seven bag parameters (as defined in [15]) with uncertainties $\rho_i = 1 \pm 1/12$, $\sigma_i = 0 \pm 1/6$, we find

$$\frac{\tau(B^+)}{\tau(B_d)}\bigg|_{\text{exp}} = 1.076 \pm 0.004 \, ,$$

$$\frac{\tau(B^+)}{\tau(B_d)}\bigg|_{\overline{\text{MS}}} = 1.078^{+0.021}_{-0.023}$$

$$= 1.078^{+0.020}_{-0.019}(\text{had.})^{+0.002}_{-0.011}(\text{scale}) \pm 0.006(\text{param.}) \, ,$$

$$\frac{\tau(B^+)}{\tau(B_d)}\bigg|_{\text{PS}} = 1.082^{+0.022}_{-0.026}$$

$$= 1.082 \pm 0.021(\text{had.})^{+0.000}_{-0.015}(\text{scale}) \pm 0.006(\text{param.}) \, ,$$

$$\frac{\tau(B^+)}{\tau(B_d)}\bigg|_{1S} = 1.082^{+0.023}_{-0.028}$$

$$= 1.082^{+0.022}_{-0.021}(\text{had.})^{+0.001}_{-0.017}(\text{scale})^{+0.007}_{-0.006}(\text{param.}) \, ,$$

$$\frac{\tau(B^+)}{\tau(B_d)}\bigg|_{\text{kin}} = 1.081^{+0.022}_{-0.027}$$

$$= 1.081 \pm 0.021(\text{had.})^{+0.001}_{-0.016}(\text{scale}) \pm 0.006(\text{param.}) \, ,$$

$$(6.5.14)$$

as our predictions for the lifetime ratio—showing excellent agreement with the experimental value and very good consistency between different mass schemes. The biggest contributions to the total uncertainty are still from the hadronic matrix elements, specifically from ϵ_1 with ± 0.015 and σ_3 with ± 0.013. In the future, they can be reduced with an independent determination of the dimension-six bag parameters and a sum-rule determination of the dimension-seven bag parameters.

We also update the prediction for the lifetime ratio $\tau(B_s)/\tau(B_d)$ in the $\overline{\text{MS}}$ scheme, by using Eq. 117 from [13]:

$$\frac{\tau(B_s)}{\tau(B_d)}\bigg|_{\text{exp}} = 0.994 \pm 0.004 \, ,$$

$$\frac{\tau(B_s)}{\tau(B_d)}\bigg|_{\overline{\text{MS}}} = 0.9994 \pm 0.0025$$

$$= 0.9994 \pm 0.0014(\text{had.}) \pm 0.0006(\text{scale}) \pm 0.0020(1/m_b^4) \, ,$$

$$(6.5.15)$$

where we have added an uncertainty estimate for the spectator effects at order $1/m_b^4$ which have not been considered in [13]. With respect to our determination in Chap. 3, the difference between the theory prediction and the experimental value for $\tau(B_s)/\tau(B_d)$ is reduced from 2.5 to 1.1 σ.

6.6 Matrix Elements for Charm and the $D^+ - D^0$ Lifetime Ratio

The HQET sum rule analysis can easily be adapted to the charm sector. It is common to quote the matrix elements for the charm sector at the scale 3 GeV instead of the charm quark mass (see [70–72]), and we adopt that convention for ease of comparison. Consequently we also use 3 GeV as the central matching scale. In the error analysis it is varied between 2 and 4 GeV. To account for the lower value of charm quark mass we assume that the uncertainty due to power corrections is 0.03 instead of 0.01 for the bottom sector. Otherwise, we use the same analysis strategy as in the bottom sector which is outlined in Sect. 6.4.1.

6.6.1 Matrix Elements for D Mixing

The latest lattice QCD study [72] for D mixing only gives results for the matrix elements and not for the bag parameters. We do the same here and obtain, using the value of the D meson decay constant from Appendix G.2,

$$
\begin{aligned}
\left\langle Q_1(3\,\text{GeV}) \right\rangle / \text{GeV}^4 &= 0.265^{+0.024}_{-0.021} \\
&= 0.265^{+0.006}_{-0.010}(\text{s.r.})^{+0.019}_{-0.014}(\text{matching})^{+0.013}_{-0.012}(f_D), \\
-\left\langle Q_2(3\,\text{GeV}) \right\rangle / \text{GeV}^4 &= 0.502^{+0.124}_{-0.092} \\
&= 0.502^{+0.094}_{-0.078}(\text{s.r.})^{+0.076}_{-0.044}(\text{matching})^{+0.024}_{-0.023}(f_D), \\
\left\langle Q_3(3\,\text{GeV}) \right\rangle / \text{GeV}^4 &= 0.135^{+0.037}_{-0.029} \\
&= 0.135^{+0.031}_{-0.026}(\text{s.r.})^{+0.019}_{-0.010}(\text{matching})^{+0.006}_{-0.006}(f_D), \\
\left\langle Q_4(3\,\text{GeV}) \right\rangle / \text{GeV}^4 &= 0.792^{+0.175}_{-0.122} \\
&= 0.792^{+0.116}_{-0.093}(\text{s.r.})^{+0.125}_{-0.070}(\text{matching})^{+0.038}_{-0.037}(f_D), \\
\left\langle Q_5(3\,\text{GeV}) \right\rangle / \text{GeV}^4 &= 0.340^{+0.060}_{-0.039} \\
&= 0.340^{+0.027}_{-0.021}(\text{s.r.})^{+0.051}_{-0.029}(\text{matching})^{+0.016}_{-0.016}(f_D).
\end{aligned}
\tag{6.6.1}
$$

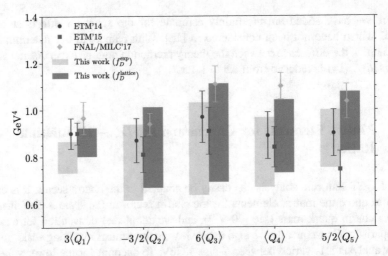

Fig. 6.7 Comparison of our results for the $\Delta C = 2$ matrix elements at the scale 3 GeV to the lattice values of ETM'14 [70], ETM'15 [71] and FNAL/MILC'17 [72]. The values for the matrix elements of the ETM collaboration are extracted from Fig. 16 of [72]

The relative uncertainties in the charm sector are consistently larger than those in the bottom sector because of larger perturbative corrections due to a larger value of α_s at the smaller scales and larger power corrections. This effect is most pronounced for Q_2, Q_4 and Q_5 where the relative uncertainty is larger by a factor of order two. In the matrix elements we have an additional uncertainty from the value of the decay constant which is added in quadrature.

We compare our results to those from the lattice in Fig. 6.7. There is a consistent hierarchy with decreasing values from the results of the FNAL/MILC collaboration [72], those of the ETM collaboration [70, 71] and ours. The only exception is the value of $\langle Q_5 \rangle$ from [71] which lies below ours. If we use the lattice average [47] for the decay constant $f_D^{\text{lattice}} = (211.9 \pm 1.1)$ MeV in place of the experimental average $f_D^{\text{exp}} = (203.7 \pm 4.8)$ MeV [47], we find very good agreement between our results and those of ETM and the remaining differences with respect to the FNAL/MILC results are comfortably below $2\,\sigma$. We prefer the experimental average of the decay constant since it is in significantly better agreement with recent sum rule results [67, 68, 73, 74]. On the other hand, using the lattice value yields a more meaningful comparison with the lattice results since the quantities we determine with the sum rule are the bag parameters and the decay constant cancels out in the comparison if the same value is used on both sides. We therefore conclude that our sum rule results for the non-factorizable contributions to the bag parameters are in good agreement with lattice simulations. An investigation of the differences in the numerical values of the decay constant is beyond the scope of this work.

6.6.2   Matrix Elements for D Lifetimes and $\tau(D^+)/\tau(D^0)$

Our results for the $\Delta C = 0$ bag parameters are

$$\overline{B}_1(3\,\text{GeV}) = 0.902^{+0.077}_{-0.051}$$
$$= 0.902^{+0.018}_{-0.018}(\text{sum rule})^{+0.075}_{-0.048}(\text{matching})\,,$$
$$\overline{B}_2(3\,\text{GeV}) = 0.739^{+0.124}_{-0.073}$$
$$= 0.739^{+0.015}_{-0.015}(\text{sum rule})^{+0.123}_{-0.072}(\text{matching})\,,$$
$$\overline{\epsilon}_1(3\,\text{GeV}) = -0.132^{+0.041}_{-0.046} \qquad\qquad (6.6.2)$$
$$= -0.132^{+0.025}_{-0.026}(\text{sum rule})^{+0.033}_{-0.038}(\text{matching})\,,$$
$$\overline{\epsilon}_2(3\,\text{GeV}) = -0.005^{+0.032}_{-0.032}$$
$$= -0.005^{+0.011}_{-0.012}(\text{sum rule})^{+0.030}_{-0.030}(\text{matching})\,.$$

While the uncertainties in $B_{1,2}$ are similar to those in the B sector we find that those in $\epsilon_{1,2}$ are larger by about 50%. The latter ones are dominated by the non-factorizable power correction and the intrinsic sum rule errors which are both based on somewhat ad-hoc estimates. As such, our values for the uncertainties of $\epsilon_{1,2}$ should be taken with a grain of salt and lattice results for the $\Delta C = 0$ bag parameters could provide an important consistency check. As an alternative, it should also be possible to reduce the dominant error due to non-factorizable $1/m_c$ corrections by performing the operator matching up to the order $1/m_c$ and determine the matrix elements of the subleading HQET operators using sum rules.

We update our result for the D meson lifetime ratio from [15] using the dimension-six bag parameters (Eq. 6.6.2) and the VSA for the dimension-seven bag parameters, with uncertainties $\rho_i = 1 \pm 1/12$, $\sigma_i = 0 \pm 1/6$. We have converted the $\overline{\text{MS}}$ value of the charm quark mass to the PS mass at $\mu_f = 1\,\text{GeV}$ and the 1S mass at four-loop accuracy using RunDec. The kinetic mass at the scale $1\,\text{GeV}$ is determined with two-loop accuracy using an unpublished version of the QQbar_Threshold code [75, 76]. The central value for the scales μ_1 and μ_0 is fixed to $1.5\,\text{GeV}$ for all mass schemes and varied between 1 and 3 GeV. We find

$$\left.\frac{\tau(D^+)}{\tau(D^0)}\right|_{\text{exp}} = 2.536 \pm 0.019\,,$$

$$\left.\frac{\tau(D^+)}{\tau(D^0)}\right|_{\overline{\text{MS}}} = 2.61^{+0.72}_{-0.77}$$
$$= 2.61^{+0.70}_{-0.66}(\text{had.})^{+0.12}_{-0.38}(\text{scale}) \pm 0.09(\text{param.})\,,$$

$$\left.\frac{\tau(D^+)}{\tau(D^0)}\right|_{\text{PS}} = 2.70^{+0.74}_{-0.82}$$

$$= 2.70^{+0.72}_{-0.68}(\text{had.})^{+0.11}_{-0.45}(\text{scale}) \pm 0.10(\text{param.}),$$

$$\left.\frac{\tau(D^+)}{\tau(D^0)}\right|_{1S} = 2.56^{+0.81}_{-0.99}$$

$$= 2.56^{+0.78}_{-0.74}(\text{had.})^{+0.22}_{-0.65}(\text{scale}) \pm 0.10(\text{param.}), \qquad (6.6.3)$$

$$\left.\frac{\tau(D^+)}{\tau(D^0)}\right|_{\text{kin}} = 2.53^{+0.72}_{-0.76}$$

$$= 2.53^{+0.70}_{-0.66}(\text{had.})^{+0.13}_{-0.37}(\text{scale}) \pm 0.10(\text{param.}),$$

which is in very good agreement. The various mass schemes are all consistent and we again take the PS result as our preferred value. The dominant sources of uncertainties are the bag parameters ϵ_1 and σ_3 which both contribute ± 0.5 to the error budget of the lifetime ratio. Both errors can be reduced in the future with a lattice determination of the dimensions-six matrix elements and a sum-rule determination of the dimension-seven bag parameters, respectively. In the PS scheme, the radiative corrections are of the order $+27\%$, and the power corrections of the order -34%, which indicate good convergence behaviour. We therefore conclude that the HQE provides a good description of the lifetime ratio $\tau(D^+)/\tau(D^0)$.

6.7 Summary

We have determined the matrix elements of the dimension-six $\Delta F = 0, 2$ operators for the bottom and charm sector using HQET sum rules. Our findings for the $\Delta F = 2$ matrix elements are in good agreement with recent lattice [2–4, 70–72] and sum rule [9] results. Our $\Delta F = 0$ results are the first state-of-the-art values for the matrix elements required for B and D meson lifetime ratios. The uncertainties in our analyses for the bag parameters are similar to those of recent lattice determinations in the B sector and somewhat larger in the D sector. This suggests that the uncertainty of the $\Delta C = 0$ matrix elements could be reduced by a lattice simulation. In most cases, the dominant errors in our approach stem from the matching of QCD to HQET operators, see Appendix G.2. These errors could be reduced substantially by performing the matching calculation at NNLO. First steps towards this goal have recently been taken in [57, 61]. Consequently, in the future, sum rules will continue to be competitive with lattice simulations in the determination of four quark operators.

Our predictions for the mixing observables and lifetime ratios in the B sector are in good agreement with the experimental averages as summarized in Figs. 6.8 and 6.9. In particular, the small tensions that follow from using the FNAL/MILC results [4] for the matrix elements are not confirmed by our results. We note that the predictions based on matrix elements from sum rules and from lattice simulations are compatible and lead to overall uncertainties of the same size. Taking the naive average of the bag parameters, the relative uncertainties of the mass and decay rate difference are only

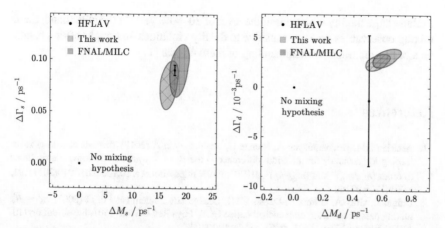

Fig. 6.8 Comparison of our predictions for the mass and decay rate difference in the B_s (left) and B_d (right) system with the present experimental averages (error bars). We also show the results obtained with the lattice results of [4] for $f_{B_q}^2 \overline{B}_{Q_i}$ and the matrix element $\langle R_0 \rangle$ and the values given in Appendix G.2 for the other input parameters. The PS mass scheme for the bottom quarks has been used in both cases

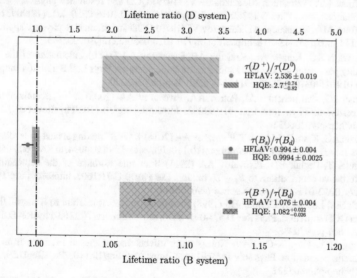

Fig. 6.9 Comparison of our predictions for the lifetime ratios of heavy mesons with the present experimental averages

reduced by about 9% and 6% respectively, because other sources of uncertainties (e.g. the matrix elements of dimension-seven operators) are dominant.

We find that the experimental value for the lifetime ratio $\tau(D^+)/\tau(D^0)$ can be reproduced within the HQE. This is a strong indication that the HQE does not break down in the charm sector. However, due to sizeable hadronic uncertainties, we cannot exclude large duality violations at the level of 20–30% yet. On the other hand, the D mixing observables are very sensitive to duality violations and might offer a handle on a better quantitative understanding of these effects [17].

References

1. Asatrian HM, Hovhannisyan A, Nierste U, Yeghiazaryan A (2017) Towards next-to-next-to-leading-log accuracy for the width difference in the $B_s - \bar{B}_s$ system: fermionic contributions to order $(m_c/m_b)^0$ and $(m_c/m_b)^1$. JHEP 10:191. https://doi.org/10.1007/JHEP10(2017)191, arXiv:1709.02160
2. Dalgic E, Gray A, Gamiz E, Davies CTH, Lepage GP, Shigemitsu J et al (2007) $B_s^0 - \bar{B}_s^0$ mixing parameters from unquenched lattice QCD. Phys Rev D 76:011501. https://doi.org/10.1103/PhysRevD.76.011501, arXiv:hep-lat/0610104
3. ETM collaboration, Carrasco N et al (2014) B-physics from $N_f = 2$ tmQCD: the Standard Model and beyond. JHEP 03:016. https://doi.org/10.1007/JHEP03(2014)016, arXiv:1308.1851
4. Fermilab Lattice, MILC collaboration, Bazavov A et al (2016) $B_{(s)}^0$-mixing matrix elements from lattice QCD for the Standard Model and beyond. Phys Rev D 93:113016. https://doi.org/10.1103/PhysRevD.93.113016, arXiv:1602.03560
5. Shifman MA, Vainshtein AI, Zakharov VI (1979) QCD and resonance physics. Theoretical foundations. Nucl Phys B 147:385–447. https://doi.org/10.1016/0550-3213(79)90022-1
6. Shifman MA, Vainshtein AI, Zakharov VI (1979) QCD and resonance physics: applications. Nucl Phys B 147:448–518. https://doi.org/10.1016/0550-3213(79)90023-3
7. Chetyrkin KG, Kataev AL, Krasulin AB, Pivovarov AA (1986) Calculation of the K^0-$\bar{K}^0$ mixing parameter via the QCD sum rules at finite energies. Phys Lett B 174:104. https://doi.org/10.1016/0370-2693(86)91137-8, arXiv:hep-ph/0103230
8. Korner JG, Onishchenko AI, Petrov AA, Pivovarov AA (2003) $B^0 - \bar{B}^0$ mixing beyond factorization. Phys Rev Lett 91:192002. https://doi.org/10.1103/PhysRevLett.91.192002, arXiv:hep-ph/0306032
9. Grozin AG, Klein R, Mannel T, Pivovarov AA (2016) $B^0 - \bar{B}^0$ mixing at next-to-leading order. Phys Rev D 94:034024. https://doi.org/10.1103/PhysRevD.94.034024, arXiv:1606.06054
10. Mannel T, Pecjak BD, Pivovarov AA (2011) Sum rule estimate of the subleading non-perturbative contributions to $B_s - \bar{B}_s$ mixing. Eur Phys J C 71:1607. https://doi.org/10.1140/epjc/s10052-011-1607-4, arXiv:hep-ph/0703244
11. Baek MS, Lee J, Liu C, Song HS (1998) Four quark operators relevant to B meson lifetimes from QCD sum rules. Phys Rev D 57:4091–4096. https://doi.org/10.1103/PhysRevD.57.4091, arXiv:hep-ph/9709386
12. Cheng H-Y, Yang K-C (1999) Nonspectator effects and B meson lifetimes from a field theoretic calculation. Phys Rev D 59:014011. https://doi.org/10.1103/PhysRevD.59.014011, arXiv:hep-ph/9805222
13. Lenz A (2015) Lifetimes and heavy quark expansion. Int J Mod Phys A 30:1543005. https://doi.org/10.1142/S0217751X15430058, arXiv:1405.3601
14. Becirevic D (2001) Theoretical progress in describing the B meson lifetimes. PoS HEP2001 098. arXiv:hep-ph/0110124
15. Lenz A, Rauh T (2013) D-meson lifetimes within the heavy quark expansion. Phys Rev D 88:034004. https://doi.org/10.1103/PhysRevD.88.034004, arXiv:1305.3588
16. Bobrowski M, Lenz A, Riedl J, Rohrwild J (2010) How large can the SM contribution to CP violation in $D^0 - \bar{D}^0$ mixing be? JHEP 03:009. https://doi.org/10.1007/JHEP03(2010)009, arXiv:1002.4794
17. Bigi IIY, Uraltsev NG (2001) $D^0 - \bar{D}^0$ oscillations as a probe of quark hadron duality. Nucl Phys B 592:92–106. https://doi.org/10.1016/S0550-3213(00)00604-0, arXiv:hep-ph/0005089

18. Georgi H (1992) D-$\bar{D}$ mixing in heavy quark effective field theory. Phys Lett B 297:353–357. https://doi.org/10.1016/0370-2693(92)91274-D, arXiv:hep-ph/9209291

19. Ohl T, Ricciardi G, Simmons EH (1993) D-anti-D mixing in heavy quark effective field theory: the sequel. Nucl Phys B 403:605–632. https://doi.org/10.1016/0550-3213(93)90364-U, arXiv:hep-ph/9301212

20. Bobrowski M, Lenz A, Rauh T (2012) Short distance $D^0 - \bar{D}^0$ mixing. In: Proceedings, 5th International workshop on charm physics (Charm 2012): Honolulu, Hawaii, USA, May 14–17, 2012. arXiv:1208.6438

21. Flynn JM, Hernandez OF, Hill BR (1991) Renormalization of four fermion operators determining B anti-B mixing on the lattice. Phys Rev D 43:3709–3714. https://doi.org/10.1103/PhysRevD.43.3709

22. Buchalla G (1997) Renormalization of $\Delta B = 2$ transitions in the static limit beyond leading logarithms. Phys Lett B 395:364–368. https://doi.org/10.1016/S0370-2693(97)00043-9, arXiv:hep-ph/9608232

23. Ciuchini M, Franco E, Gimenez V (1996) Next-to-leading order renormalization of the $\Delta B = 2$ operators in the static theory. Phys Lett B 388:167–172. https://doi.org/10.1016/0370-2693(96)01131-8, arXiv:hep-ph/9608204

24. Collins JC (1986) Renormalization. In: Cambridge monographs on mathematical physics, vol 26. Cambridge University Press, Cambridge

25. Buras AJ, Weisz PH (1990) QCD nonleading corrections to weak decays in dimensional regularization and 't Hooft-Veltman schemes. Nucl Phys B 333:66–99. https://doi.org/10.1016/0550-3213(90)90223-Z

26. Herrlich S, Nierste U (1995) Evanescent operators, scheme dependences and double insertions. Nucl Phys B 455:39–58. https://doi.org/10.1016/0550-3213(95)00474-7, arXiv:hep-ph/9412375

27. Beneke M, Buchalla G, Greub C, Lenz A, Nierste U (1999) Next-to-leading order QCD corrections to the lifetime difference of B_s mesons. Phys Lett B 459:631–640. https://doi.org/10.1016/S0370-2693(99)00684-X, arXiv:hep-ph/9808385

28. Beneke M, Buchalla G, Dunietz I (1996) Width difference in the $B_s - \bar{B}_s$ system. Phys Rev D 54:4419–4431. https://doi.org/10.1103/PhysRevD.54.4419, https://doi.org/10.1103/PhysRevD.83.119902, arXiv:hep-ph/9605259

29. Nogueira P (1993) Automatic Feynman graph generation. J Comput Phys 105:279–289. https://doi.org/10.1006/jcph.1993.1074

30. Jamin M, Lautenbacher ME (1993) TRACER: version 1.1: a mathematica package for γ algebra in arbitrary dimensions. Comput Phys Commun 74:265–288. https://doi.org/10.1016/0010-4655(93)90097-V

31. Patel HH (2015) Package-X: a mathematica package for the analytic calculation of one-loop integrals. Comput Phys Commun 197:276–290. https://doi.org/10.1016/j.cpc.2015.08.017, arXiv:1503.01469

32. Patel HH (2017) Package-X 2.0: a mathematica package for the analytic calculation of one-loop integrals. Comput Phys Commun 218:66–70. https://doi.org/10.1016/j.cpc.2017.04.015, arXiv:1612.00009

33. Lenz A, Nierste U (2007) Theoretical update of $B_s - \bar{B}_s$ mixing. JHEP 06:072. https://doi.org/10.1088/1126-6708/2007/06/072, arXiv:hep-ph/0612167

34. Eichten E, Hill BR (1990) An effective field theory for the calculation of matrix elements involving heavy quarks. Phys Lett B 234:511–516. https://doi.org/10.1016/0370-2693(90)92049-O

35. Smirnov AV (2008) Algorithm FIRE-Feynman Integral REduction. JHEP 10:107. https://doi.org/10.1088/1126-6708/2008/10/107, arXiv:0807.3243

36. Smirnov AV, Smirnov VA (2013) FIRE4, LiteRed and accompanying tools to solve integration by parts relations. Comput Phys Commun 184:2820–2827. https://doi.org/10.1016/j.cpc.2013.06.016, arXiv:1302.5885

37. Smirnov AV (2015) FIRE5: a C++ implementation of Feynman Integral REduction. Comput Phys Commun 189:182–191. https://doi.org/10.1016/j.cpc.2014.11.024, arXiv:1408.2372

38. Chetyrkin KG, Tkachov FV (1981) Integration by parts: the algorithm to calculate beta functions in 4 loops. Nucl Phys B 192:159–204. https://doi.org/10.1016/0550-3213(81)90199-1
39. Laporta S (2000) High precision calculation of multiloop Feynman integrals by difference equations. Int J Mod Phys A 15:5087–5159. https://doi.org/10.1016/S0217-751X(00)00215-7, https://doi.org/10.1142/S0217751X00002157, arXiv:hep-ph/0102033
40. Grozin AG, Lee RN (2009) Three-loop HQET vertex diagrams for B0-anti-B0 mixing. JHEP 02:047. https://doi.org/10.1088/1126-6708/2009/02/047, arXiv:0812.4522
41. Ball P, Braun VM (1994) Next-to-leading order corrections to meson masses in the heavy quark effective theory. Phys Rev D 49:2472–2489. https://doi.org/10.1103/PhysRevD.49.2472, arXiv:hep-ph/9307291
42. Broadhurst DJ, Grozin AG (1992) Operator product expansion in static quark effective field theory: large perturbative correction. Phys Lett B 274:421–427. https://doi.org/10.1016/0370-2693(92)92009-6, arXiv:hep-ph/9908363
43. Bagan E, Ball P, Braun VM, Dosch HG (1992) QCD sum rules in the effective heavy quark theory. Phys Lett B 278:457–464. https://doi.org/10.1016/0370-2693(92)90585-R
44. Neubert M (1992) Heavy meson form-factors from QCD sum rules. Phys Rev D 45:2451–2466. https://doi.org/10.1103/PhysRevD.45.2451
45. Chetyrkin KG, Kuhn JH, Steinhauser M (2000) RunDec: a mathematica package for running and decoupling of the strong coupling and quark masses. Comput Phys Commun 133:43–65. https://doi.org/10.1016/S0010-4655(00)00155-7, arXiv:hep-ph/0004189
46. Herren F, Steinhauser M (2018) Version 3 of RunDec and CRunDec. Comput Phys Commun 224:333–345. https://doi.org/10.1016/j.cpc.2017.11.014, arXiv:1703.03751
47. Particle Data Group collaboration, Patrignani C et al (2016) Review of particle physics. Chin Phys C 40:100001. https://doi.org/10.1088/1674-1137/40/10/100001
48. Beneke M, Maier A, Piclum J, Rauh T (2015) The bottom-quark mass from non-relativistic sum rules at NNNLO. Nucl Phys B 891:42–72. https://doi.org/10.1016/j.nuclphysb.2014.12.001, arXiv:1411.3132
49. Beneke M, Maier A, Piclum J, Rauh T (2016) NNNLO determination of the bottom-quark mass from non-relativistic sum rules. PoS RADCOR2015 035. https://doi.org/10.22323/1.235.0035, arXiv:1601.02949
50. Baikov PA, Chetyrkin KG, Kühn JH (2017) Five-loop running of the QCD coupling constant. Phys Rev Lett 118:082002. https://doi.org/10.1103/PhysRevLett.118.082002, arXiv:1606.08659
51. Herzog F, Ruijl B, Ueda T, Vermaseren JAM, Vogt A (2017) The five-loop beta function of Yang-Mills theory with fermions. JHEP 02:090. https://doi.org/10.1007/JHEP02(2017)090, arXiv:1701.01404
52. Luthe T, Maier A, Marquard P, Schroder Y (2017) Complete renormalization of QCD at five loops. JHEP 03:020. https://doi.org/10.1007/JHEP03(2017)020, arXiv:1701.07068
53. Luthe T, Maier A, Marquard P, Schroder Y (2017) The five-loop Beta function for a general gauge group and anomalous dimensions beyond Feynman gauge. JHEP 10:166. https://doi.org/10.1007/JHEP10(2017)166, arXiv:1709.07718
54. Chetyrkin KG, Falcioni G, Herzog F, Vermaseren JAM (2017) Five-loop renormalisation of QCD in covariant gauges. JHEP 10:179. https://doi.org/10.1007/JHEP12(2017)006, https://doi.org/10.3204/PUBDB-2018-02123, https://doi.org/10.1007/JHEP10(2017)179, arXiv:1709.08541
55. Aoki S et al (2017) Review of lattice results concerning low-energy particle physics. Eur Phys J C 77:112. https://doi.org/10.1140/epjc/s10052-016-4509-7, arXiv:1607.00299
56. HFLAV collaboration (2017) B lifetime and oscillation parameters, Summer 2017. http://www.slac.stanford.edu/xorg/hflav/osc/summer_2017/
57. Grozin AG, Mannel T, Pivovarov AA (2017) Towards a next-to-next-to-leading order analysis of matching in $B^0 - \bar{B}^0$ mixing. Phys Rev D 96:074032. https://doi.org/10.1103/PhysRevD.96.074032, arXiv:1706.05910
58. Beneke M (1998) A quark mass definition adequate for threshold problems. Phys Lett B 434:115–125. https://doi.org/10.1016/S0370-2693(98)00741-2, arXiv:hep-ph/9804241

59. Hoang AH, Ligeti Z, Manohar AV (1999) B decay and the Upsilon mass. Phys Rev Lett 82:277–280. https://doi.org/10.1103/PhysRevLett.82.277, arXiv:hep-ph/9809423

60. Bigi IIY, Shifman MA, Uraltsev N, Vainshtein AI (1997) High power n of m_b in beauty widths and $n = 5 \to \infty$ limit. Phys Rev D 56:4017–4030. https://doi.org/10.1103/PhysRevD.56.4017, arXiv:hep-ph/9704245

61. Grozin AG, Mannel T, Pivovarov AA (2018) $B^0 - \bar{B}^0$ mixing: matching to HQET at NNLO. Phys Rev D 98:054020. https://doi.org/10.1103/PhysRevD.98.054020, arXiv:1806.00253

62. Laplace S, Ligeti Z, Nir Y, Perez G (2002) Implications of the CP asymmetry in semileptonic B decay. Phys Rev D 65:094040. https://doi.org/10.1103/PhysRevD.65.094040, arXiv:hep-ph/0202010

63. Beneke M, Buchalla G, Greub C, Lenz A, Nierste U (2002) The $B^+ - B_d^0$ lifetime difference beyond leading logarithms. Nucl Phys B 639:389–407. https://doi.org/10.1016/S0550-3213(02)00561-8, arXiv:hep-ph/0202106

64. Ciuchini M, Franco E, Lubicz V, Mescia F (2002) Next-to-leading order QCD corrections to spectator effects in lifetimes of beauty hadrons. Nucl Phys B 625:211–238. https://doi.org/10.1016/S0550-3213(02)00006-8, arXiv:hep-ph/0110375

65. Franco E, Lubicz V, Mescia F, Tarantino C (2002) Lifetime ratios of beauty hadrons at the next-to-leading order in QCD. Nucl Phys B 633:212–236. https://doi.org/10.1016/S0550-3213(02)00262-6, arXiv:hep-ph/0203089

66. Gabbiani F, Onishchenko AI, Petrov AA (2004) Spectator effects and lifetimes of heavy hadrons. Phys Rev D 70:094031. https://doi.org/10.1103/PhysRevD.70.094031, arXiv:hep-ph/0407004

67. Lucha W, Melikhov D, Simula S (2011) OPE, charm-quark mass, and decay constants of D and D_s mesons from QCD sum rules. Phys Lett B 701:82–88. https://doi.org/10.1016/j.physletb.2011.05.031, arXiv:1101.5986

68. Gelhausen P, Khodjamirian A, Pivovarov AA, Rosenthal D (2013) Decay constants of heavy-light vector mesons from QCD sum rules. Phys Rev D 88:014015. https://doi.org/10.1103/PhysRevD.88.014015, https://doi.org/10.1103/PhysRevD.91.099901, https://doi.org/10.1103/PhysRevD.89.099901, arXiv:1305.5432

69. UKQCD collaboration, Di Pierro M, Sachrajda CT (1998) A lattice study of spectator effects in inclusive decays of B mesons. Nucl Phys B 534:373–391. https://doi.org/10.1016/S0550-3213(98)00580-X, arXiv:hep-lat/9805028

70. Carrasco N et al (2014) $D^0 - \bar{D}^0$ mixing in the standard model and beyond from $N_f = 2$ twisted mass QCD. Phys Rev D 90:014502. https://doi.org/10.1103/PhysRevD.90.014502, arXiv:1403.7302

71. ETM collaboration, Carrasco N, Dimopoulos P, Frezzotti R, Lubicz V, Rossi GC, Simula S et al (2015) $\Delta S = 2$ and $\Delta C = 2$ bag parameters in the standard model and beyond from $N_f = 2 + 1 + 1$ twisted-mass lattice QCD. Phys Rev D 92:034516. https://doi.org/10.1103/PhysRevD.92.034516, arXiv:1505.06639

72. Bazavov A et al (2018) Short-distance matrix elements for D^0-meson mixing for $N_f = 2 + 1$ lattice QCD. Phys Rev D 97:034513. https://doi.org/10.1103/PhysRevD.97.034513, arXiv:1706.04622

73. Wang Z-G (2015) Analysis of the masses and decay constants of the heavy-light mesons with QCD sum rules. Eur Phys J C 75:427. https://doi.org/10.1140/epjc/s10052-015-3653-9, arXiv:1506.01993

74. Narison S (2013) A fresh look into $m_{c,b}$ and precise $f_{D(s),B(s)}$ from heavy-light QCD spectral sum rules. Phys Lett B 718:1321–1333. https://doi.org/10.1016/j.physletb.2012.10.057, arXiv:1209.2023

75. Beneke M, Kiyo Y, Maier A, Piclum J (2016) Near-threshold production of heavy quarks with QQbar_threshold. Comput Phys Commun 209:96–115. https://doi.org/10.1016/j.cpc.2016.07.026, arXiv:1605.03010

76. Beneke M, Maier A, Rauh T, Ruiz-Femenia P (2018) Non-resonant and electroweak NNLO correction to the e^+e^- top anti-top threshold. JHEP 02:125. https://doi.org/10.1007/JHEP02(2018)125, arXiv:1711.10429

Chapter 7
One Constraint to Kill Them All?

7.1 Introduction

As we have discussed in Sect. 1.3.4, there are many intriguing anomalies in the b quark sector, for which a variety of possible classes of new physics model have been proposed. However, any new $b\bar{s}$-coupling immediately gives tree-level contributions to B_s mixing, which is strongly constrained by experiment. For many years, the SM prediction for ΔM_s (see Eq. 7.2.1) has perfectly agreed with the experimental measurement. Recently, there has been an updated average provided by FLAG of the important input parameter $f_{B_s}\hat{B}$, and this increases the SM prediction by around 10% to a value considerably above experiment. In this chapter, we investigate the consequences of this updated theory prediction. We will find that the allowed parameter space for Z' and leptoquark models that can explain the B anomalies is severely reduced by this change. Remarkably, for Z' models the upper bound on the Z' mass approaches dangerously close to the energy scales already probed by the LHC. We finally identify some directions for model building in order to alleviate the tension with B_s mixing. More details of the SM prediction and error breakdown, along with a critical discussion of the theoretical uncertainties can be found in Appendix H.

7.2 B_s Mixing in the SM

Having discussed in detail the theory of B_s mixing in Sect. 2.4.2, we move straight into our updated prediction of the mass difference.

Two commonly used SM predictions of ΔM_s are

$$
\begin{aligned}
\Delta M_s^{\text{SM,2011}} &= (17.3 \pm 2.6)\,\text{ps}^{-1}, \\
\Delta M_s^{\text{SM,2015}} &= (18.3 \pm 2.7)\,\text{ps}^{-1},
\end{aligned}
\tag{7.2.1}
$$

© Springer Nature Switzerland AG 2019
M. J. Kirk, *Charming New Physics in Beautiful Processes?*,
Springer Theses, https://doi.org/10.1007/978-3-030-19197-9_7

given in [1, 2] respectively, which both give very good agreement with the experimental measurement [3]:

$$\Delta M_s^{\text{exp}} = (17.757 \pm 0.021)\,\text{ps}^{-1}\,.\tag{7.2.2}$$

In 2016 Fermilab/MILC presented a new calculation [4], which gave considerably larger values for the non-perturbative input $f_{B_s}\hat{B}$, resulting in values around $20\,\text{ps}^{-1}$ for the mass difference [4–7] which are much larger than experiment. An independent confirmation of these large values would of course be desirable; a first step in that direction is our work in Chap. 6 which is in agreement with Fermilab/MILC for the bag parameters.

Using the most recent numerical inputs (as listed in Appendix H.1) we predict the mass difference of the neutral B_s mesons to be[1]

$$\boxed{\Delta M_s^{\text{SM,2017}} = (20.01 \pm 1.25)\,\text{ps}^{-1}\,.}\tag{7.2.3}$$

As in previous calculations, the dominant uncertainty in this result comes from the lattice predictions for the non-perturbative parameters B and f_{B_s}, giving a relative error of 5.8%, while the next-to-leading contribution comes from the uncertainty in the CKM elements. A detailed discussion of the error budget is given in Appendix H.2.

The updated central value for the mass difference in Eq. 7.2.3 is $1.8\,\sigma$ above the experimental one given in Eq. 7.2.2, which has profound implications for NP models that predict sizeable positive contributions to B_s mixing. As the SM prediction depends strongly on the non-perturbative input as well as the values of the CKM elements, we take our values from the relevant groups in those fields, FLAG and CKMfitter respectively. In general, the presence of BSM physics can affect the determination of the CKM elements, and hence the SM prediction of ΔM can in general differ from the one we use—see e.g. the case of a fourth chiral fermion generation [8]. In the following, we will assume that NP effects do not invoke sizeable shifts in the CKM elements. Further discussion of the lattice and CKM dependencies can be found in Appendices H.3 and H.4.

7.3 B_s Mixing Beyond the SM

To determine the allowed space for NP effects in B_s mixing we compare the experimental measurement of the mass difference with the SM prediction plus some NP component:

---

[1]A more conservative determination of the SM value of the mass difference using only tree-level inputs for the CKM parameters can be found in Eq. H.4.6.

$$\Delta M_s^{\text{exp}} = 2 \left| M_{12}^{\text{SM}} + M_{12}^{\text{NP}} \right| = \Delta M_s^{\text{SM}} \left| 1 + \frac{M_{12}^{\text{NP}}}{M_{12}^{\text{SM}}} \right| . \tag{7.3.1}$$

A simple estimate shows that the change in the SM prediction from Eqs. 7.2.1 to 7.2.3 can have a drastic impact on the size of the allowed BSM effects on B_s mixing. For a generic NP model we can introduce the parameterisation

$$\frac{\Delta M_s^{\text{exp}}}{\Delta M_s^{\text{SM}}} = \left| 1 + \frac{\kappa}{\Lambda_{\text{NP}}^2} \right| , \tag{7.3.2}$$

where Λ_{NP} denotes the mass scale of the NP mediator and κ is a dimensionful quantity which encodes NP couplings and the SM contribution. If $\kappa > 0$, which is often the case in BSM scenarios considered in the literature,[2] the 2σ bound on Λ_{NP} scales like

$$\frac{\Lambda_{\text{NP}}^{2017}}{\Lambda_{\text{NP}}^{2015}} = \sqrt{ \frac{ \frac{\Delta M_s^{\text{exp}}}{\left(\Delta M_s^{\text{SM}} - 2\delta \Delta M_s^{\text{SM}} \right)_{2015}} - 1 }{ \frac{\Delta M_s^{\text{exp}}}{\left(\Delta M_s^{\text{SM}} - 2\delta \Delta M_s^{\text{SM}} \right)_{2017}} - 1 } } \simeq 5.2 , \tag{7.3.3}$$

where $\delta \Delta M_s^{\text{SM}}$ denotes the 1σ error of the SM prediction. Hence in models where $\kappa > 0$, the limit on the mass of the NP mediators is strengthened by a factor 5. On the other hand, if the tension between the SM prediction and ΔM_s^{exp} increases in the future, a NP contribution with $\kappa < 0$ would be required in order to accommodate the discrepancy.

A typical example where $\kappa > 0$ is that of a purely left-handed (LH) vector-current operator, which arises from the exchange of a single mediator featuring real couplings (see Sect. 7.3.1 for an example). In such a case, the short distance contribution to B_s mixing is described by the effective Lagrangian

$$\mathcal{L}_{\Delta B=2}^{\text{NP}} = -\frac{4 G_F}{\sqrt{2}} \left(V_{tb} V_{ts}^* \right)^2 C_{bs}^{LL} \left(\bar{s}_L \gamma_\mu b_L \right)^2 + \text{h.c.} , \tag{7.3.4}$$

where C_{bs}^{LL} is a Wilson coefficient to be matched with UV models, and which enters Eq. 7.3.1 as

$$\frac{\Delta M_s^{\text{exp}}}{\Delta M_s^{\text{SM}}} = \left| 1 + \frac{C_{bs}^{LL}}{R_{\text{SM}}^{\text{loop}}} \right| , \tag{7.3.5}$$

where

$$R_{\text{SM}}^{\text{loop}} = \frac{\sqrt{2} G_F M_W^2 \hat{\eta}_{2B} S_0(x_t)}{16\pi^2} = 1.3397 \times 10^{-3} . \tag{7.3.6}$$

---

[2] See for example the papers by Blanke and Buras [5, 9] where they show that so-called CMFV models always lead to $\kappa > 0$.

In the rest of this chapter, we will show how the updated bound from ΔM_s impacts the parameter space of simplified models (with $\kappa > 0$) that have been put forth to the explain the recent discrepancies in semileptonic B physics data (Sect. 7.3.1), and then discuss some model-building directions in order to achieve $\kappa < 0$ (Sect. 7.3.2).

7.3.1 Impact of B_s Mixing on NP Models for B-Anomalies

A useful application of the refined SM prediction in Eq. 7.2.3 is in the context of the recent hints of LFU violation in both neutral and charged current semileptonic B meson decays. Focussing first on neutral current anomalies, the main observables are the LFU violating ratios (which we defined in Eq. 1.3.2), together with the angular distributions of $B \to K^{(*)}\mu^+\mu^-$ [10–19] and the branching ratios of hadronic $b \to s\mu^+\mu^-$ decays [10, 11, 20]. As hinted at by various recent global fits [21–26] (and in order to simplify the discussion), we assume NP contributions only in purely LH vector currents involving muons. The generalisation to different type of operators is straightforward. The effective Lagrangian for semileptonic $b \to s\mu^+\mu^-$ transitions contains the terms

$$\mathcal{L}^{\mathrm{NP}}_{b\to s\mu\mu} \supset \frac{4G_F}{\sqrt{2}} V_{tb}V_{ts}^* \left(\Delta C_9^\mu \mathcal{O}_9^\mu + \Delta C_{10}^\mu \mathcal{O}_{10}^\mu\right) + \mathrm{h.c.}, \qquad (7.3.7)$$

with

$$\mathcal{O}_9^\mu = \frac{\alpha}{4\pi}(\bar{s}_L\gamma_\mu b_L)(\bar{\mu}\gamma^\mu\mu), \qquad (7.3.8)$$

$$\mathcal{O}_{10}^\mu = \frac{\alpha}{4\pi}(\bar{s}_L\gamma_\mu b_L)(\bar{\mu}\gamma^\mu\gamma_5\mu). \qquad (7.3.9)$$

Assuming purely LH currents and real Wilson coefficients the best-fit of R_K and R_{K^*} yields (from e.g. [22]): $\mathrm{Re}(\Delta C_9^\mu) = -\mathrm{Re}(\Delta C_{10}^\mu) \in [-0.81, -0.48]$ ([−1.00, −0.32]) at 1σ (2σ). Including also the data on $B \to K^{(*)}\mu^+\mu^-$ angular distributions and other $b \to s\mu^+\mu^-$ observables[3] improves the statistical significance of the fit, but does not necessarily imply larger deviations of $\mathrm{Re}(\Delta C_9^\mu)$ from zero (see e.g. [21]). In the following we will limit ourselves to the R_K and R_{K^*} observables and denote this benchmark as "$R_{K^{(*)}}$".

7.3.1.1 Z′

A paradigmatic NP model for explaining the B-anomalies in neutral currents is that of a Z' dominantly coupled via LH currents. Here we focus only on the part of the

---

[3]These include for instance $\mathcal{B}(B_s \to \mu^+\mu^-)$ which is particularly constraining in the case of pseudo-scalar mediated quark transitions (see e.g. [27]).

Lagrangian relevant for $b \rightarrow s\mu^+\mu^-$ transitions and B_s mixing, namely

$$\mathcal{L}_{Z'} = \frac{1}{2}M_{Z'}^2(Z'_\mu)^2 + \left(\lambda_{ij}^Q \bar{d}_L^i \gamma^\mu d_L^j + \lambda_{\alpha\beta}^L \bar{\ell}_L^\alpha \gamma^\mu \ell_L^\beta\right)Z'_\mu, \qquad (7.3.10)$$

where d^i and ℓ^α denote down quark and charged lepton mass eigenstates, and $\lambda^{Q,L}$ are hermitian matrices in flavour space. Of course, any full-fledged (i.e. $SU(2)_L \times U(1)_Y$ gauge invariant and anomaly free) Z' model attempting an explanation of $R_{K^{(*)}}$ via LH currents can be mapped into Eq. 7.3.10. After integrating out the Z' at tree level, we obtain the effective Lagrangian

$$\mathcal{L}_{Z'}^{\text{eff}} = -\frac{1}{2M_{Z'}^2}\left(\lambda_{ij}^Q \bar{d}_L^i \gamma_\mu d_L^j + \lambda_{\alpha\beta}^L \bar{\ell}_L^\alpha \gamma_\mu \ell_L^\beta\right)^2$$

$$\supset -\frac{1}{2M_{Z'}^2}\left[(\lambda_{23}^Q)^2 \left(\bar{s}_L \gamma_\mu b_L\right)^2 + 2\lambda_{23}^Q \lambda_{22}^L (\bar{s}_L \gamma_\mu b_L)(\bar{\mu}_L \gamma^\mu \mu_L) + \text{h.c.}\right].$$

Matching with Eqs. 7.3.4 and 7.3.7 we get

$$\Delta C_9^\mu = -\Delta C_{10}^\mu = -\frac{\pi}{\sqrt{2}G_F M_{Z'}^2 \alpha}\left(\frac{\lambda_{23}^Q \lambda_{22}^L}{V_{tb}V_{ts}^*}\right), \qquad (7.3.11)$$

and

$$C_{bs}^{LL} = \frac{\hat{\eta}^{LL}(M_{Z'})}{4\sqrt{2}G_F M_{Z'}^2}\left(\frac{\lambda_{23}^Q}{V_{tb}V_{ts}^*}\right)^2, \qquad (7.3.12)$$

where $\hat{\eta}^{LL}(M_{Z'})$ encodes the RG running down to the bottom mass scale using NLO anomalous dimensions [28, 29]—e.g. for $M_{Z'} \in [1\text{--}10]\,\text{TeV}$ we find $\hat{\eta}^{LL}(M_{Z'}) \in [0.79, 0.75]$.

Here we consider the case of a real coupling λ_{23}^Q, so that $C_{bs}^{LL} > 0$ and $\Delta C_9^\mu = -\Delta C_{10}^\mu$ is also real. This assumption is consistent with the fact that nearly all of the groups performing global fits [21–26, 30–35] (see however [36] for an exception) have so far assumed real Wilson coefficients in Eq. 7.3.7, and also follows the standard approach adopted in the literature for the Z' models aiming at an explanation of the $b \rightarrow s\mu^+\mu^-$ anomalies (for an incomplete list, see [37–66]). In fact, complex Z' couplings can arise via fermion mixing, but are subject to additional constraints from CP-violating observables (see Sect. 7.3.2).

The impact of the improved SM calculation of B_s mixing on the parameter space of the Z' explanation of $R_{K^{(*)}}$ is displayed in Fig. 7.1 for the reference value $\lambda_{22}^L = 1$.[4] Note that the old SM determination, $\Delta M_s^{\text{SM,2015}}$, allowed for values of M'_Z

---

[4]For $M_{Z'} \lesssim 1$ TeV the coupling λ_{22}^L is bounded by the $Z \rightarrow 4\mu$ measurement at LHC and by neutrino trident production [67]. See for instance Fig. 1 in [68] for a recent analysis.

Fig. 7.1 Bounds from B_s mixing on the parameter space of the simplified Z' model of Eq. 7.3.10, for real λ_{23}^Q and $\lambda_{22}^L = 1$. The blue and red shaded areas correspond respectively to the 2σ exclusions from $\Delta M_s^{\mathrm{SM,2015}}$ and $\Delta M_s^{\mathrm{SM,2017}}$, while the dashed (solid) black curves encompass the 1σ (2σ) best-fit region from $R_{K^{(*)}}$

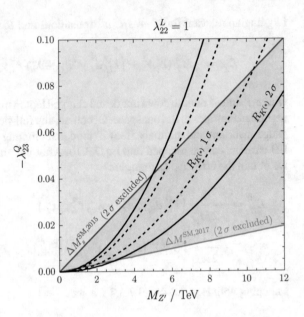

up to $\approx 10\,\mathrm{TeV}$ in order to explain $R_{K^{(*)}}$ at 1σ. In contrast, our updated $\Delta M_s^{\mathrm{SM,2017}}$ now implies $M_Z' \lesssim 2\,\mathrm{TeV}$. Remarkably, even for $\lambda_{22}^L = \sqrt{4\pi}$, which saturates the perturbative unitarity bound [69, 70], we find that the updated limit from B_s mixing requires $M_Z' \lesssim 8\,\mathrm{TeV}$ for the 1σ explanation of $R_{K^{(*)}}$. However, whether a Z' of mass a few TeV is ruled out or not by direct searches at LHC depends on the details of the Z' model. For instance, the stringent constraints from di-lepton searches [71] are relaxed in models where the Z' couples mainly third generation fermions (as e.g. in [65]). This notwithstanding, the updated limit from B_s mixing cuts dramatically into the parameter space allowed for a Z' explanation of the $b \to s\mu^+\mu^-$ anomalies, with important implications for the sensitivity of LHC direct searches and future colliders [72] in discovering or ruling out models of this kind.

7.3.1.2   Leptoquarks

Another popular class of simplified models which has been proposed in order to address the $b \to s\mu^+\mu^-$ anomalies consists of leptoquark mediators (see e.g. [73–90]).[5] Although B_s mixing is generated at one loop [91, 92],[5] and hence the constraints are expected to be milder compared to the Z' case, the connection with the anomalies

---

[5]The scalar leptoquark model proposed in [85] is a notable exception.

is more direct due to the structure of the leptoquark couplings. For instance, let us consider the scalar leptoquark $S_3 \sim (\bar{3}, 3)_{1/3}$,[6] with the Lagrangian

$$\mathcal{L}_{S_3} = -M_{S_3}^2 |S_3^a|^2 + y_{i\alpha}^{QL} \overline{Q^c}^i (\epsilon \sigma^a) L^\alpha S_3^a + \text{h.c.}, \tag{7.3.13}$$

where σ^a (for $a = 1, 2, 3$) are the Pauli matrices, $\epsilon = i\sigma^2$, and $Q^i = (V_{ji}^* u_L^j \ d_L^i)^T$ and $L^\alpha = (v_L^\alpha \ell_L^\alpha)^T$ are the quark and lepton doublet representations respectively (V being the CKM matrix). The contribution to the Wilson coefficients in Eq. 7.3.7 is present at tree level and reads

$$\Delta C_9^\mu = -\Delta C_{10}^\mu = \frac{\pi}{\sqrt{2} G_F M_{S_3}^2 \alpha} \left(\frac{y_{32}^{QL} y_{22}^{QL*}}{V_{tb} V_{ts}^*} \right), \tag{7.3.14}$$

while that to B_s mixing in Eq. 7.3.4 is induced at one loop [94]

$$C_{bs}^{LL} = \frac{\hat{\eta}^{LL}(M_{S_3})}{4\sqrt{2} G_F M_{S_3}^2} \frac{5}{64\pi^2} \left(\frac{y_{3\alpha}^{QL} y_{2\alpha}^{QL*}}{V_{tb} V_{ts}^*} \right)^2, \tag{7.3.15}$$

where the sum over the leptonic index $\alpha = 1, 2, 3$ is understood. In order to compare the two observables we consider in Fig. 7.2 the case in which only the couplings $y_{32}^{QL} y_{22}^{QL*}$ (namely those directly connected to $R_{K^{(*)}}$) contribute to B_s mixing, and further assume real couplings so that we can use the results of global fits which apply to real $\Delta C_9^\mu = -\Delta C_{10}^\mu$.

The bound on M_{S_3} from B_s mixing is strengthened by a factor of 5 due to our new determination of ΔM_s, which requires $M_{S_3} \lesssim 22$ TeV to explain $R_{K^{(*)}}$ at 1σ (see Fig. 7.2).

On the other hand, in flavour models predicting a hierarchical structure for the leptoquark couplings one expects $y_{i3}^{QL} \gg y_{i2}^{QL}$, so that the dominant contribution to ΔM_s is given by $y_{33}^{QL} y_{23}^{QL*}$. For example, $y_{i3}^{QL}/y_{i2}^{QL} \sim \sqrt{m_\tau/m_\mu} \approx 4$ in the partial compositeness framework of [74], so that the upper bound on M_{S_3} is strengthened by a factor $y_{33}^{QL} y_{23}^{QL*}/y_{32}^{QL} y_{22}^{QL*} \sim 16$. The latter can then easily approach the limits from LHC direct searches which imply $M_{S_3} \gtrsim 900$ GeV, e.g. for a QCD pair-produced S_3 dominantly coupled to third generation fermions [95].

---

[6]Similar considerations apply to the vector leptoquarks $U_1^\mu \sim (3, 1)_{2/3}$ and $U_3^\mu \sim (3, 3)_{2/3}$, which also provide a good fit for $R_{K^{(*)}}$. The case of massive vectors is however subtler, since the calculability of loop observables depends upon the UV completion (for a recent discussion, see e.g. [93]).

Fig. 7.2 Bounds from B_s
mixing on the parameter
space of the scalar
leptoquark model of
Eq. 7.3.13, for real $y_{32}^{QL} y_{22}^{QL*}$
couplings. Meaning of
shaded areas and curves as in
Fig. 7.1

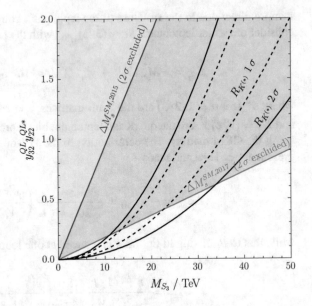

7.3.1.3   Combined $R_{K^{(*)}}$ and $R_{D^{(*)}}$ Explanations

Another set of intriguing anomalies in B physics data are those related to the LFU
violating ratios $R_{D^{(*)}} \equiv \mathcal{B}(B \to D^{(*)} \tau \bar{\nu})/\mathcal{B}(B \to D^{(*)} \ell \bar{\nu})$ (here, $\ell = e, \mu$), which
are observed to be larger than the SM [96–98]. Notably, in this case NP competes
with a tree-level SM charged current, thus requiring a sizeably larger effect compared
to that needed to explain the neutral current anomalies. The conditions under which
a combined explanation of $R_{K^{(*)}}$ and $R_{D^{(*)}}$ can be obtained, while being compatible
with a plethora of other indirect constraints (such as e.g. those pointed out in [99,
100]), have been recently reassessed at the EFT level in [101]. Regarding B_s mixing,
dimensional analysis (see e.g. Eq. (6) in [101]) shows that models without some
additional dynamical suppression (compared to semileptonic operators) are already
severely constrained by the old ΔM_s value. For instance, solutions based on a vector
triplet $V' \sim (\mathbf{1}, \mathbf{3})_0$ [102], where B_s mixing arises at tree level, are in serious tension
with data unless one invokes e.g. a percent level cancellation from extra contributions
[101]. The updated value of ΔM_s in Eq. 7.2.3 only increases the tuning required. On
the other hand, leptoquark solutions (e.g. the vector $U_1^\mu \sim (\mathbf{3}, \mathbf{1})_{2/3}$) comply better
with the bound due to the fact that B_s mixing arises at one loop, but the contribution
to ΔM_s should be addressed in specific UV models whenever calculable [88].

7.3.2 Model Building Directions for $\Delta M_s^{NP} < 0$

Given the fact that $\Delta M_s^{SM} > \Delta M_s^{exp}$ at about 2σ, it is interesting to consider possible ways to obtain a negative NP contribution to ΔM_s, thus relaxing the tension between the SM and the experimental measurement.

Sticking to the simplified models of Sect. 7.3.1 (Z' and leptoquarks coupled only to LH currents), an obvious solution to achieve $C_{bs}^{LL} < 0$ is to allow for complex couplings (see Eqs. 7.3.11 and 7.3.15). For instance, in Z' models this could happen as a consequence of fermion mixing if the Z' does not couple universally in the gauge-current basis. A similar mechanism could occur for vector leptoquarks arising from a spontaneously broken gauge theory, while scalar-leptoquark couplings to SM fermions are in general complex even before transforming to the mass basis.

Extra phases in the couplings are constrained by CP-violating observables, that we discuss now. In order to quantify the allowed parameter space for a generic, complex coefficient C_{bs}^{LL} in Eq. 7.3.4, we parameterise NP effects in B_s mixing via

$$\frac{M_{12}^{SM + NP}}{M_{12}^{SM}} \equiv |\Delta| \, e^{i\phi_\Delta} , \tag{7.3.16}$$

where

$$|\Delta| = \left| 1 + \frac{C_{bs}^{LL}}{R_{SM}^{loop}} \right| , \qquad \phi_\Delta = \arg\left(1 + \frac{C_{bs}^{LL}}{R_{SM}^{loop}} \right) . \tag{7.3.17}$$

The former is constrained by $\Delta M_s^{exp}/\Delta M_s^{SM} = |\Delta|$, while the latter by the mixing-induced CP asymmetry [2, 103][7]

$$A_{CP}^{mix}(B_s \to J/\psi\phi) = \sin(\phi_\Delta - 2\beta_s) , \tag{7.3.18}$$

where $A_{CP}^{mix} = -0.021 \pm 0.031$ [3], $\beta_s = 0.01852 \pm 0.00032$ [104], and we neglected penguin contributions [2]. The combined 2σ constraints on the Wilson coefficient C_{bs}^{LL} are displayed in Fig. 7.3.

For $\arg(C_{bs}^{LL}) = 0$ we recover the 2σ bound $\left| C_{bs}^{LL} \right| / R_{SM}^{loop} \lesssim 0.014$, which corresponds to the case discussed in Sect. 7.3.1 where we assumed a nearly real C_{bs}^{LL} (up to a small imaginary part due to V_{ts}). On the other hand, a non-zero phase of C_{bs}^{LL} allows relaxation of the bound from ΔM_s, or even accommodation of ΔM_s at 1σ (region between the red dashed lines in Fig. 7.3), and compatibility with the 2σ allowed region from A_{CP}^{mix} (blue shaded region in Fig. 7.3). For $\arg(C_{bs}^{LL}) \approx \pi$ values of $\left| C_{bs}^{LL} \right| / R_{SM}^{loop}$ as high as 0.21 are allowed at 2σ, relaxing the bound on the modulus of the Wilson coefficient by a factor 15 with respect to the $\arg(C_{bs}^{LL}) = 0$

---

[7]The semileptonic CP asymmetries for flavour-specific decays, a_{sl}^s, cannot generate competitive constraints as the experimental errors are currently still too large [2].

Fig. 7.3 Combined constraints on the complex Wilson coefficient C_{bs}^{LL}. The blue shaded area is the 2σ allowed region from A_{CP}^{mix}, while the solid (dashed) red curves enclose the 2σ (1σ) allowed regions from $\Delta M_s^{SM,2017}$

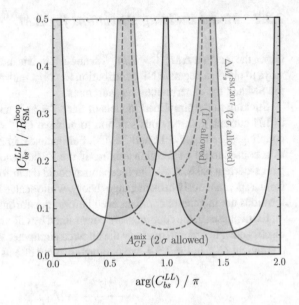

case. Note, however, that the limit $\arg(C_{bs}^{LL}) = \pi$ corresponds to a nearly imaginary $\Delta C_9^\mu = -\Delta C_{10}^\mu$ which likely spoil the fit of $R_{K^{(*)}}$, as the interference with the SM contribution would be strongly suppressed. Nevertheless, it would be interesting to perform a global fit of $R_{K^{(*)}}$ together with ΔM_s and A_{CP}^{mix} while allowing for non-zero values of the phase in order to see whether a better agreement with the data can be obtained. Non-zero weak phases can potentially reveal themselves via their contribution to triple product CP asymmetries in $B \to K^{(*)}\mu^+\mu^-$ angular distributions [36].

An alternative way to achieve a negative contribution for ΔM_s^{NP} is to go beyond the simplified models of Sect. 7.3.1 and consider generalised chirality structures. Let us choose, for definiteness, the case of a Z' coupled both to LH and RH down quark currents

$$\mathcal{L}_{Z'} \supset \frac{1}{2}M_{Z'}^2(Z'_\mu)^2 + \left(\lambda_{ij}^Q \, \bar{d}_L^i \gamma^\mu d_L^j + \lambda_{ij}^d \, \bar{d}_R^i \gamma^\mu d_R^j\right) Z'_\mu. \tag{7.3.19}$$

Upon integrating out the Z' one obtains

$$\mathcal{L}_{Z'}^{eff} \supset -\frac{1}{2M_{Z'}^2}\Big[\; (\lambda_{23}^Q)^2 \left(\bar{s}_L \gamma_\mu b_L\right)^2 + (\lambda_{23}^d)^2 \left(\bar{s}_R \gamma_\mu b_R\right)^2$$
$$+ 2\lambda_{23}^Q \lambda_{23}^d (\bar{s}_L \gamma_\mu b_L)(\bar{s}_R \gamma_\mu b_R) + \text{h.c.}\Big]. \tag{7.3.20}$$

The left-right (LR) vector operator can clearly have either sign, even for real couplings. Moreover, since it gets enhanced by renormalisation-group effects compared to LL and RR vector operators [105], it can dominate the contribution to ΔM_s^{NP}.

Note, however, that λ_{23}^d contributes to $R_{K^{(*)}}$ via RH quark currents whose presence is disfavoured by global fits, since they break the approximate relation $R_K \approx R_{K^*}$ that is observed experimentally (see e.g. [23]). Hence a careful study would also be required in this case in order to assess the simultaneous explanation of $R_{K^{(*)}}$ and ΔM_s.

7.4 Summary

In this chapter we made an updated SM prediction for the B_s mixing observable ΔM_s (Eq. 7.2.3) using the most recent values for the input parameters, in particular new results from the lattice averaging group FLAG. Our update shifts the central value of the SM theory prediction upwards and away from experiment by 13%, while reducing the theory uncertainty compared to the previous SM determination by a factor of two—which when combined result in a $1.8\,\sigma$ discrepancy between experiment and theory.

We further discussed an important application of the ΔM_s update for NP models aimed at explaining the recent anomalies in semileptonic B_s decays. The latter typically predict a positive shift in the NP contribution to ΔM_s, thus making the discrepancy with respect to the experimental value even worse. We have shown generically that whenever the NP contribution to ΔM_s is positive, the upper limit on the mass of the NP mediators that must be invoked to explain any of the anomalies is reduced by a factor of five (for a given size of couplings) compared to using the 2015 SM calculation for ΔM_s.

In particular, we considered two representative examples of NP models featuring purely LH current and real couplings—that of a Z' with the minimal couplings needed to explain $R_{K^{(*)}}$ anomalies, and a scalar $(SU(2)_L$ triplet) leptoquark model. For the Z' case we get an upper bound on the Z' mass of 2 TeV (for unit Z' coupling to muons, see Fig. 7.1), an energy scale that is already being probed by direct searches at LHC. On the other hand, the bounds on leptoquark models from B_s mixing are generally milder, as the contributions to mixing are loop suppressed. For instance, taking only the contribution of the couplings needed to fit $R_{K^{(*)}}$ for the evaluation of ΔM_s we find that the upper bound on the scalar leptoquark mass is brought down to about 20 TeV (see Fig. 7.2). This limit gets strengthened, however, in flavour models predicting a hierarchical structure of the leptoquark couplings to SM fermions and can then easily approach the region probed by the LHC. Solving the deviations in $R_{D^{(*)}}$ in addition to $R_{K^{(*)}}$ implies very severe bounds from B_s mixing as well, since the overall scale of NP must be lowered compared to the case of only neutral current anomalies.

Given the current status of a higher theory value for ΔM_s compared to experiment, we also have looked at possible ways in which NP can provide a negative contribution that lessens the tension. A non-zero phase in the NP couplings is one such way, and we have shown how extra constraints from the CP-violating observable A_{CP}^{mix} in $B_s \to J/\psi\phi$ decays cuts out parameter space where otherwise a significant NP contribution could be present. However, a large phase can potentially worsen the fit for $R_{K^{(*)}}$, increasing the need for a combined global fit of ΔM_s, $A_{\mathrm{CP}}^{\mathrm{mix}}$ and $R_{K^{(*)}}$. Another possibility is to consider NP models with a generalised chirality structure. In particular, $\Delta B = 2$ LR vector operators, which are renormalisation-group enhanced, can accommodate any sign for ΔM_s^{NP}, even for real couplings. Large contributions from RH currents are however disfavoured by the $R_{K^{(*)}}$ fit, and again a more careful analysis is needed.

References

1. Lenz A, Nierste U (2011) Numerical updates of lifetimes and mixing parameters of B mesons. In: CKM unitarity triangle. Proceedings, 6th International Workshop, CKM 2010, Warwick, UK, September 6–10, 2010. arXiv:1102.4274, http://inspirehep.net/record/890169/files/arXiv:1102.4274.pdf
2. Artuso M, Borissov G, Lenz A (2016) CP violation in the B_s^0 system. Rev Mod Phys 88:045002. https://doi.org/10.1103/RevModPhys.88.045002, arXiv:1511.09466
3. HFLAV collaboration, Amhis Y et al (2017) Averages of b-hadron, c-hadron, and τ-lepton properties as of summer 2016. Eur Phys J C 77:895. https://doi.org/10.1140/epjc/s10052-017-5058-4, arXiv:1612.07233
4. Fermilab Lattice, MILC collaboration, Bazavov A et al (2016) $B_{(s)}^0$-mixing matrix elements from lattice QCD for the Standard Model and beyond. Phys Rev D 93:113016. https://doi.org/10.1103/PhysRevD.93.113016, arXiv:1602.03560
5. Blanke M, Buras AJ (2016) Universal Unitarity Triangle 2016 and the tension between $\Delta M_{s,d}$ and ε_K in CMFV models. Eur Phys J C 76:197. https://doi.org/10.1140/epjc/s10052-016-4044-6, arXiv:1602.04020
6. Buras AJ, De Fazio F (2016) 331 models facing the tensions in $\Delta F = 2$ processes with the impact on ε'/ε, $B_s \to \mu^+\mu^-$ and $B \to K^*\mu^+\mu^-$. JHEP 08:115. https://doi.org/10.1007/JHEP08(2016)115, arXiv: 1604.02344
7. Altmannshofer W, Gori S, Robinson DJ, Tuckler D (2018) The flavor-locked flavorful two higgs doublet model. JHEP 03:129. https://doi.org/10.1007/JHEP03(2018)129, arXiv:1712.01847
8. Bobrowski M, Lenz A, Riedl J, Rohrwild J (2009) How much space is left for a new family of fermions? Phys Rev D 79:113006. https://doi.org/10.1103/PhysRevD.79.113006, arXiv:0902.4883
9. Blanke M, Buras AJ (2007) Lower bounds on $\Delta M_{s,d}$ from constrained minimal flavour violation. JHEP 05:061. https://doi.org/10.1088/1126-6708/2007/05/061, arXiv: hep-ph/0610037
10. LHCb collaboration, Aaij R et al (2015) Angular analysis and differential branching fraction of the decay $B_s^0 \to \phi\mu^+\mu^-$. JHEP 09:179. https://doi.org/10.1007/JHEP09(2015)179, arXiv: 1506.08777
11. CMS collaboration, Khachatryan V et al (2016) Angular analysis of the decay $B^0 \to K^{*0}\mu^+\mu^-$ from pp collisions at $\sqrt{s} = 8$ TeV. Phys Lett B 753:424–448. https://doi.org/10.1016/j.physletb.2015.12.020, arXiv:1507.08126
12. BaBar collaboration, Lees JP et al (2016) Measurement of angular asymmetries in the decays $B \to K^*\ell^+\ell^-$. Phys Rev D 93:052015. https://doi.org/10.1103/PhysRevD.93.052015, arXiv:1508.07960
13. Belle collaboration, Wei JT et al (2009) Measurement of the differential branching fraction and forward-backword asymmetry for $B \to K^{(*)}\ell^+\ell^-$. Phys Rev Lett 103:171801. https://doi.org/10.1103/PhysRevLett.103.171801, arXiv:0904.0770

14. CDF collaboration, Aaltonen T et al (2012) Measurements of the angular distributions in the decays $B \to K^{(*)} \mu^+ \mu^-$ at CDF. Phys Rev Lett 108:081807. https://doi.org/10.1103/PhysRevLett.108.081807, arXiv:1108.0695

15. LHCb collaboration, Aaij R et al (2016) Angular analysis of the $B^0 \to K^{*0} \mu^+ \mu^-$ decay using 3 fb^{-1} of integrated luminosity. JHEP 02:104. https://doi.org/10.1007/JHEP02(2016)104, arXiv:1512.04442

16. Belle collaboration, Abdesselam A et al (2016) Angular analysis of $B^0 \to K^*(892)^0 \ell^+ \ell^-$. In: Proceedings, LHCSki 2016—a first discussion of 13 TeV results: Obergurgl, Austria, April 10–15, 2016. arXiv: 1604.04042, https://inspirehep.net/record/1446979/files/arXiv:1604.04042.pdf

17. Belle collaboration, Wehle S et al (2017) Lepton-flavor-dependent angular analysis of $B \to K^* \ell^+ \ell^-$. Phys Rev Lett 118:111801. https://doi.org/10.1103/PhysRevLett.118.111801, arXiv:1612.05014

18. ATLAS collaboration, Aaboud M et al (2018) Angular analysis of $B_d^0 \to K^* \mu^+ \mu^-$ decays in pp collisions at $\sqrt{s} = 8$ TeV with the ATLAS detector. JHEP 10:047. https://doi.org/10.1007/JHEP10(2018)047, arXiv: 1805.04000

19. CMS collaboration, Sirunyan AM et al (2018) Measurement of angular parameters from the decay $B^0 \to K^{*0} \mu^+ \mu^-$ in proton-proton collisions at $\sqrt{s} = 8$ TeV. Phys Lett B 781:517–541. https://doi.org/10.1016/j.physletb.2018.04.030, arXiv:1710.02846

20. LHCb collaboration, Aaij R et al (2014) Differential branching fractions and isospin asymmetries of $B \to K^{(*)} \mu^+ \mu^-$ decays. JHEP 06:133. https://doi.org/10.1007/JHEP06(2014)133, arXiv:1403.8044

21. Capdevila B, Crivellin A, Descotes-Genon S, Matias J, Virto J (2018) Patterns of new physics in $b \to s \ell^+ \ell^-$ transitions in the light of recent data. JHEP 01:093. https://doi.org/10.1007/JHEP01(2018)093, arXiv: 1704.05340

22. Altmannshofer W, Stangl P, Straub DM (2017) Interpreting hints for lepton flavor universality violation. Phys Rev D 96:055008. https://doi.org/10.1103/PhysRevD.96.055008, arXiv:1704.05435

23. D'Amico G, Nardecchia M, Panci P, Sannino F, Strumia A, Torre R et al (2017) Flavour anomalies after the R_{K^*} measurement. JHEP 09:010. https://doi.org/10.1007/JHEP09(2017)010, arXiv: 1704.05438

24. Alok AK, Bhattacharya B, Datta A, Kumar D, Kumar J, London D (2017) New physics in $b \to s \mu^+ \mu^-$ after the measurement of R_{K^*}. Phys Rev D 96:095009. https://doi.org/10.1103/PhysRevD.96.095009, arXiv:1704.07397

25. Geng L-S, Grinstein B, Jäger S, Martin Camalich J, Ren X-L, Shi R-X (2017) Towards the discovery of new physics with lepton-universality ratios of $b \to s \ell \ell$ decays. Phys Rev D 96:093006. https://doi.org/10.1103/PhysRevD.96.093006, arXiv:1704.05446

26. Ciuchini M, Coutinho AM, Fedele M, Franco E, Paul A, Silvestrini L et al (2017) On flavourful easter eggs for new physics hunger and lepton flavour universality violation. Eur Phys J C 77:688. https://doi.org/10.1140/epjc/s10052-017-5270-2, arXiv:1704.05447

27. Golowich E, Hewett J, Pakvasa S, Petrov AA, Yeghiyan GK (2011) Relating B_s mixing and $B_s \to \mu^+ \mu^-$ with new physics. Phys Rev D 83:114017. https://doi.org/10.1103/PhysRevD.83.114017, arXiv:1102.0009

28. Ciuchini M, Franco E, Lubicz V, Martinelli G, Scimemi I, Silvestrini L (1998) Next-to-leading order QCD corrections to $\Delta F = 2$ effective Hamiltonians. Nucl Phys B 523:501–525. https://doi.org/10.1016/S0550-3213(98)00161-8, arXiv:hep-ph/9711402

29. Buras AJ, Misiak M, Urban J (2000) Two loop QCD anomalous dimensions of flavor changing four quark operators within and beyond the standard model. Nucl Phys B 586:397–426. https://doi.org/10.1016/S0550-3213(00)00437-5, arXiv:hep-ph/0005183

30. Descotes-Genon S, Matias J, Virto J (2013) Understanding the $B \to K^* \mu^+ \mu^-$ anomaly. Phys Rev D 88:074002. https://doi.org/10.1103/PhysRevD.88.074002, arXiv:1307.5683

31. Beaujean F, Bobeth C, van Dyk D (2017) Comprehensive Bayesian analysis of rare (semi)leptonic and radiative B decays. Eur Phys J C 74:2897. https://doi.org/10.1140/epjc/s10052-014-2897-0, https://doi.org/10.1140/epjc/s10052-014-3179-6, arXiv:1310.2478

32. Altmannshofer W, Straub DM (2015) New physics in $b \to s$ transitions after LHC run 1. Eur Phys J C 75:382. https://doi.org/10.1140/epjc/s10052-015-3602-7, arXiv:1411.3161
33. Descotes-Genon S, Hofer L, Matias J, Virto J (2016) Global analysis of $b \to s\ell\ell$ anomalies. JHEP 06:092. https://doi.org/10.1007/JHEP06(2016)092, arXiv: 1510.04239
34. Hurth T, Mahmoudi F, Neshatpour S (2016) On the anomalies in the latest LHCb data. Nucl Phys B 909:737–777. https://doi.org/10.1016/j.nuclphysb.2016.05.022, arXiv:1603.00865
35. Altmannshofer W, Niehoff C, Stangl P, Straub DM (2017) Status of the $B \to K^*\mu^+\mu^-$ anomaly after Moriond 2017. Eur Phys J C 77:377. https://doi.org/10.1140/epjc/s10052-017-4952-0, arXiv:1703.09189
36. Alok AK, Bhattacharya B, Kumar D, Kumar J, London D, Sankar SU (2017) New physics in $b \to s\mu^+\mu^-$: Distinguishing models through CP-violating effects. Phys Rev D 96:015034. https://doi.org/10.1103/PhysRevD.96.015034, arXiv:1703.09247
37. Buras AJ, Girrbach J (2013) Left-handed Z' and Z FCNC quark couplings facing new $b \to s\mu^+\mu^-$ data. JHEP 12:009. https://doi.org/10.1007/JHEP12(2013)009, arXiv:1309.2466
38. Gauld R, Goertz F, Haisch U (2014) An explicit Z'-boson explanation of the $B \to K^*\mu^+\mu^-$ anomaly. JHEP 01:069. https://doi.org/10.1007/JHEP01(2014)069, arXiv:1310.1082
39. Buras AJ, De Fazio F, Girrbach J (2014) 331 models facing new $b \to s\mu^+\mu^-$ data. JHEP 02:112. https://doi.org/10.1007/JHEP02(2014)112, arXiv:1311.6729
40. Altmannshofer W, Gori S, Pospelov M, Yavin I (2014) Quark flavor transitions in $L_\mu - L_\tau$ models. Phys Rev D 89:095033. https://doi.org/10.1103/PhysRevD.89.095033, arXiv:1403.1269
41. Crivellin A, D'Ambrosio G, Heeck J (2015) Explaining $h \to \mu^\pm\tau^\mp$, $B \to K^*\mu^+\mu^-$ and $B \to K\mu^+\mu^-/B \to Ke^+e^-$ in a two-Higgs-doublet model with gauged $L_\mu - L_\tau$. Phys Rev Lett 114:151801. https://doi.org/10.1103/PhysRevLett.114.151801, arXiv:1501.00993
42. Crivellin A, D'Ambrosio G, Heeck J (2015) Addressing the LHC flavor anomalies with horizontal gauge symmetries. Phys Rev D 91:075006. https://doi.org/10.1103/PhysRevD.91.075006, arXiv:1503.03477
43. Celis A, Fuentes-Martin J, Jung M, Serôdio H (2015) Family nonuniversal Z' models with protected flavor-changing interactions. Phys Rev D 92:015007. https://doi.org/10.1103/PhysRevD.92.015007, arXiv:1505.03079
44. Belanger G, Delaunay C, Westhoff S (2015) A dark matter relic from muon anomalies. Phys Rev D 92:055021. https://doi.org/10.1103/PhysRevD.92.055021, arXiv:1507.06660
45. Falkowski A, Nardecchia M, Ziegler R (2015) Lepton flavor non-universality in B-meson decays from a U(2) flavor model. JHEP 11:173. https://doi.org/10.1007/JHEP11(2015)173, arXiv: 1509.01249
46. Carmona A, Goertz F (2016) Lepton flavor and nonuniversality from minimal composite higgs setups. Phys Rev Lett 116:251801. https://doi.org/10.1103/PhysRevLett.116.251801, arXiv:1510.07658
47. Allanach B, Queiroz FS, Strumia A, Sun S (2016) Z? models for the LHCb and $g - 2$ muon anomalies:Phys Rev D 93:055045. https://doi.org/10.1103/PhysRevD.93.055045, https://doi.org/10.1103/PhysRevD.95.119902, arXiv:1511.07447
48. Chiang C-W, He X-G, Valencia G (2016) Z? model for b?s?$\bar{?}$ flavor anomalies. Phys Rev D 93:074003. https://doi.org/10.1103/PhysRevD.93.074003, arXiv:1601.07328
49. Boucenna SM, Celis A, Fuentes-Martin J, Vicente A, Virto J (2016) Non-abelian gauge extensions for B-decay anomalies. Phys Lett B 760:214–219. https://doi.org/10.1016/j.physletb.2016.06.067, arXiv:1604.03088
50. Megias E, Panico G, Pujolas O, Quiros M (2016) A Natural origin for the LHCb anomalies. JHEP 09:118. https://doi.org/10.1007/JHEP09(2016)118, arXiv:1608.02362
51. Boucenna SM, Celis A, Fuentes-Martin J, Vicente A, Virto J (2016) Phenomenology of an $SU(2) \times SU(2) \times U(1)$ model with lepton-flavour non-universality. JHEP 12:059. https://doi.org/10.1007/JHEP12(2016)059, arXiv: 1608.01349
52. Altmannshofer W, Gori S, Profumo S, Queiroz FS (2016) Explaining dark matter and B decay anomalies with an $L_\mu - L_\tau$ model. JHEP 12:106. https://doi.org/10.1007/JHEP12(2016)106, arXiv: 1609.04026

53. Crivellin A, Fuentes-Martin J, Greljo A, Isidori G (2017) Lepton flavor non-universality in B decays from Dynamical Yukawas. Phys Lett B 766:77–85. https://doi.org/10.1016/j.physletb.2016.12.057, arXiv:1611.02703

54. Garcia Garcia I (2017) LHCb anomalies from a natural perspective. JHEP 03:040. https://doi.org/10.1007/JHEP03(2017)040, arXiv: 1611.03507

55. Bhatia D, Chakraborty S, Dighe A (2017) Neutrino mixing and R_K anomaly in U(1)$_X$ models: a bottom-up approach. JHEP 03:117. https://doi.org/10.1007/JHEP03(2017)117, arXiv:1701.05825

56. Cline JM, Cornell JM, London D, Watanabe R (2017) Hidden sector explanation of B-decay and cosmic ray anomalies. Phys Rev D 95:095015. https://doi.org/10.1103/PhysRevD.95.095015, arXiv:1702.00395

57. Baek S (2018) Dark matter contribution to $b \to s\mu^+\mu^-$ anomaly in local $U(1)_{L_\mu-L_\tau}$ model. Phys Lett B 781:376–382. https://doi.org/10.1016/j.physletb.2018.04.012, arXiv:1707.04573

58. Cline JM, Martin Camalich J (2017) B decay anomalies from nonabelian local horizontal symmetry. Phys Rev D 96:055036. https://doi.org/10.1103/PhysRevD.96.055036, arXiv:1706.08510

59. Di Chiara S, Fowlie A, Fraser S, Marzo C, Marzola L, Raidal M et al (2017) Minimal flavor-changing Z' models and muon $g - 2$ after the R_{K^*} measurement. Nucl Phys B 923:245–257. https://doi.org/10.1016/j.nuclphysb.2017.08.003, arXiv:1704.06200

60. Kamenik JF, Soreq Y, Zupan J (2018) Lepton flavor universality violation without new sources of quark flavor violation. Phys Rev D 97:035002. https://doi.org/10.1103/PhysRevD.97.035002, arXiv:1704.06005

61. Ko P, Omura Y, Shigekami Y, Yu C (2017) LHCb anomaly and B physics in flavored Z' models with flavored Higgs doublets. Phys Rev D 95:115040. https://doi.org/10.1103/PhysRevD.95.115040, arXiv:1702.08666

62. Ko P, Nomura T, Okada H (2017) Explaining $B \to K^{(*)}\ell^+\ell^-$ anomaly by radiatively induced coupling in $U(1)_{\mu-\tau}$ gauge symmetry. Phys Rev D 95:111701. https://doi.org/10.1103/PhysRevD.95.111701, arXiv:1702.02699

63. Alonso R, Cox P, Han C, Yanagida TT (2017) Anomaly-free local horizontal symmetry and anomaly-full rare B-decays. Phys Rev D 96:071701. https://doi.org/10.1103/PhysRevD.96.071701, arXiv:1704.08158

64. Ellis J, Fairbairn M, Tunney P (2018) Anomaly-free models for flavour anomalies. Eur Phys J C 78:238. https://doi.org/10.1140/epjc/s10052-018-5725-0, arXiv:1705.03447

65. Alonso R, Cox P, Han C, Yanagida TT (2017) Flavoured $B - L$ Local symmetry and anomalous rare B decays. Phys Lett B 774:643–648. https://doi.org/10.1016/j.physletb.2017.10.027, arXiv:1705.03858

66. Carmona A, Goertz F (2018) Recent **B** physics anomalies: a first hint for compositeness? Eur Phys J C 78:979. https://doi.org/10.1140/epjc/s10052-018-6437-1, arXiv:1712.02536

67. Altmannshofer W, Gori S, Pospelov M, Yavin I (2014) Neutrino trident production: a powerful probe of new physics with neutrino beams. Phys Rev Lett 113:091801. https://doi.org/10.1103/PhysRevLett.113.091801, arXiv:1406.2332

68. Falkowski A, King SF, Perdomo E, Pierre M (2018) Flavourful Z' portal for vector-like neutrino Dark Matter and $R_{K^{(*)}}$. JHEP 08:061. https://doi.org/10.1007/JHEP08(2018)061, arXiv: 1803.04430

69. Di Luzio L, Nardecchia M (2017) What is the scale of new physics behind the B-flavour anomalies?. Eur Phys J C 77:536. https://doi.org/10.1140/epjc/s10052-017-5118-9, arXiv:1706.01868

70. Di Luzio L, Kamenik JF, Nardecchia M (2017) Implications of perturbative unitarity for scalar di-boson resonance searches at LHC. Eur Phys J C 77:30. https://doi.org/10.1140/epjc/s10052-017-4594-2, arXiv:1604.05746

71. ATLAS collaboration, Aaboud M et al (2017) Search for new high-mass phenomena in the dilepton final state using 36 fb^{-1} of proton-proton collision data at $\sqrt{s} = $ 13 TeV with the ATLAS detector. JHEP 10:182. https://doi.org/10.1007/JHEP10(2017)182, arXiv:1707.02424

72. Allanach BC, Gripaios B, You T (2018) The case for future hadron colliders from $B \to K^{(*)} \mu^+ \mu^-$ decays. JHEP 03:021. https://doi.org/10.1007/JHEP03(2018)021, arXiv: 1710.06363

73. Hiller G, Schmaltz M (2014) R_K and future $b \to s\ell\ell$ physics beyond the standard model opportunities. Phys Rev D 90:054014. https://doi.org/10.1103/PhysRevD.90.054014, arXiv:1408.1627

74. Gripaios B, Nardecchia M, Renner SA (2015) Composite leptoquarks and anomalies in B-meson decays. JHEP 05:006. https://doi.org/10.1007/JHEP05(2015)006, arXiv:1412.1791

75. de Medeiros Varzielas I, Hiller G (2015) Clues for flavor from rare lepton and quark decays. JHEP 06:072. https://doi.org/10.1007/JHEP06(2015)072, arXiv: 1503.01084

76. Bečirević D, Fajfer S, Košnik N (2015) Lepton flavor nonuniversality in $b \to s\mu^+\mu^-$ processes. Phys Rev D 92:014016. https://doi.org/10.1103/PhysRevD.92.014016, arXiv:1503.09024

77. Alonso R, Grinstein B, Martin Camalich J (2015) Lepton universality violation and lepton flavor conservation in B-meson decays. JHEP 10:184. https://doi.org/10.1007/JHEP10(2015)184, arXiv: 1505.05164

78. Bauer M, Neubert M (2016) Minimal Leptoquark Explanation for the $R_{D^{(*)}}$, R_K and $(g - 2)_g$ anomalies. Phys Rev Lett 116:141802. https://doi.org/10.1103/PhysRevLett.116.141802, arXiv:1511.01900

79. Fajfer S, Košnik N (2016) Vector leptoquark resolution of R_K and $R_{D^{(*)}}$ puzzles. Phys Lett B 755:270–274. https://doi.org/10.1016/j.physletb.2016.02.018, arXiv:1511.06024

80. Barbieri R, Isidori G, Pattori A, Senia F (2016) Anomalies in B-decays and $U(2)$ flavour symmetry. Eur Phys J C 76:67. https://doi.org/10.1140/epjc/s10052-016-3905-3, arXiv:1512.01560

81. Bečirević D, Košnik N, Sumensari O, Zukanovich Funchal R (2016) Palatable leptoquark scenarios for lepton flavor violation in exclusive $b \to s\ell_1\ell_2$ modes. JHEP 11:035. https://doi.org/10.1007/JHEP11(2016)035, arXiv: 1608.07583

82. Bečirević D, Fajfer S, Košnik N, Sumensari O (2016) Leptoquark model to explain the B-physics anomalies, R_K and R_D. Phys Rev D 94:115021. https://doi.org/10.1103/PhysRevD. 94.115021, arXiv:1608.08501

83. Crivellin A, Müller D, Ota T (2017) Simultaneous explanation of $R(D^{(*)})$ and $b \to s\mu^+\mu^-$: the last scalar leptoquarks standing. JHEP 09:040. https://doi.org/10.1007/JHEP09(2017)040, arXiv: 1703.09226

84. Hiller G, Nisandzic I (2017) R_K and R_{K^*} beyond the standard model. Phys Rev D 96:035003. https://doi.org/10.1103/PhysRevD.96.035003, arXiv:1704.05444

85. Bečirević D, Sumensari O (2017) A leptoquark model to accommodate $R_K^{\mathrm{exp}} < R_K^{\mathrm{SM}}$ and $R_{K^*}^{\mathrm{exp}} < R_{K^*}^{\mathrm{SM}}$. JHEP 08:104. https://doi.org/10.1007/JHEP08(2017)104, arXiv: 1704.05835

86. Doršner I, Fajfer S, Faroughy DA, Košnik N, The role of the S_3 GUT leptoquark in flavor universality and collider searches. arXiv:1706.07779

87. Assad N, Fornal B, Grinstein B (2018) Baryon number and lepton universality violation in leptoquark and diquark models. Phys Lett B 777:324–331. https://doi.org/10.1016/j.physletb. 2017.12.042, arXiv:1708.06350

88. Di Luzio L, Greljo A, Nardecchia M (2017) Gauge leptoquark as the origin of B-physics anomalies. Phys Rev D 96:115011. https://doi.org/10.1103/PhysRevD.96.115011, arXiv:1708.08450

89. Calibbi L, Crivellin A, Li T (2018) Model of vector leptoquarks in view of the B-physics anomalies. Phys Rev D 98:115002. https://doi.org/10.1103/PhysRevD.98.115002, arXiv:1709.00692

90. Bordone M, Cornella C, Fuentes-Martin J, Isidori G (2018) A three-site gauge model for flavor hierarchies and flavor anomalies. Phys Lett B 779:317–323. https://doi.org/10.1016/j. physletb.2018.02.011, arXiv:1712.01368

91. Davidson S, Bailey DC, Campbell BA (1994) Model independent constraints on lepto-quarks from rare processes. Z Phys C 61:613–644. https://doi.org/10.1007/BF01552629, arXiv:hep-ph/9309310

92. Doršner I, Fajfer S, Greljo A, Kamenik JF, Košnik N (2016) Physics of leptoquarks in precision experiments and at particle colliders. Phys Rept 641:1–68. https://doi.org/10.1016/j.physrep. 2016.06.001, arXiv:1603.04993

93. Biggio C, Bordone M, Di Luzio L, Ridolfi G (2016) Massive vectors and loop observables: the $g - 2$ case. JHEP 10:002. https://doi.org/10.1007/JHEP10(2016)002, arXiv: 1607.07621

94. Bobeth C, Buras AJ (2018) Leptoquarks meet ε'/ε and rare Kaon processes. JHEP 02:101. https://doi.org/10.1007/JHEP02(2018)101, arXiv: 1712.01295

95. CMS collaboration, Sirunyan AM et al (2017) Search for third-generation scalar leptoquarks and heavy right-handed neutrinos in final states with two tau leptons and two jets in proton-proton collisions at $\sqrt{s} = 13$ TeV. JHEP 07:121. https://doi.org/10.1007/JHEP07(2017)121, arXiv: 1703.03995

96. BaBar collaboration, Lees JP et al (2013) Measurement of an excess of $\bar{B} \to D^{(*)}\tau^-\bar{\nu}_\tau$ decays and implications for charged Higgs bosons. Phys Rev D 88:072012. https://doi.org/10.1103/PhysRevD.88.072012. arXiv:1303.0571

97. LHCb collaboration, Aaij R et al (2015) Measurement of the ratio of branching fractions $\mathcal{B}(\bar{B}^0 \to D^{*+}\tau^-\bar{\nu}_\tau)/\mathcal{B}(\bar{B}^0 \to D^{*+}\mu^-\bar{\nu}_\mu)$, Phys Rev Lett 115:111803. https://doi.org/10.1103/PhysRevLett.115.159901, https://doi.org/10.1103/PhysRevLett.115.111803, arXiv: 1506.08614

98. Belle collaboration, Hirose S et al (2017) Measurement of the τ lepton polarization and $R(D^*)$ in the decay $\bar{B} \to D^*\tau^-\bar{\nu}_\tau$. Phys Rev Lett 118:211801. https://doi.org/10.1103/PhysRevLett. 118.211801, arXiv:1612.00529

99. Feruglio F, Paradisi P, Pattori A (2017) Revisiting lepton flavor universality in B decays. Phys Rev Lett 118:011801. https://doi.org/10.1103/PhysRevLett.118.011801, arXiv:1606.00524

100. Feruglio F, Paradisi P, Pattori A (2017) On the importance of electroweak corrections for B anomalies. JHEP 09:061. https://doi.org/10.1007/JHEP09(2017)061, arXiv: 1705.00929

101. Buttazzo D, Greljo A, Isidori G, Marzocca D (2017) B-physics anomalies: a guide to combined explanations. JHEP 11:044. https://doi.org/10.1007/JHEP11(2017)044, arXiv: 1706.07808

102. Greljo A, Isidori G, Marzocca D (2015) On the breaking of lepton flavor universality in B decays. JHEP 07:142. https://doi.org/10.1007/JHEP07(2015)142, arXiv: 1506.01705

103. Lenz A, Nierste U (2007) Theoretical update of $B_s - \bar{B}_s$ mixing. JHEP 06:072. https://doi.org/10.1088/1126-6708/2007/06/072, arXiv:hep-ph/0612167

104. CKMfitter collaboration (2016) ICHEP 2016 results. http://ckmfitter.in2p3.fr/www/results/plots_ichep16/num/ckmEval_results_ichep16.html

105. Buras AJ, Jäger S, Urban J (2001) Master formulae for $\Delta F = 2$ NLO QCD factors in the standard model and beyond. Nucl Phys B 605:600–624. https://doi.org/10.1016/S0550-3213(01)00207-3, arXiv:hep-ph/0102316

Chapter 8
Conclusions

The aim of this thesis has been to focus on certain observables within particle physics, and to understand how they may lead us either to greater heights in confirming the Standard Model, or show us the way to as yet unknown physics. We have chosen on flavour physics as our target as it holds the key to many rare processes that are ideal for our purpose. The structure of the SM means that a great many flavour observables are suppressed by various means, and so any new physics that can lift the suppression should either stand out easily or have very strong bounds placed upon it.

Within this thesis we have approached our task from several angles, in an attempt to elucidate all we can. We have looked at the basic underlying assumptions of the way we calculate, so that we might be sure if we can distinguish a failure of the tool from genuine new effects; we have improved the accuracy of calculations by improving the determinations of crucial inputs; and studied directly a few models of BSM physics and the limitations imposed by the current and future precision of experimental and theoretical results.

Following on from Chap. 1 where we introduced the Standard Model, flavour physics, and the problems facing theory today, in Chap. 2 we outlined many of the standard tools of flavour physics used to perform calculations and make predictions. As part of that, we focussed in depth on two simple calculations—the matching of the SM to the weak effective theory and the leading order calculation of the mass difference ΔM_s arising from B_s mixing, both to illuminate the tools we had described at the start of the chapter, and as a helpful guide to those who follow by including details that are so often "left as an exercise to the reader".

In Chap. 3, we studied one of the underlying assumptions of the HQE, quark-hadron duality. The HQE is an important technique for calculating in our effective theory, where the b quark mass is far removed from most other relevant scales, but it cannot account for any violation of the duality between inclusive quark level calculations and exclusive studies of fully identified hadronic decay modes. As the rest of the thesis relies on the HQE for precision calculations and for identifying possible NP effects in flavour observables, it is of vital importance that we can trust

© Springer Nature Switzerland AG 2019
M. J. Kirk, *Charming New Physics in Beautiful Processes?*,
Springer Theses, https://doi.org/10.1007/978-3-030-19197-9_8

our tools–we must be sure that our phenomenological predictions correctly reflect the underlying theory.

We tested quark-hadron duality by constructing a simple phenomenological model of how duality violations could occur, which we presented in Eqs. 3.2.10–3.2.12. Using this simple model we studied how the theory prediction for mixing observables $\Delta\Gamma$ and a_{sl} would be affected. Due to cancellations, the ratio $\Delta\Gamma_s/\Delta M_s$ is one of the highest precision mixing observables, and so we used that to bound our simple model. Comparing theory and experiment, we showed that within our model, duality could be violated by no more than 20–25% in the $c\bar{c}$ decay channel (see Eq. 3.2.14). We next examined a generalised version of our model and saw (for example in Fig. 3.2) how large duality violating effects in $\Delta\Gamma_s$ could be hidden in a_{sl}^s. The general model allowed us to make more profound statements about possible future measurements of the currently unmeasured quantities a_{sl}^s, a_{sl}^d, and $\Delta\Gamma_d$. Taking duality to be violated by the maximum allowed by current results, we showed in Fig. 3.3 what size of measurement by a future experiment could only be explained by new physics rather than a breakdown of the HQE. In the spirit of looking forward, we took recent lattice results of greatly improved precision and briefly examined how they would affect our conclusions, and how far the limits might conceivably be pushed. Our brief adventure saw the possibility of constraining duality violation at close to the 5% level.

Since the structure of the HQE and processes that contribute to the lifetime ratio $\tau(B_s)/\tau(B_d)$ are similar to those in $\Delta\Gamma$, we took a slight detour to see whether that observable could be affected by duality violations. The results are shown in Fig. 3.5—currently we cannot make any further statement beyond what we have already learned, but future scenarios could be very enlightening depending on the direction experiment takes.

While in the B sector the agreement between theory and experiment for mixing allows us to bound duality violations, in the D sector we saw that opposite is true. Currently it is not clear how well the HQE works for the simple reason that $1/m_b \ll 1/m_c$, and the inclusive calculation of ΔM and $\Delta\Gamma$ for D mixing seems to totally disagree with experiment, at least with the current level of calculated corrections. However a small duality violation, of roughly the same size as currently allowed in B_s mesons can bring the HQE calculation into agreement by breaking the extremely effective GIM cancellation that arises in the standard calculation.

What we learned from this chapter is that while current precision does place limits on violations of quark-hadron duality, there are a few areas where swift progress could be made. The major uncertainty in most of our calculations comes from the matrix elements of dimension-six and dimension-seven operators, and future improvement there would be very meaningful—our work in Chap. 6 is an example of what can be done here.

In Chap. 4 we study the problem of dark matter that was first discussed in Sect. 1.3.1. As we explained there, the existence of dark matter has been known for many decades now, but our knowledge of the nature of dark matter is effectively still limited to its abundance in the universe and the interactions with our current detectors. With so much unknown about the dark sector, it is tempting to make use of effective field theories as these can be sufficiently general as to assume almost nothing about the undiscovered particles. Yet there is an issue with these—while valid at low scales, they break down once experiments start to probe close to the scale associated with the new physics. This is a problem as the search for dark matter is currently proceeding from three directions: direct detection, indirect detection and collider searches. Since there are some theoretical reasons to believe that at least part of the dark sector should be close to the energy scale of the LHC, an EFT approach cannot be used to include all current searches for dark matter.

Once this became clear to the community, so-called simplified models gained popularity as a tool for dark matter investigations. In these models, a minimal set of dark sector particles are kept as dynamical degrees of freedom, allowing the correct behaviour at LHC energy scales to be reproduced. As we have seen in this thesis, the flavour sector of the SM has a non-trivial structure, and so the question arises of how to proceed with any interaction with dark matter. In our work, we chose to go beyond the MFV hypothesis to study a dark sector with a coloured mediator and three dark sector particles (of which the lightest will be DM) with complex (in the sense of not simple) couplings to the quark sector. Since we know that the current theory predictions for mixing in the up type sector are not in agreement with the experimental results, we looked at dark matter with couplings to right handed up type quarks to see how it might contribute.

We approached it using an MCMC to deal with our large parameter space and the many constraints. MCMC techniques not only allowed us to probe the relatively high dimensional parameter space efficiently, but also to get a statistical handle on the allowed regions that we found. Looking at Fig. 4.14, we see that the MCMC tells us that for dark matter coupling predominantly to top quarks, both the mediator and dark matter are, at $2\,\sigma$, forbidden from having masses below $\mathcal{O}(1\,\text{TeV})$. On the contrary, for predominant charm quark coupling we can go lighter in mass for both particles and, by allowing a small mass splitting between the flavour triplet dark particles, push this bound down to almost 100 GeV. We note that collider constraints can rule out most of this low mass range if we allow roughly equal coupling to all quarks, as the LHC searches for stop dijet production are constraining at this mass range. There are few searches (aimed at e.g. scharm production) which would constrain the charming dark matter scenario, so we maintain our result that allowing coupling to multiple up type quarks can allow what might otherwise be considered excluded parameter space.

In the SM there are many rare B processes that are highly suppressed, which makes them ideal probes of BSM physics. We introduced in Sect. 1.3.4 how rare B decays have thrown up intriguing signs of new physics. The first signs of these NP effects were in $B \rightarrow K\ell\bar{\ell}$ decays and were interpreted as lepton flavour universal effects in the Wilson coefficient C_9. In the SM, around half of the effect in this operator is generated by a 1-loop diagram involving $b \rightarrow c\bar{c}s$ transitions. As such, it seemed plausible that a relatively small NP contribution here could account for the observed deviations.

We studied this possibility in Chap. 5, since these four quark operators show up in the B_s mixing and lifetime calculations, as well as the radiative decay mode $B \rightarrow X_s\gamma$. All these observables are well measured experimentally, and are under decent theoretical control—a correlated effect between all these could provide a distinct signature. We looked at operators of the form $(\bar{s}b)(\bar{c}c)$ with the full set of possible Dirac structures, and used the aforementioned observables to constrain the NP Wilson coefficients. As part of our calculation, we computed the leading order mixing of our chosen effective operators into the full set of $\Delta B = -\Delta S = 1$ operators.

We use our computation to show how, if these effective operators are generated at the weak scale or beyond, strong renormalisation-group effects can enhance the impact on the semileptonic decays $b \rightarrow s\ell\ell$ while leaving radiative B decay largely unaffected. Meanwhile, if they are generated by low scale NP the resulting effect in C_9 has a non-trivial dependence on q^2 (the invariant mass of the lepton pair), which is contrary to the prevailing wisdom in much of the literature. As a result of our study, we showed that there are certain contributions within this scenario where the overall fit between theory and experiment can be improved in the SM for our chosen observables. This can be seen from Fig. 5.3, where we have marked with a gold star points in parameter space where the C_9 shift can be accommodated alongside improvements in at least some of our observables. An obvious caveat to our results in this chapter is the non universality shown in $R_{K^{(*)}}$, as our loop level contribution to the rare decay anomalies must be lepton flavour universal without some extra model-dependent NP.

As a final study in that chapter, we examined how future prospects for the mixing and lifetime observables would alter our conclusions, and allow us to discriminate between different BSM effects.

As we saw in Chap. 3, precision flavour calculations are heavily reliant on lattice calculations of the matrix elements of effective operators, and advances in this area can bring a wealth of benefits. In Chap. 6, we used sum rule techniques to calculate the bag parameters for the dimensions-six operators contributing to mixing and lifetime calculations, in both the bottom and charm sector. These types of calculation are typically not as precise as lattice, but for the bag parameters we have a unique situation where competitive results are possible. Since these inputs are of vital importance to theory, for both precision calculations and for constraining new physics (as we saw in Chap. 7), an independent determination is vital. Sum rules have completely different systematics to lattice calculations, and so they can provide this independent checks of lattice results.

The technical aspect of our work came in three parts—we formulated our sum rules at low scales using HQET, then calculated the anomalous dimensions of the operators so as to RG run them up to the b mass scale, and then matched from HQET to QCD at 1-loop. Our main results are given in Eqs. 6.4.2, 6.5.13, 6.6.1 and 6.6.2. The $\Delta B = 2$ results show (see Fig. 6.4) a competitive precision compared to current lattice calculations—this is achievable because we have formulated the sum rule for the quantity $\Delta B = B - 1$, i.e. the deviation from the VSA, and so the overall error is reduced for the full bag parameter result. For $\Delta C = 2$ (Fig. 6.4), our uncertainties are still slightly behind lattice, as the lower charm scale increases the size of the perturbative corrections.

For the lifetime operators ($\Delta F = 0$), our work is of even more significance. The operators needed for B lifetime calculations have not been updated since 2001 and so our result (seen in Fig. 6.6) stands as the only state-of-the-art determination available today. In the calculation of D meson lifetimes, up until now there has only been the VSA to rely upon as no lattice study has ever been done. Our new result allows the first real test of the HQE for D lifetimes, which is shown in the top section of Fig. 6.9. The central result for the lifetime ratio $\tau(D^+)/\tau(D^0)$ agrees very well with experiment, which is a major improvement, but there is still significant room for improvement in both.

With our new results many of the dominant uncertainties are now due to dimension-seven operators, which are currently undetermined. Our sum rules method can be extended to these subleading contributions, and could allow a substantial reduction in theoretical uncertainties.

In recent years, it has become clear that something is going on in a variety of rare $b \to s\ell\ell$ decays, and that NP might well be the explanation. The most likely effective theory scenario is one in which new physics couples to the SM in form of the operator $(\bar{s}\gamma_\mu P_L b)(\bar{\mu}\gamma^\mu \mu)$. Since the quark part of this operator matches the "square root" of the SM operator that generates B_s mixing, many of the proposed explanations have profound implications for the theoretical prediction for mixing, and so in Chap. 7 we looked at how these models are affected by changes in the SM prediction. Following on from our initial look in Chap. 3, we take the most recent set of lattice averages and consider them seriously in terms of their effect on ΔM_s.

Our first result was the updated calculation for ΔM_s, which is given in Eq. 7.2.3. It shows a $1.8\,\sigma$ deviation from the experimental result, as a result of both an enlarged central value and a greater than 50% reduction in the uncertainty. The lattice result we first discussed in Chap. 3 plays a large role in this, but there is also a major improvement in the precision of V_{cb} from the CKM fitting community. Our next result is that this updated SM prediction strengthens the mass constraints by a factor of five for any NP model that provides a positive contribution to ΔM_s, and we demonstrate this explicitly for Z' models in Fig. 7.1 and leptoquarks in Fig. 7.2. This means that a minimal Z' that doesn't increase the tension in the mixing sector to unreasonable levels cannot also explain the flavour anomalies unless it is lighter than around 2–3 TeV.

In light of these tight bounds, we thought about how a NP model could reduce the tension in mixing alongside the tension in b decays. We presented a solution where the Z' off-diagonal couplings are complex, allowing for a negative contribution to ΔM_s, and showed a CP-violating observable that would come into play. An alternative direction was also pursued, where allowing multiple chiralities in the quark coupling might allow for reduced tension in the mass difference, and we discussed that RG effects could play an important role here. Our final thought was that assuming all the anomalies, in both decays and mixing, hold up under further inspection, a careful study of these possibilities would be something worth pursuing. Particularly for the mixing result, a confirmation of our calculation (by further lattice groups confirming the large FNAL/MILC results for the four quark matrix elements, as well as a definite solution of the V_{cb} puzzle) would give further confidence in the extraordinary strength of the bounds we found.

Over the course of this thesis, we have aimed to study how flavour physics observables can be used to test the Standard Model as well as being used as a search space for beyond the Standard Model effects. Following our initial look at some underlying assumptions, we identified areas where progress could be made, and worked towards these in the rest of the thesis, following interesting hints when they appeared. We hope that the steps laid out here have provided a useful improvement in the areas of B mixing and lifetimes, as well as searches for new physics. We are optimistic that the study of flavour physics and the lines of enquiry within this thesis will continue to be fruitful in the years ahead.

Appendix A
Fierz Transformations

Fierz identities are often considered in the context of particular relationships between pairs of fermion bilinears—probably the most well known is

$$(\gamma^\mu P_L)_{ij}(\gamma_\mu P_L)_{kl} = -(\gamma^\mu P_L)_{il}(\gamma_\nu P_L)_{kj}$$

which is often useful in flavour physics calculations and also in e.g. the calculation of the MSW [1, 2] effect for neutrino propagation in matter.

However they can instead be considered (perhaps more correctly) as a set of identities that can be found for any complete set of square matrices, which arise simply as a consequence of completeness relations over this basis set.

In general, if we have some matrix vector space which is spanned by the basis set $\{M^i\}$, then we can write identities of the form

$$M^i \otimes M^j = M^k \otimes M^l .$$

In this way, the colour rearrangement identity

$$t^a_{ij} t^a_{kl} = \frac{1}{2}\left(\delta_{il}\delta_{kj} - \frac{1}{N_c}\delta_{ij}\delta_{kl}\right) \tag{A.0.1}$$

can be viewed as a Fierz identity, since the space of 3×3 matrices is spanned by the set $\{t^a\}$ of generators of $SU(3)$ plus the identity matrix.

A useful introduction to generalised Fierz identities and how to calculate them is [3]. We will use the notation of that paper (itself taken from Takahashi [4]) in the rest of this appendix, where we list some identities that are used in this thesis. It is important to note that when these identities are used in conjunction with fermionic operators rather than spinors, an extra minus sign will arise from the anti-commuting nature of the operators.

$$(\gamma^\mu P_L)[\gamma_\mu P_L] = -(\gamma^\mu P_L)[\gamma_\mu P_L) \tag{A.0.2}$$

© Springer Nature Switzerland AG 2019
M. J. Kirk, *Charming New Physics in Beautiful Processes?*,
Springer Theses, https://doi.org/10.1007/978-3-030-19197-9

$$(\gamma^\mu \gamma^\nu \gamma^\lambda P_L)[\gamma_\lambda \gamma_\nu \gamma_\mu P_L] = -4(\gamma^\mu P_L)[\gamma_\mu P_L) \qquad (A.0.3)$$

$$(P_L)[P_L] = \frac{1}{2}(P_L)[P_L) + \frac{1}{8}(\sigma^{\mu\nu} P_L)[\sigma_{\mu\nu} P_L) \qquad (A.0.4)$$

$$(P_L)[P_R] = \frac{1}{2}(\gamma^\mu P_R)[\gamma_\mu P_L) \qquad (A.0.5)$$

$$(\sigma_{\mu\nu} P_L)[\sigma^{\mu\nu} P_L] = 6(P_L)[P_L) - \frac{1}{2}(\sigma^{\mu\nu} P_L)[\sigma_{\mu\nu} P_L) \qquad (A.0.6)$$

$$(\sigma_{\mu\nu} P_L)[\sigma^{\mu\nu} P_R] = 0 \qquad (A.0.7)$$

$$(\sigma_{\alpha\beta} \sigma_{\mu\nu} P_L)[\sigma^{\mu\nu} \sigma^{\alpha\beta} P_L] = 72(P_L)[P_L) + 2(\sigma^{\mu\nu} P_L)[\sigma_{\mu\nu} P_L) \qquad (A.0.8)$$

References

1. Mikheyev SP, Smirnov A Yu (1985) Resonance amplification of oscillations in matter and spectroscopy of solar neutrinos. Sov J Nucl Phys 42:913–917
2. Wolfenstein L (1978) Neutrino oscillations in matter. Phys Rev D 17:2369–2374. https://doi.org/10.1103/PhysRevD.17.2369
3. Nishi CC (2005) Simple derivation of general Fierz-like identities. Am J Phys 73:1160–1163. https://doi.org/10.1119/1.2074087, arXiv:hep-ph/0412245
4. Takahashi Y (1986) The Fierz identities. In: Ezawa H, Kamefuchi S (eds) Progress in quantum field theory. North-Holland, pp 121–132

Appendix B
Feynman Rules

Our SM Feynman rules can be found by taking those in [1] and setting $\eta = \eta_s = \eta' = \eta_e = -1$ and $\eta_Z = \eta_\theta = \eta_Y = 1$. (Note that in that work all particles are considered incoming in Feynman diagrams.) These match the conventions of [2].

In Table B.1, we show the Feynman rules for heavy quarks in HQET.

© Springer Nature Switzerland AG 2019
M. J. Kirk, *Charming New Physics in Beautiful Processes?*,
Springer Theses, https://doi.org/10.1007/978-3-030-19197-9

Table B.1 HQET Feynman rules

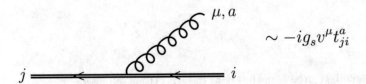

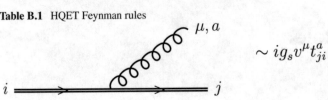

$$\sim ig_s v^\mu t^a_{ji}$$

$$\sim -ig_s v^\mu t^a_{ji}$$

$$\sim \frac{i(1+\not v)}{2v\cdot\tilde p}$$

$$p = mv + \tilde p$$

$$\sim \frac{i(1-\not v)}{2v\cdot\tilde p}$$

$$p = mv + \tilde p$$

References

1. Romao JC, Silva JP (2012) A resource for signs and Feynman diagrams of the standard model. Int J Mod Phys A 27:1230025. https://doi.org/10.1142/S0217751X12300256, arXiv:1209.6213
2. Peskin ME, Schroeder DV (1995) An introduction to quantum field theory. Addison-Wesley, Reading, USA

Appendix C
Four Quark Matrix Elements

In phenomenological calculations matrix elements of four quark operators often arise, and so knowing these matrix elements is of vital importance to being able to make theoretical predictions. However these are non-perturbative objects and so calculating them is non-trivial. Nowadays, lattice QCD has progressed to the point of being able to provide precision determinations, but historically other techniques were used. In particular the Vacuum Saturation Approximation (VSA) is a simple way of analytically approximating the result, and current data shows it even holds numerically at the 10–20% level [1, 2]. The VSA approximates the four quark matrix elements as a product of decay to vacuum operators, and so gives us an estimate in terms of meson decay constants plus various colour factors. Below we show an example of a VSA calculation, and then we list the results for different operators.

Note that we show results in this appendix for B_s–$\bar{B}_s$ mixing and B_s lifetime calculations—the generalisation to B_d, D, etc. mesons can be found by making the obvious substitutions.

VSA Calculation Example

If we take the operator $\mathcal{O}_1 = (\bar{b}^i \gamma_\mu P_L s^i)(\bar{b}^j \gamma^\mu P_L s^j)$, there are four ways to contract this with an incoming B_s and outgoing $\bar{B}_s$ state—two for the b quarks and two for the s quarks. Making use of the Fierz identities stated in Eqs. A.0.1 and A.0.2, we can write the matrix element $\langle \bar{B}_s | \mathcal{O}_1 | B_s \rangle \equiv \langle \mathcal{O}_1 \rangle$ as

$$
\begin{aligned}
\langle \mathcal{O}_1 \rangle &= \langle \bar{B}_s | (\bar{b}^i \gamma^\mu P_L s^i)(\bar{b}^j \gamma_\mu P_L s^j) | B_s \rangle \\
&= 2 \Big[\ \left(\langle \bar{B}_s | \bar{b}^i \gamma^\mu P_L s^i \rangle \right) \left(\bar{b}^j \gamma_\mu P_L s^j | B_s \rangle \right) \\
&\quad + \left(\langle \bar{B}_s | \bar{b}^i \gamma^\mu P_L s^j \rangle \right) \left(\bar{b}^j \gamma_\mu P_L s^i | B_s \rangle \right) \Big] \\
&= 2 \Big[\left(1 + \frac{1}{N_c} \right) \left(\langle \bar{B}_s | \bar{b}^i \gamma^\mu P_L s^i \rangle \right) \left(\bar{b}^j \gamma_\mu P_L s^j | B_s \rangle \right) \\
&\quad + 2 \left(\langle \bar{B}_s | \bar{b}^i \gamma^\mu P_L t^a_{ij} s^j \rangle \right) \left(\bar{b}^k \gamma_\mu P_L t^a_{kl} s^l | B_s \rangle \right) \Big],
\end{aligned}
$$

© Springer Nature Switzerland AG 2019
M. J. Kirk, *Charming New Physics in Beautiful Processes?*,
Springer Theses, https://doi.org/10.1007/978-3-030-19197-9

where the round brackets fix the operator contraction. (The factor of two is the same as seen in Eq. 2.4.18.) We then take the completeness relation $I = \sum_X |X\rangle\langle X|$ and insert it into the middle of the four quark operator:

$$\langle \mathcal{O}_1 \rangle = 2 \sum_X \left[\left(1 + \frac{1}{N_c} \right) \langle \bar{B}_s | \bar{b}^i \gamma^\mu P_L s^i | X \rangle \langle X | \bar{b}^j \gamma_\mu P_L s^j | B_s \rangle \right.$$
$$\left. + 2 \langle \bar{B}_s | \bar{b}^i \gamma^\mu P_L t_{ij}^a s^j | X \rangle \langle X | \bar{b}^k \gamma_\mu P_L t_{kl}^a s^l | B_s \rangle \right].$$

The VSA amounts to assuming the vacuum state dominates this sum, and so we can write

$$\langle \mathcal{O}_1 \rangle \overset{\text{VSA}}{=} 2 \left[\left(1 + \frac{1}{N_c} \right) \langle \bar{B}_s | \bar{b}^i \gamma^\mu P_L s^i | 0 \rangle \langle 0 | \bar{b}^j \gamma_\mu P_L s^j | B_s \rangle \right.$$
$$\left. + 2 \langle \bar{B}_s | \bar{b}^i \gamma^\mu P_L t_{ij}^a s^j | 0 \rangle \langle 0 | \bar{b}^k \gamma_\mu P_L t_{kl}^a s^l | B_s \rangle \right].$$

The second term goes to zero as the colour octet operator cannot annihilate a colour singlet meson, while for the first term we can use the definitions of the meson decay constant

$$\left. \begin{matrix} \langle 0 | \bar{b} \gamma_\mu s | B_s(p) \rangle = 0 \\ \langle 0 | \bar{b} \gamma_\mu \gamma^5 s | B_s(p) \rangle = i p_\mu f_{B_s} \end{matrix} \right\} \Rightarrow \langle 0 | \bar{b} \gamma_\mu P_{L,R} s | B_s(p) \rangle = \mp \frac{i}{2} p_\mu f_{B_s} \qquad \text{(C.0.1)}$$

to give our result

$$\langle \mathcal{O}_1 \rangle \overset{\text{VSA}}{=} 2 \left(1 + \frac{1}{N_c} \right) \left(-\frac{i}{2} p^\mu f_{B_s} \right)^\dagger \left(-\frac{i}{2} p_\mu f_{B_s} \right)$$
$$= \frac{1}{2} \left(1 + \frac{1}{N_c} \right) M_{B_s}^2 f_{B_s}^2 .$$

Since the VSA is only an approximation, we introduce an extra correction factor, known as the bag parameter,[1] to account for deviations from this behaviour and write the matrix element as

$$\langle \bar{B}_s | \mathcal{O}_1 | B_s \rangle = \frac{2}{3} f_{B_s}^2 M_{B_s}^2 B_1 .$$

$\Delta F = 2$ Operators

For $\Delta F = 2$ operators, we can use Fierz relations to reduce the set of possible dimension-six operators which can arise to a minimal basis set, which are shown below:

---

[1] The term "bag" dates back to the MIT bag model of hadrons.

$$\mathcal{O}_1 = (\bar{b}^\alpha \gamma^\mu P_L s^\alpha)(\bar{b}^\beta \gamma_\mu P_L s^\beta),$$

$$\mathcal{O}_2 = (\bar{b}^\alpha P_L s^\alpha)(\bar{b}^\beta P_L s^\beta),$$

$$\mathcal{O}_3 = (\bar{b}^\alpha P_L s^\beta)(\bar{b}^\beta P_L s^\alpha),$$

$$\mathcal{O}_4 = (\bar{b}^\alpha P_L s^\alpha)(\bar{b}^\beta P_R s^\beta),$$

$$\mathcal{O}_5 = (\bar{b}^\alpha P_L s^\beta)(\bar{b}^\beta P_R s^\alpha),$$

$$\tilde{\mathcal{O}}_1 = (\bar{b}^\alpha \gamma^\mu P_R s^\alpha)(\bar{b}^\beta \gamma_\mu P_R s^\beta),$$

$$\tilde{\mathcal{O}}_2 = (\bar{b}^\alpha P_R s^\alpha)(\bar{b}^\beta P_R s^\beta),$$

$$\tilde{\mathcal{O}}_3 = (\bar{b}^\alpha P_R s^\beta)(\bar{b}^\beta P_R s^\alpha).$$

This minimal set was first derived in the context of supersymmetric extensions of the SM [3], and so is often referred to as the "SUSY basis". Since parity is a symmetry of QCD, $\langle \tilde{\mathcal{O}} \rangle = \langle \mathcal{O} \rangle$ holds as long as we don't consider electroweak corrections to the matrix elements. At dimension seven, there are several more operators that arise—see [4] for details.

The matrix elements of the SUSY basis can be written in the form (notation taken from [5])

$$\langle \mathcal{O}_1 \rangle = c_1 f_{B_s}^2 M_{B_s}^2 B_1,$$

$$\langle \mathcal{O}_i \rangle = c_i f_{B_s}^2 M_{B_s}^2 \left(\frac{M_{B_s}}{m_b + m_s} \right)^2 B_i \quad (i = 2, 3),$$

$$\langle \mathcal{O}_i \rangle = c_i f_{B_s}^2 M_{B_s}^2 \left[\left(\frac{M_{B_s}}{m_b + m_s} \right)^2 + d_i \right] B_i \quad (i = 4, 5),$$

(C.0.2)

where $c_i = \{2/3, -5/12, 1/12, 1/2, 1/6\}$, $d_4 = 1/6$, and $d_5 = 3/2$. These prefactors can be derived in the same way as in the above example, making use of Eqs. A.0.4 and A.0.5, as well as the following two further relations involving the decay constant. We can show

$$\langle 0 | \bar{b} P_{L,R} s | B_s(p) \rangle = \pm \frac{i}{2} \frac{M_{B_s}}{m_b + m_s} f_{B_s}$$

by contracting the meson momentum p^μ with our result in Eq. C.0.1 (see Sect. 4 of [6]); we must also have

$$\langle 0 | \bar{b} \sigma^{\mu\nu} s | B_s(p) \rangle = 0$$

since we cannot construct an antisymmetric Lorentz tensor out of a single 4-momentum.

$\Delta F = 0$ Operators

For $\Delta F = 0$ operators, the standard set considered (which are those which arise in the SM) is

$$Q_1 = (\bar{b}\gamma^\mu P_L q)(\bar{q}\gamma_\mu P_L b)$$
$$Q_2 = (\bar{b} P_L q)(\bar{q} P_L b)$$
$$T_1 = (\bar{b}\gamma^\mu t^a P_L q)(\bar{q}\gamma_\mu t^a P_L b)$$
$$T_2 = (\bar{b} t^a P_L q)(\bar{q} t^a P_L b)$$

which have matrix elements

$$\langle Q_i \rangle = A_i f_{B_s}^2 M_{B_s}^2 B_i$$
$$\langle T_i \rangle = A_i f_{B_s}^2 M_{B_s}^2 \epsilon_i \tag{C.0.3}$$

where $B_i = 1$ and $\epsilon_i = 0$ in the VSA and

$$A_1 = \frac{1}{4}, \quad A_2 = \frac{1}{4}\left(\frac{M_{B_s}}{m_b + m_s}\right)^2. \tag{C.0.4}$$

References

1. Aoki S et al (2017) Review of lattice results concerning low-energy particle physics. Eur Phys J C 77:112. https://doi.org/10.1140/epjc/s10052-016-4509-7, arXiv:1607.00299
2. FLAG collaboration. http://flag.unibe.ch/MainPage
3. Gabbiani F, Gabrielli E, Masiero A, Silvestrini L (1996) A Complete analysis of FCNC and CP constraints in general SUSY extensions of the standard model. Nucl Phys B 477:321–352. https://doi.org/10.1016/0550-3213(96)00390-2, arXiv:hep-ph/9604387
4. Beneke M, Buchalla G, Dunietz I (1996) Width difference in the $B_s - \bar{B}_s$ system. Phys Rev D 54:4419–4431. https://doi.org/10.1103/PhysRevD.54.4419, https://doi.org/10.1103/PhysRevD.83.119902, arXiv:hep-ph/9605259
5. Fermilab Lattice, MILC collaboration, Bazavov A et al (2016) $B_{(s)}^0$-mixing matrix elements from lattice QCD for the standard model and beyond. Phys. Rev. D 93:113016. https://doi.org/10.1103/PhysRevD.93.113016, arXiv:1602.03560
6. Tanedo P (2009) One-Loop MSSM predictions for $B_{s,d} \to \ell^+\ell^-$ at low tan β, Master's thesis, Durham University. http://etheses.dur.ac.uk/2040/. (See also https://www.physics.uci.edu/~tanedo/files/documents/TanedoMSc.pdf for a non-scanned version)

Appendix D
Additional Information from "Quark-Hadron Duality"

In this appendix we present more information on the work in Chap. 3.

In Table D.1, we show the error breakdown for our updated lifetime ratio calculation in Eq. 3.2.32, which has an error smaller than the current experimental measurement by a factor of four, and the previous SM calculation by a factor two.

In Table D.2, we present the inputs for our "aggressive" SM calculation in Sect. 3.3.

In Tables D.3, D.4, D.5, D.6, D.7, D.8, D.9 and D.10 we present the error breakdown for our updated predictions.

Finally in Appendix D.2 we show the proof of the inequality used in Sect. 3.4.

D.1 Inputs and Detailed View of Uncertainties

Table D.1 Error breakdown for the SM prediction for $\tau(B_s)/\tau(B_d)$ given in Eq. 3.2.32

Parameter	Error contribution (%)
$\delta(\epsilon_1)$	0.0710
$\delta(\epsilon_2)$	0.0510
$\delta(f_{B_s})$	0.0290
$\delta(\mu_G^2(B_s)/\mu_G^2(B_d))$	0.0280
$\mu_\pi^2(B_s) - \mu_\pi^2(B_d))$	0.0230
$\delta(f_{B_d})$	0.0230
$\delta(c_3)$	0.0230
$\delta(\mu)$	0.0160
$\delta(B_1)$	0.0140
$\delta(\mu_G^2(B_d))$	0.0130

(continued)

© Springer Nature Switzerland AG 2019
M. J. Kirk, *Charming New Physics in Beautiful Processes?*,
Springer Theses, https://doi.org/10.1007/978-3-030-19197-9

Table D.1 (continued)

Parameter	Error contribution (%)
$\delta(B_2)$	0.0100
$\delta(c_G)$	0.007
$\delta(m_b)$	0.004
$\delta(|V_{cb}|)$	0.003
$\delta(m_c)$	0.001
$\delta(\tau_{B_s})$	<0.001
$\delta(M_{B_s})$	<0.001
$\delta(M_{B_d})$	<0.001
$\delta(|V_{us}|)$	<0.001
$\delta(\gamma)$	<0.001
$\delta(|V_{ub}/V_{cb}|)$	<0.001
$\sum \delta$	0.108

Table D.2 Input parameters used in the "aggressive" determination of Sect. 3.3, compared to the previous SM calculation

Parameter	This work	ABL 2015 [1]
$f_{B_s}\sqrt{B}$	$(0.223 \pm 0.007)\,\text{GeV}$	$(0.215 \pm 0.015)\,\text{GeV}$
$f_{B_d}\sqrt{B}$	$(0.185 \pm 0.008)\,\text{GeV}$	$(0.175 \pm 0.012)\,\text{GeV}$
$B_3/B\ (B_s)$	1.15 ± 0.16	1.07 ± 0.06
$B_3/B\ (B_d)$	1.17 ± 0.24	1.04 ± 0.12
$\tilde{B}_{R_0}/B\ (B_s)$	0.54 ± 0.55	1.00 ± 0.3
$\tilde{B}_{R_0}/B\ (B_d)$	0.35 ± 0.80	1.00 ± 0.3
$\tilde{B}_{R_1}/B\ (B_s)$	1.61 ± 0.10	1.71 ± 0.26
$\tilde{B}_{R_1}/B\ (B_d)$	1.72 ± 0.15	1.71 ± 0.26
$\tilde{B}_{\tilde{R}_1}/B\ (B_s)$	1.223 ± 0.093	1.27 ± 0.16
$\tilde{B}_{\tilde{R}_1}/B\ (B_d)$	1.31 ± 0.14	1.27 ± 0.16
$|V_{cb}|$	$0.04180^{+0.00033}_{-0.00068}$	$0.04117^{+0.00090}_{-0.00114}$
$|V_{ub}/V_{cb}|$	0.0889 ± 0.0019	0.0862 ± 0.0044
γ	$1.170^{+0.015}_{-0.035}$	$1.171^{+0.017}_{-0.038}$
$|V_{us}|$	$0.22542^{+0.00042}_{-0.00031}$	$0.22548^{+0.00068}_{-0.00034}$

Table D.3 Error breakdown for the mass difference ΔM_s

Parameter	Error contribution	
	This work (%)	ABL 2015 [1] (%)
$\delta(f_{B_s}\sqrt{B})$	0.0635	0.139
$\delta(|V_{cb}|)$	0.0240	0.049
$\delta(m_t)$	0.0066	0.007
$\delta(\Lambda_{QCD})$	0.0013	0.001
$\delta(\gamma)$	0.0009	0.001
$\delta(m_b)$	0.0005	<0.001
$\delta(|V_{ub}/V_{cb}|)$	0.0004	0.001
$\sum \delta$	0.0682	0.148

Table D.4 Error breakdown for the mass difference ΔM_d

Parameter	Error contribution	
	This work (%)	ABL 2015 [1] (%)
$\delta(f_{B_d}\sqrt{B})$	0.0872	0.137
$\delta(|V_{cb}|)$	0.0240	0.049
$\delta(m_t)$	0.0066	0.001
$\delta(\Lambda_{QCD})$	0.0013	0.0
$\delta(\gamma)$	0.0208	0.002
$\delta(m_b)$	0.0005	0.0
$\delta(|V_{ub}/V_{cb}|)$	0.0001	0.0
$\sum \delta$	0.0931	0.148

Table D.5 Error breakdown for the width difference $\Delta\Gamma_s$

Parameter	Error contribution	
	This work (%)	ABL 2015 [1] (%)
$\delta(\mu)$	0.0889	0.084
$\delta(f_{B_s})$	0.0635	0.139
$\delta(B_{R_2})$	0.0604	0.148
$\delta(B_3)$	0.0539	0.021
$\delta(B_{R_0})$	0.0301	0.021
$\delta(|V_{cb}|)$	0.0240	0.049
$\delta(\bar{z})$	0.0109	0.011
$\delta(m_b)$	0.0080	0.008
$\delta(B_{\tilde{R}_1})$	0.0038	0.007
$\delta(m_s)$	0.0024	0.001
$\delta(B_{R_3})$	0.0023	0.002
$\delta(B_{R_1})$	0.0018	0.005
$\delta(\gamma)$	0.0010	0.001
$\delta(\Lambda_{QCD})$	0.0010	0.001
$\delta(|V_{ub}/V_{cb}|)$	0.0004	0.001
$\delta(m_t)$	0	<0.001
$\sum\delta$	0.1421	0.228

Table D.6 Error breakdown for the width difference $\Delta\Gamma_d$

Parameter	Error contribution	
	This work (%)	ABL 2015 [1] (%)
$\delta(\mu)$	0.0929	0.079
$\delta(f_{B_d})$	0.0872	0.137
$\delta(B_3)$	0.0809	0.04
$\delta(B_{R_2})$	0.0623	0.144
$\delta(B_{R_0})$	0.0533	0.025
$\delta(|V_{cb}|)$	0.0240	0.049
$\delta(\gamma)$	0.0233	0.002
$\delta(\bar{z})$	0.0109	0.011
$\delta(m_b)$	0.0076	0.008
$\delta(B_{R_3})$	0.0023	0.005
$\delta(\Lambda_{QCD})$	0.0009	0.001
$\delta(|V_{ub}/V_{cb}|)$	0.0008	0.001
$\delta(B_{\tilde{R}_1})$	0.0	0.0
$\delta(m_d)$	—	—
$\delta(B_{R_1})$	0.0	0.0
$\sum\delta$	0.175	0.227

Table D.7 Error breakdown for the ratio $\Delta\Gamma_s/\Delta M_s$

Parameter	Error contribution (%)
$\delta(\mu)$	0.0889
$\delta(B_{R_2})$	0.0604
$\delta(B_3)$	0.0539
$\delta(B_{R_0})$	0.0301
$\delta(\bar{z})$	0.0109
$\delta(m_b)$	0.0080
$\delta(m_t)$	0.0066
$\delta(\tilde{B}_{R_1})$	0.0038
$\delta(m_s)$	0.0024
$\delta(\Lambda_{QCD})$	0.0023
$\delta(B_{R_3})$	0.0023
$\delta(B_{R_1})$	0.0018
$\delta(\gamma)$	0.0001
$\delta(|V_{ub}/V_{cb}|)$	0
$\delta(|V_{cb}|)$	0
$\sum \delta$	0.125

Table D.8 Error breakdown for the ratio $\Delta\Gamma_d/\Delta M_d$

Parameter	Error contribution (%)
$\delta(\mu)$	0.0929
$\delta(B_3)$	0.0809
$\delta(B_{R_2})$	0.0623
$\delta(B_{R_0})$	0.0533
$\delta(\bar{z})$	0.0109
$\delta(m_b)$	0.0076
$\delta(m_t)$	0.0066
$\delta(\gamma)$	0.0025
$\delta(B_{R_3})$	0.0023
$\delta(\Lambda_{QCD})$	0.0022
$\delta(|V_{ub}/V_{cb}|)$	0.0009
$\delta(\tilde{B}_{R_1})$	0.0
$\delta(m_d)$	0.0
$\delta(B_{R_1})$	0.0
$\delta(|V_{cb}|)$	0.0
$\sum \delta$	0.149

Table D.9 Error breakdown for a_{sl}^s

Parameter	Error contribution (%)
$\delta(\mu)$	0.0946
$\delta(\bar{z})$	0.0463
$\delta(V_{ub}/V_{cb})$	0.0211
$\delta(\gamma)$	0.0118
$\delta(B_{R3})$	0.0106
$\delta(m_b)$	0.0101
$\delta(m_t)$	0.0066
$\delta(B_S)$	0.0078
$\delta(\Lambda_{QCD})$	0.0053
$\delta(B_{R2})$	0.0039
$\delta(\tilde{B}_{R1})$	0.0030
$\delta(B_{R0})$	0.0026
$\delta(m_s)$	0.0021
$\delta(B_{R1})$	0.0002
$\delta(V_{cb})$	0
$\sum \delta$	0.1098

Table D.10 Error breakdown for a_{sl}^d

Parameter	Error contribution (%)
$\delta(\mu)$	0.0937
$\delta(\bar{z})$	0.0487
$\delta(|V_{ub}/V_{cb}|)$	0.0215
$\delta(m_b)$	0.0129
$\delta(B_3)$	0.0123
$\delta(B_{R3})$	0.0115
$\delta(\gamma)$	0.0105
$\delta(m_t)$	0.0066
$\delta(\Lambda_{QCD})$	0.0054
$\delta(B_{R0})$	0.0049
$\delta(B_{R2})$	0.0042
$\delta(\tilde{B}_{R1})$	0.0
$\delta(m_d)$	0.0
$\delta(B_{R1})$	0.0
$\delta(|V_{cb}|)$	0.0
$\sum \delta$	0.111

D.2   Proof of $\Delta\Gamma \leq 2|\Gamma_{12}|$

In the B system we get the very simple expression Eq. 2.4.16 for the mixing observables ΔM and $\Delta\Gamma$ in terms of M_{12} and Γ_{12}, by expanding in the small parameter $\Delta\Gamma/\Delta M$. In the D system $\Delta\Gamma$ and ΔM are of the same order and so one would have to exactly solve the two defining equations. One can find however, the inequality $\Delta\Gamma \leq 2|\Gamma_{12}|$, which gives us the opportunity to calculate only Γ_{12} and to give an upper bound on $\Delta\Gamma$.

We start with the two fundamental equations for the mixing observables:

$$(\Delta M)^2 - \frac{1}{4}(\Delta\Gamma)^2 = 4|M_{12}|^2 - |\Gamma_{12}|^2, \tag{D.2.1}$$

$$\Delta M \Delta\Gamma = 4|M_{12}||\Gamma_{12}|\cos\phi_{12}. \tag{D.2.2}$$

Next we eliminate ΔM by substituting Eq. D.2.2 into Eq. D.2.1, and then solve for $|M_{12}|$.

$$\frac{16|M_{12}|^2|\Gamma_{12}|^2\cos^2\phi_{12}}{(\Delta\Gamma)^2} - \frac{1}{4}(\Delta\Gamma)^2 = 4|M_{12}|^2 - |\Gamma_{12}|^2$$

$$|M_{12}|^2\left(\frac{16|\Gamma_{12}|^2\cos^2\phi_{12}}{(\Delta\Gamma)^2} - 4\right) = \frac{1}{4}(\Delta\Gamma)^2 - |\Gamma_{12}|^2 \tag{D.2.3}$$

$$|M_{12}|^2 = \frac{\frac{1}{4}(\Delta\Gamma)^2 - |\Gamma_{12}|^2}{\left(\frac{16|\Gamma_{12}|^2\cos^2\phi_{12}}{(\Delta\Gamma)^2} - 4\right)}.$$

Since $|M_{12}|^2 \geq 0$, we can say that the numerator and denominator on the right hand side of the last line of Eq. D.2.3 must have the same sign.

First, assume both terms are ≥ 0:

$$\frac{1}{4}(\Delta\Gamma)^2 - |\Gamma_{12}|^2 \geq 0 \quad \text{and} \quad \frac{16|\Gamma 12|^2\cos^2\phi_{12}}{(\Delta\Gamma)^2} - 4 \geq 0 \tag{D.2.4}$$

$$\implies (\Delta\Gamma)^2 \geq 4|\Gamma_{12}|^2 \quad \text{and} \quad (\Delta\Gamma)^2 \leq 4\Gamma_{12}^2\cos^2\phi_{12}. \tag{D.2.5}$$

These inequalities are only consistent in the case $\cos^2\phi_{12} = 1$ and $\Delta\Gamma = 2|\Gamma_{12}|$.

Now, assume both terms are ≤ 0:

$$\frac{1}{4}(\Delta\Gamma)^2 - |\Gamma_{12}|^2 \leq 0 \quad \text{and} \quad \frac{16|\Gamma 12|^2\cos^2\phi_{12}}{(\Delta\Gamma)^2} - 4 \leq 0 \tag{D.2.6}$$

$$\implies (\Delta\Gamma)^2 \leq 4|\Gamma_{12}|^2 \quad \text{and} \quad 4\Gamma_{12}^2\cos^2\phi_{12} \leq (\Delta\Gamma)^2. \tag{D.2.7}$$

As $0 \leq \cos^2 \phi_{12} \leq 1$, these inequalities are consistent for either a) $\cos^2 \phi_{12} = 1$ which gives $\Delta\Gamma = 2|\Gamma_{12}|$ or b) $2|\Gamma_{12}||\cos \phi_{12}| \leq \Delta\Gamma \leq 2|\Gamma_{12}|$.

We see that for either assumption the inequality $\Delta\Gamma \leq 2|\Gamma_{12}|$ holds. A similar line of reasoning shows that the inequality $\Delta M \leq 2|M_{12}|$ also holds.

Reference

1. Artuso M, Borissov G, Lenz A (2016) CP violation in the B_s^0 system. Rev Mod Phys 88:045002. https://doi.org/10.1103/RevModPhys.88.045002, arXiv:1511.09466

Appendix E
Additional Information from "Charming Dark Matter"

In this appendix we present more information on the work in Chap. 4.

In Appendix E.1 we explicitly give the Wilson coefficients that arise when considering contributions to rare D decays, in Appendix E.2 we give more information on our technique for calculating constraints from direct detection experiments, and in Appendix E.3 we show some of the Feynman diagrams that contribute to collider signatures for dark matter.

E.1 Rare Decays

The non-zero Wilson coefficients arise from electroweak penguins (shown in Fig. E.1) and (neglecting Z penguins since the small momentum transfer means they amount to an $\mathcal{O}(1)\%$ correction) we find

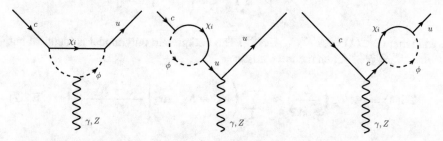

Fig. E.1 The DMFV model contribution to the effective operators governing rare decays of charm mesons, including explicit self-energy corrections to the external quark legs as explained in the text. The γ, Z couple to a lepton pair

© Springer Nature Switzerland AG 2019
M. J. Kirk, *Charming New Physics in Beautiful Processes?*,
Springer Theses, https://doi.org/10.1007/978-3-030-19197-9

$$C_7' = \sum_i \frac{\lambda_{1i}\lambda_{2i}^*}{6\sqrt{2}G_F} \Big[\ C_1(m_c^2, q^2, 0, m_{\chi_i}^2, m_\phi^2, m_\phi^2) + C_{11}(m_c^2, q^2, 0, m_{\chi_i}^2, m_\phi^2, m_\phi^2) \quad \text{(E.1.1)}$$

$$+ C_{12}(m_c^2, q^2, 0, m_{\chi_i}^2, m_\phi^2, m_\phi^2) \Big],$$

$$C_9' = \sum_i \frac{\lambda_{1i}\lambda_{2i}^*}{3\sqrt{2}G_F q^2} \Big[\ B_1(m_c^2, m_{\chi_i}^2, m_\phi^2) + 2C_{00}(m_c^2, q^2, 0, m_{\chi_i}^2, m_\phi^2, m_\phi^2)$$

$$+ m_c^2 \Big\{ \ C_1(m_c^2, q^2, 0, m_{\chi_i}^2, m_\phi^2, m_\phi^2) + C_{11}(m_c^2, q^2, 0, m_{\chi_i}^2, m_\phi^2, m_\phi^2)$$
$$\text{(E.1.2)}$$
$$+ C_{12}(m_c^2, q^2, 0, m_{\chi_i}^2, m_\phi^2, m_\phi^2) \Big\} \Big],$$

where B and C are loop functions using LoopTools [1] notation.

E.2   Direct Detection

E.2.1   LUX

For situations where we have both a measured event count N_k^{obs} (binned into energy bins labelled by k), and theoretical background N_k^{bck}, we can use the *likelihood ratio test*, a method based on a hypothesis test between a background only, and background+signal model, with likelihoods $\mathcal{L}$, $\mathcal{L}_{\text{bck}}$ respectively [2].

The likelihood of observing the data, D, assuming a particular set of parameters $\{\lambda\}$, is denoted $\mathcal{L}(D|\{\lambda\})$. The likelihood of each bin is a Poisson distribution $\text{Poiss}(N^{\text{obs}}, N^{\text{th}}(\lambda))$ where N_k^{th} are the predicted number of signal events (including background):

$$\mathcal{L}(N^{\text{obs}}|\{\lambda\}) = \prod_k \frac{\left(N_k^{\text{th}}\right)^{N_k^{\text{obs}}}}{N_k^{\text{obs}}!} \exp\left[-N_k^{\text{th}}\right] \qquad \text{(E.2.1)}$$

and where $N^{\text{th}}(\lambda) = N^{\text{DM}}(\lambda) + N^{\text{bck}}$. The background only model is identical but with $N^{\text{th}} = N^{\text{bck}}$. Then the test statistic

$$\text{TS}(\lambda) = -2\log\left(\frac{\mathcal{L}}{\mathcal{L}_{\text{bck}}}\right) \approx 2\sum_k \left(N_k^{\text{th}} - N_k^{\text{obs}} \log\left[\frac{N_k^{\text{th}} + N_k^{\text{bck}}}{N_k^{\text{bck}}}\right]\right) \qquad \text{(E.2.2)}$$

follows a χ^2 distribution—the cumulative probability density function of $\chi^2(x)$ represents the probability that we observe the data given the model parameters λ. The value of x such that $\chi^2(x) = C$ (i.e. the $C\%$ confidence limit) depends on the number of parameters $\{\lambda\}$—for only one parameter for example one can look up that $\chi^2(2.71) = 0.9$, which means that the 90% confidence bounds on λ are given by $\text{TS}(\lambda) = 2.71$.

E.2.2   CDMSlite

For CDMSlite, we use a conservative method based on the statement that the 90% confidence limit is such that *there is a probability of 0.9 that if the model were true, then the experiment would have measured more events (n) than have been measured* (n_{obs}). Using the Poisson distribution this probability is

$$P(n > n_{obs}|\mu) = \sum_{n=n_{obs}}^{\infty} \frac{\mu^n}{n!} \exp(-\mu) \approx \int_{n_{obs}}^{\infty} \frac{1}{\sqrt{2\pi\mu}} \exp\left(-\frac{(t-\mu)^2}{2\mu}\right) dt = 0.9$$

(E.2.3)

and in the limit $n_{obs} \gg 1$, this can be approximated by

$$P(n > n_{obs}|\mu) = \frac{1}{2}\left(\text{Erfc}\left(\frac{n_{obs} - \mu}{\sqrt{2\mu}}\right)\right) = 0.9.$$
(E.2.4)

This equation is numerically solvable for μ giving a required signal $\mu = 109^{+51}_{-50}$, 88 ± 14, 635 ± 37 and 207 ± 20 events for energy bins 1–4 respectively. This is conservative since a large portion of the measured events are background, and the resulting limits are slightly weaker than those given by the CDMSlite collaboration.

E.3   Feynman Diagrams for Collider Searches

E.3.1   Monojet Processes

The dominant diagrams contributing to the pure monojet process are shown here. Each processes scales as $\sigma \propto (\lambda\lambda^\dagger)\alpha_s$ and can become extremely large for large λ. The cross section is dominated by the diagrams containing a heavy ϕ resonance.

E.3.2   Dijet Processes

The dominant processes contributing to the production of on-shell ϕ, which decay $\phi \to q_i \chi_j$ producing a dijet signal, are shown in Figs. E.2 and E.3. In monojet analyses, this provides a subdominant contribution compared with pure monojet processes (Figs. E.4 and E.5) in most of the parameter space.

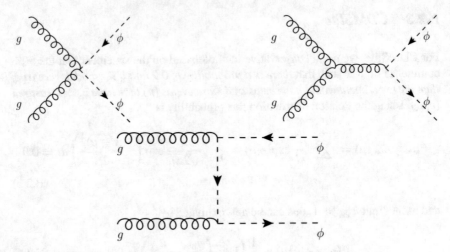

Fig. E.2 Gluon fusion dijet processes $\sigma \propto \alpha_s^2$

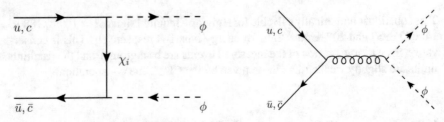

Fig. E.3 The left (right) process has $\sigma \propto (\lambda\lambda^\dagger)^2 (\alpha_s^2)$ and so the dominance depends on the size of the new couplings—for couplings which are large enough to be excluded it is usually the left diagram which dominates

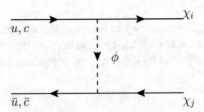

Fig. E.4 The above diagram must include initial/final state radiation from external legs or internal bremsstrahlung from the mediator. The contribution is roughly equal amongst these emissions

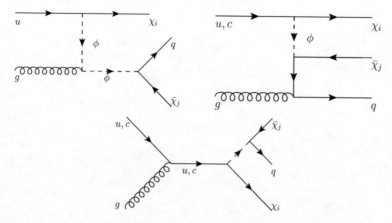

Fig. E.5 The s-channel ϕ resonance is responsible for top left and bottom dominating over top right, and the additional enhancement due to the gluon pdf over Fig. E.4 makes these the overall dominant monojet contribution. For very heavy mediators top left is suppressed due to the two propagators

References

1. Hahn T, Perez-Victoria M (1999) Automatized one loop calculations in four-dimensions and D-dimensions. Comput Phys Commun 118:153–165. https://doi.org/10.1016/S0010-4655(98)00173-8, arXiv:hep-ph/9807565
2. Cirelli M, Del Nobile E, Panci P (2013) Tools for model-independent bounds in direct dark matter searches. JCAP 1310:019. https://doi.org/10.1088/1475-7516/2013/10/019, arXiv:1307.5955

Appendix F
Additional Information from "Charming New Physics in Rare B_s Decays and Mixing?"

In this appendix we provide additional information on the technical aspects of our results for the anomalous dimensions entering in the RGE (Eq. 5.6.1) in Chap. 5.

A set of Wilson coefficients that contains C_7, C_9, and $C^c_{1...4}$ and is closed under renormalisation necessarily also contains four QCD-penguin coefficients C_{P_i} multiplying the operators $P_{3...6}$ (we define them as in [1]) and the chromodipole coefficient C_{8g}, resulting in an 11×11 anomalous dimension matrix γ. If the rescaled semileptonic operator $\tilde{Q}_9(\mu) = (4\pi/\alpha_s(\mu))Q_{9V}(\mu)$ is used, then to leading order $\gamma_{ij}(\mu) = \alpha_s(\mu)/(4\pi)\gamma^{(0)}_{ij}$, with constant $\gamma^{(0)}_{ij}$. As is well known, this matrix is already scheme-dependent at LO [2–4]. A scheme-independent matrix $\gamma^{\text{eff}(0)}$ can be achieved by replacing C_7 and C_8 by the scheme-independent combinations

$$C^{\text{eff}}_7 = C_7 + \sum_i y_i C_i, \tag{F.0.1}$$

$$C^{\text{eff}}_8 = C_8 + \sum_i z_i C_i, \tag{F.0.2}$$

where

$$\langle s\gamma|Q_i|b\rangle = y_i \langle s\gamma|Q_{7\gamma}|b\rangle, \tag{F.0.3}$$

$$\langle sg|Q_i|b\rangle = z_i \langle s\gamma|Q_{8g}|b\rangle, \tag{F.0.4}$$

to lowest order and the sums run over all four-quark operators. We find that y_i and z_i vanish for $Q^c_{1...4}$, leaving only the known coefficients $y_{P_i} = (-1/3, -4/9, -20/3, -80/9)_i$ and $z_{P_i} = (1, -1/6, 20, -10/3)_i$ $(i = 3 \ldots 6)$ [1]. The BSM correction ΔC^{eff}_9 in Eqs. 5.3.2 and 5.3.7 coincides with the (BSM correction to the) coefficient C_9 of Q_{9V} to LL accuracy.

Many of the elements of $\gamma^{\text{eff}(0)}$ are known [2, 3, 5–11], except for $\gamma^{\text{eff}(0)}_{Q^c_i Q_{7\gamma}}$, $\gamma^{\text{eff}(0)}_{Q^c_i Q_{8g}}$, $\gamma^{\text{eff}(0)}_{Q^c_i P_j}$, and $\gamma^{\text{eff}(0)}_{Q^c_i \tilde{Q}_9}$, for $i = 3, 4$. The latter can be read off from the logarithmic terms in Eq. 5.3.2, and the mixing into P_i follows from substituting gauge coupling and

© Springer Nature Switzerland AG 2019
M. J. Kirk, *Charming New Physics in Beautiful Processes?*,
Springer Theses, https://doi.org/10.1007/978-3-030-19197-9

colour factors in the diagram shown on the left of Fig. 5.1. This gives

$$\gamma^{(0)}_{Q^c_i \tilde{Q}_9} = \left(-\frac{8}{3}, -\frac{8}{9}, \frac{4}{3}, \frac{4}{9}\right)_i , \quad \gamma^{(0)}_{Q^c_i P_4} = \left(0, \frac{4}{3}, 0, -\frac{2}{3}\right)_i ,$$

for $i = 1, 2, 3, 4$, with the mixing into $C_{P_{3,5,6}}$ vanishing.

The leading mixing into C_7^{eff} arises at two loops [12–15] and is the technically most challenging aspect of our results. Our calculation employs the 1PI (off-shell) formalism and the method of [16] for computing UV divergences, which involves an infrared-regulator mass and the appearance of a set of gauge-non-invariant counterterms. The result is

$$\gamma^{\text{eff}(0)}_{Q^c_i Q_7} = \left(0, \frac{416}{81}, 0, \frac{224}{81}\right)_i \quad (i = 1, 2, 3, 4) .$$

Our stated results for $i = 1, 2$ agree with the results in [1, 9], which constitutes a cross-check of our calculation.

We have not obtained the 2-loop mixing of $C^c_{3,4}$ into C_{8g} and set these anomalous dimension elements to zero. For the case of $C^c_{1,2}$ where this mixing is known, the impact of neglecting $\gamma^{\text{eff}(0)}_{i8}$ on $\Delta C_7^{\text{eff}}(\mu)$ is small (the only change being $-0.19\Delta C_2 \rightarrow -0.18\Delta C_2$ in Eq. 5.3.6). We expect a similarly small error in the case of $\Delta C_{3,4}$.

References

1. Chetyrkin KG, Misiak M, Munz M (1997) Weak radiative B meson decay beyond leading logarithms. Phys Lett B 400:206–219. https://doi.org/10.1016/S0370-2693(97)00324-9, arXiv:hep-ph/9612313
2. Ciuchini M, Franco E, Martinelli G, Reina L, Silvestrini L (1993) Scheme independence of the effective Hamiltonian for $b \rightarrow s\gamma$ and $b \rightarrow sg$ decays. Phys Lett B 316:127–136. https://doi.org/10.1016/0370-2693(93)90668-8, arXiv:hep-ph/9307364
3. Ciuchini M, Franco E, Martinelli G, Reina L, Silvestrini L (1994) $b \rightarrow s\gamma$ and $b \rightarrow sg$: a theoretical reappraisal. Phys Lett B 334:137–144. https://doi.org/10.1016/0370-2693(94)90602-5, arXiv:hep-ph/9406239
4. Ciuchini M, Franco E, Reina L, Silvestrini L (1994) Leading order QCD corrections to $b \rightarrow s\gamma$ and $b \rightarrow sg$ decays in three regularization schemes. Nucl Phys B 421:41–64. https://doi.org/10.1016/0550-3213(94)90223-2, arXiv:hep-ph/9311357
5. Gaillard MK, Lee BW (1974) $\Delta I = 1/2$ Rule for nonleptonic decays in asymptotically free field theories. Phys Rev Lett 33:108. https://doi.org/10.1103/PhysRevLett.33.108
6. Altarelli G, Maiani L (1974) Octet enhancement of nonleptonic weak interactions in asymptotically free gauge theories. Phys Lett B 52:351–354. https://doi.org/10.1016/0370-2693(74)90060-4
7. Gilman FJ, Wise MB (1979) Effective Hamiltonian for $\Delta S = 1$ weak nonleptonic decays in the six quark model. Phys Rev D 20:2392. https://doi.org/10.1103/PhysRevD.20.2392
8. Shifman MA, Vainshtein AI, Zakharov VI (1977) Nonleptonic decays of K mesons and hyperons. Sov Phys JETP 45:670
9. Gilman FJ, Wise MB (1980) $K \rightarrow \pi e^+ e^-$ in the Six Quark Model. Phys Rev D 21:3150. https://doi.org/10.1103/PhysRevD.21.3150

10. Guberina B, Peccei RD (1980) Quantum Chromodynamic Effects and CP Violation in the Kobayashi-Maskawa Model. Nucl Phys B 163:289–311. https://doi.org/10.1016/0550-3213(80)90404-6

11. Buras AJ, Misiak M, Urban J (2000) Two loop QCD anomalous dimensions of flavor changing four quark operators within and beyond the standard model. Nucl Phys B 586:397–426. https://doi.org/10.1016/S0550-3213(00)00437-5, arXiv:hep-ph/0005183

12. Bertolini S, Borzumati F, Masiero A (1987) QCD enhancement of radiative B decays. Phys Rev Lett 59:180. https://doi.org/10.1103/PhysRevLett.59.180

13. Grinstein B, Springer RP, Wise MB (1988) Effective Hamiltonian for weak radiative B meson decay. Phys Lett B 202:138–144. https://doi.org/10.1016/0370-2693(88)90868-4

14. Grinstein B, Springer RP, Wise MB (1990) Strong interaction effects in weak radiative $\bar{B}$ meson decay. Nucl Phys B 339:269–309. https://doi.org/10.1016/0550-3213(90)90350-M

15. Misiak M (1991) QCD corrected effective Hamiltonian for the $b \to s\gamma$ decay. Phys Lett B 269:161–168. https://doi.org/10.1016/0370-2693(91)91469-C

16. Chetyrkin KG, Misiak M, Munz M (1998) Beta functions and anomalous dimensions up to three loops. Nucl Phys B 518:473–494. https://doi.org/10.1016/S0550-3213(98)00122-9, arXiv:hep-ph/9711266

Appendix G
Additional Information from "Dimension-Six Matrix Elements from Sum Rules"

In this appendix we provide additional information relating to our work in Chap. 6.

In Appendix G.1 we provide our choice of basis for the evanescent operators for both the $\Delta B = 2$ and $\Delta B = 0$ sectors, along with the corresponding anomalous dimension matrices.

In Appendix G.2 we show our choice of input parameters (Table G.1) as well as a detailed breakdown of the uncertainties associated with our results for the bag parameters (Tables G.2, G.3, G.4 and G.5).

Finally in Tables G.6, G.7 G.8 and G.9 we provide a breakdown of the errors in our mixing and lifetime ratio observables.

G.1 Basis of Evanescent Operators and ADMs

G.1.1 $\Delta B = 2$ Operators

Our choice of basis for the evanescent operators is given by

$$
\begin{aligned}
E_1 &= \bar{b}_i \gamma_\mu (1 - \gamma^5) q_j \; \bar{b}_j \gamma^\mu (1 - \gamma^5) q_i - Q_1 \,, \\
E_2 &= \bar{b}_i \gamma_\mu \gamma_\nu (1 - \gamma^5) q_i \; \bar{b}_j \gamma^\mu \gamma^\nu (1 - \gamma^5) q_j - (8 - 4\epsilon) Q_2 - (8 - 8\epsilon) Q_3 \,, \\
E_3 &= \bar{b}_i \gamma_\mu \gamma_\nu (1 - \gamma^5) q_j \; \bar{b}_j \gamma^\mu \gamma^\nu (1 - \gamma^5) q_i - (8 - 8\epsilon) Q_2 - (8 - 4\epsilon) Q_3 \,, \\
E_4 &= \bar{b}_i \gamma_\mu \gamma_\nu \gamma_\rho (1 - \gamma^5) q_i \; \bar{b}_j \gamma^\mu \gamma^\nu \gamma^\rho (1 - \gamma^5) q_j - (16 - 4\epsilon) Q_1 \,, \\
E_5 &= \bar{b}_i \gamma_\mu \gamma_\nu \gamma_\rho (1 - \gamma^5) q_j \; \bar{b}_j \gamma^\mu \gamma^\nu \gamma^\rho (1 - \gamma^5) q_i - (16 - 4\epsilon)(Q_1 + E_1) \,, \\
E_6 &= \bar{b}_i \gamma_\mu (1 - \gamma^5) q_i \; \bar{b}_j \gamma^\mu (1 + \gamma^5) q_j + 2 Q_5 \,, \\
E_7 &= \bar{b}_i \gamma_\mu (1 - \gamma^5) q_j \; \bar{b}_j \gamma^\mu (1 + \gamma^5) q_i + 2 Q_4 \,, \\
E_8 &= \bar{b}_i \gamma_\mu \gamma_\nu (1 - \gamma^5) q_i \; \bar{b}_j \gamma^\mu \gamma^\nu (1 + \gamma^5) q_j - 4 Q_4 \,, \\
E_9 &= \bar{b}_i \gamma_\mu \gamma_\nu (1 - \gamma^5) q_j \; \bar{b}_j \gamma^\mu \gamma^\nu (1 + \gamma^5) q_i - 4 Q_5 \,,
\end{aligned}
$$

$$\tag{G.1.1}$$

© Springer Nature Switzerland AG 2019
M. J. Kirk, *Charming New Physics in Beautiful Processes?*,
Springer Theses, https://doi.org/10.1007/978-3-030-19197-9

for QCD and

$$
\tilde{E}_1 = \bar{h}_i^{\{(+)}\gamma_\mu(1-\gamma^5)q_j\ \bar{h}_j^{(-)\}}\gamma^\mu(1-\gamma^5)q_i - \tilde{Q}_1\,,
$$

$$
\tilde{E}_2 = \frac{1}{2}\tilde{Q}_1 + \tilde{Q}_2 + \bar{h}_i^{\{(+)}(1-\gamma^5)q_j\ \bar{h}_j^{(-)\}}(1-\gamma^5)q_i\,,
$$

$$
\tilde{E}_3 = \bar{h}_i^{\{(+)}\gamma_\mu\gamma_\nu(1-\gamma^5)q_i\ \bar{h}_j^{(-)\}}\gamma^\mu\gamma^\nu(1-\gamma^5)q_j + (4+a_1\epsilon)\tilde{Q}_1\,,
$$

$$
\tilde{E}_4 = \bar{h}_i^{\{(+)}\gamma_\mu\gamma_\nu(1-\gamma^5)q_j\ \bar{h}_j^{(-)\}}\gamma^\mu\gamma^\nu(1-\gamma^5)q_i + (4+a_1\epsilon)(\tilde{Q}_1+\tilde{E}_1)\,,
$$

$$
\tilde{E}_5 = \bar{h}_i^{\{(+)}\gamma_\mu\gamma_\nu\gamma_\rho(1-\gamma^5)q_i\ \bar{h}_j^{(-)\}}\gamma^\mu\gamma^\nu\gamma^\rho(1-\gamma^5)q_j - (16+a_2\epsilon)\tilde{Q}_1\,,\quad\text{(G.1.2)}
$$

$$
\tilde{E}_6 = \bar{h}_i^{\{(+)}\gamma_\mu\gamma_\nu\gamma_\rho(1-\gamma^5)q_j\ \bar{h}_j^{(-)\}}\gamma^\mu\gamma^\nu\gamma^\rho(1-\gamma^5)q_i - (16+a_2\epsilon)(\tilde{Q}_1+\tilde{E}_1)\,,
$$

$$
\tilde{E}_7 = \bar{h}_i^{\{(+)}\gamma_\mu(1-\gamma^5)q_i\ \bar{h}_j^{(-)\}}\gamma^\mu(1+\gamma^5)q_j + 2\tilde{Q}_5\,,
$$

$$
\tilde{E}_8 = \bar{h}_i^{\{(+)}\gamma_\mu(1-\gamma^5)q_j\ \bar{h}_j^{(-)\}}\gamma^\mu(1+\gamma^5)q_i + 2\tilde{Q}_4\,,
$$

$$
\tilde{E}_9 = \bar{h}_i^{\{(+)}\gamma_\mu\gamma_\nu(1-\gamma^5)q_i\ \bar{h}_j^{(-)\}}\gamma^\mu\gamma^\nu(1+\gamma^5)q_j - (4+a_3\epsilon)\tilde{Q}_4\,,
$$

$$
\tilde{E}_{10} = \bar{h}_i^{\{(+)}\gamma_\mu\gamma_\nu(1-\gamma^5)q_j\ \bar{h}_j^{(-)\}}\gamma^\mu\gamma^\nu(1+\gamma^5)q_i - (4+a_3\epsilon)\tilde{Q}_5\,,
$$

for HQET. It is straightforward to verify that the evanescent operators vanish in four dimensions by using the Fierz identities Eqs. A.0.2, A.0.4 and A.0.5 along with the relation

$$
\gamma_\mu\gamma_\nu\gamma_\rho = g_{\mu\nu}\gamma_\rho + g_{\nu\rho}\gamma_\mu - g_{\mu\rho}\gamma_\nu - i\epsilon_{\mu\nu\rho\lambda}\gamma^\lambda\gamma^5\,.\qquad\text{(G.1.3)}
$$

A useful strategy to simplify expressions with two Dirac matrices is to use projection identities, e.g.

$$
\bar{h}^{(\pm)}\gamma_\mu\gamma_\nu(1-\gamma^5)q = \pm\bar{h}^{(\pm)}\slashed{v}\gamma_\mu\gamma_\nu(1-\gamma^5)q\,,\qquad\text{(G.1.4)}
$$

and then reduce the number of Dirac matrices with Eq. G.1.3.

In the decomposition shown in Eq. 6.2.8 the LO QCD ADM is

$$
\gamma_{QQ}^{(0)} =
\begin{pmatrix}
\frac{6(N_c-1)}{N_c} & 0 & 0 & 0 & 0 \\
0 & -\frac{2\left(3N_c^2-4N_c-1\right)}{N_c} & \frac{4N_c-8}{N_c} & 0 & 0 \\
0 & \frac{4(N_c-2)(N_c+1)}{N_c} & \frac{2(N_c+1)^2}{N_c} & 0 & 0 \\
0 & 0 & 0 & -\frac{6(N_c^2-1)}{N_c} & 0 \\
0 & 0 & 0 & -6 & \frac{6}{N_c}
\end{pmatrix}\,,\qquad\text{(G.1.5)}
$$

$$
\gamma_{QE}^{(0)} = \begin{pmatrix}
6 & 0 & 0 & -\frac{1}{N_c} & 1 & 0 & 0 & 0 & 0 \\
0 & -\frac{1}{N_c} & 1 & 0 & 0 & 0 & 0 & 0 & 0 \\
0 & \frac{1}{2} & \frac{N_c}{2} - \frac{1}{N_c} & 0 & 0 & 0 & 0 & 0 & 0 \\
0 & 0 & 0 & 0 & 0 & 0 & 0 & -\frac{1}{N_c} & 1 \\
0 & 0 & 0 & 0 & 0 & 0 & 0 & \frac{1}{2} & \frac{N_c}{2} - \frac{1}{N_c}
\end{pmatrix} . \tag{G.1.6}
$$

In HQET we find

$$
\tilde{\gamma}_{\tilde{Q}\tilde{Q}}^{(0)} = \begin{pmatrix}
\frac{3}{N_c} - 3N_c & 0 & 0 & 0 \\
1 + \frac{1}{N_c} & -3N_c + 4 + \frac{7}{N_c} & 0 & 0 \\
0 & 0 & \frac{6}{N_c} - 3N_c & -3 \\
0 & 0 & -3 & \frac{6}{N_c} - 3N_c
\end{pmatrix} , \tag{G.1.7}
$$

$$
\tilde{\gamma}_{\tilde{Q}\tilde{E}}^{(0)} = \begin{pmatrix}
0 & 0 & 0 & 0 & -\frac{1}{4N_c} & \frac{1}{4} & 0 & 0 & 0 & 0 \\
-1 & -4 & -\frac{1}{4N_c} & \frac{1}{4} & 0 & 0 & 0 & 0 & 0 & 0 \\
0 & 0 & 0 & 0 & 0 & 0 & 0 & 0 & -\frac{1}{4N_c} & \frac{1}{4} \\
0 & 0 & 0 & 0 & 0 & 0 & 0 & 0 & \frac{1}{4} & -\frac{1}{4N_c}
\end{pmatrix} . \tag{G.1.8}
$$

Our result (Eq. G.1.5) with $N_c = 3$ differs from the results of [1, 2] because we have only used the replacements implied by the basis of evanescent operators (Eq. G.1.1) to simplify products of Dirac matrices. We can reproduce their result by applying 4-dimensional Fierz identities that relate Q_1, Q_2 and Q_3. The upper left 2×2 sub-matrix of Eq. G.1.7 agrees with [3].

G.1.2 $\Delta B = 0$ *Operators*

We define the basis of evanescent operators in QCD following [4]:

$$
\begin{aligned}
E_1^q &= \bar{b}\gamma_\mu\gamma_\nu\gamma_\rho(1 - \gamma^5)q \; \bar{q}\gamma^\rho\gamma^\nu\gamma^\mu(1 - \gamma^5)b - (4 - 8\epsilon)Q_1^q , \\
E_2^q &= \bar{b}\gamma_\mu\gamma_\nu(1 - \gamma^5)q \; \bar{q}\gamma^\nu\gamma^\mu(1 + \gamma^5)b - (4 - 8\epsilon)Q_2^q , \\
E_3^q &= \bar{b}\gamma_\mu\gamma_\nu\gamma_\rho(1 - \gamma^5)T^A q \; \bar{q}\gamma^\rho\gamma^\nu\gamma^\mu(1 - \gamma^5)T^A b - (4 - 8\epsilon)T_1^q , \\
E_4^q &= \bar{b}\gamma_\mu\gamma_\nu(1 - \gamma^5)T^A q \; \bar{q}\gamma^\nu\gamma^\mu(1 + \gamma^5)T^A b - (4 - 8\epsilon)T_2^q .
\end{aligned} \tag{G.1.9}
$$

In HQET we again introduce parameters $a_{1,2}$ to keep track of the scheme dependence.

$$\tilde{E}_1^q = \bar{h}\gamma_\mu\gamma_\nu\gamma_\rho(1-\gamma^5)q\ \bar{q}\gamma^\rho\gamma^\nu\gamma^\mu(1-\gamma^5)h - (4+a_1\epsilon)\tilde{Q}_1^q\,,$$
$$\tilde{E}_2^q = \bar{h}\gamma_\mu\gamma_\nu(1-\gamma^5)q\ \bar{q}\gamma^\nu\gamma^\mu(1+\gamma^5)h - (4+a_2\epsilon)\tilde{Q}_2^q\,,$$
$$\tilde{E}_3^q = \bar{h}\gamma_\mu\gamma_\nu\gamma_\rho(1-\gamma^5)T^A q\ \bar{q}\gamma^\rho\gamma^\nu\gamma^\mu(1-\gamma^5)T^A h - (4+a_1\epsilon)\tilde{T}_1^q\,,$$
$$\tilde{E}_4^q = \bar{h}\gamma_\mu\gamma_\nu(1-\gamma^5)T^A q\ \bar{q}\gamma^\nu\gamma^\mu(1+\gamma^5)T^A h - (4+a_2\epsilon)\tilde{T}_2^q\,. \tag{G.1.10}$$

The isospin breaking combinations of the evanescent operators are defined in analogy to Eq. 6.5.3. The LO ADM in QCD takes the form

$$\gamma_{QQ}^{(0)} = \begin{pmatrix} 0 & 0 & 12 & 0 \\ 0 & \frac{6}{N_c} - 6N_c & 0 & 0 \\ 3 - \frac{3}{N_c^2} & 0 & -\frac{12}{N_c} & 0 \\ 0 & 0 & 0 & \frac{6}{N_c} \end{pmatrix}, \tag{G.1.11}$$

$$\gamma_{QE}^{(0)} = \begin{pmatrix} 0 & 0 & -2 & 0 \\ 0 & 0 & 0 & -2 \\ \frac{1}{2N_c^2} - \frac{1}{2} & 0 & \frac{2}{N_c} - \frac{N_c}{2} & 0 \\ 0 & \frac{1}{2N_c^2} - \frac{1}{2} & 0 & \frac{2}{N_c} - \frac{N_c}{2} \end{pmatrix}. \tag{G.1.12}$$

The HQET result is given by

$$\tilde{\gamma}_{\tilde{Q}\tilde{Q}}^{(0)} = \begin{pmatrix} \frac{3}{N_c} - 3N_c & 0 & 6 & 0 \\ 0 & \frac{3}{N_c} - 3N_c & 0 & 6 \\ \frac{3}{2} - \frac{3}{2N_c^2} & 0 & -\frac{3}{N_c} & 0 \\ 0 & \frac{3}{2} - \frac{3}{2N_c^2} & 0 & -\frac{3}{N_c} \end{pmatrix}, \tag{G.1.13}$$

$$\tilde{\gamma}_{\tilde{Q}\tilde{E}}^{(0)} = \begin{pmatrix} 0 & 0 & -\frac{1}{2} & 0 \\ 0 & 0 & 0 & -\frac{1}{2} \\ \frac{1}{8N_c^2} - \frac{1}{8} & 0 & \frac{1}{2N_c} - \frac{N_c}{4} & 0 \\ 0 & \frac{1}{8N_c^2} - \frac{1}{8} & 0 & \frac{1}{2N_c} - \frac{N_c}{4} \end{pmatrix}. \tag{G.1.14}$$

Our result in Eq. G.1.11 is in agreement with [5, 6] and Eq. G.1.13 reproduces the result of [7].[2] The results in Eqs. G.1.12 and G.1.14 are new. The matching coefficients read

$$C_{Q_i\tilde{Q}_j}^{(0)} = \delta_{ij} \tag{G.1.15}$$

---

[2]Note that [7] contains a misprint that has been identified in [8].

at LO and

$$
C^{(1)}_{Q_i \tilde{Q}_j} = \begin{pmatrix}
-4L_\mu - \frac{32}{3} & \frac{16}{3} & -\frac{a_1}{4} - 3L_\mu - 13 & -2 \\
0 & 4L_\mu + \frac{16}{3} & -\frac{3}{2} & -\frac{a_2}{4} + 3L_\mu - 1 \\
-\frac{a_1}{18} - \frac{2L_\mu}{3} - \frac{26}{9} & -\frac{4}{9} & -\frac{7a_1}{24} + \frac{3L_\mu}{2} + \frac{7}{6} & -3 \\
-\frac{1}{3} & -\frac{a_2}{18} + \frac{2L_\mu}{3} - \frac{2}{9} & -\frac{1}{4} & -\frac{7a_2}{24} - \frac{3L_\mu}{2} - \frac{29}{6}
\end{pmatrix}
$$

$$(G.1.16)$$

at NLO where we have set $N_c = 3$ for brevity.

G.2 Inputs and Detailed Overview of Uncertainties

Table G.1 Input values for parameters. Note that for f_B we use the mean of f_B and f_{B^+}, while for f_D we use the "experimental" value instead of the lattice average, since the former is in significantly better agreement with sum rule results [14–17]

Parameter	Value	Source
$\overline{m}_b(\overline{m}_b)$	$\left(4.203^{+0.016}_{-0.034}\right)$ GeV	[9, 10]
$m_b^{PS}(2\,\text{GeV})$	$\left(4.532^{+0.013}_{-0.039}\right)$ GeV	[9, 10]
m_b^{1S}	$\left(4.66^{+0.04}_{-0.03}\right)$ GeV	[11]
$m_b^{kin}(1\,\text{GeV})$	(4.553 ± 0.020) GeV	[12]
$\overline{m}_c(\overline{m}_c)$	(1.279 ± 0.013) GeV	[13]
$\alpha_s(M_Z)$	0.1181 ± 0.0011	[11]
V_{us}	0.2248 ± 0.0006	[11]
V_{ub}	0.00409 ± 0.00039	[11]
V_{cb}	0.0405 ± 0.0015	[11]
γ_{CKM}	$\left(73.2^{+6.3}_{-7.0}\right)^\circ$	[11]
f_B	(189 ± 4) MeV	[11]
f_{B_s}	(227.2 ± 3.4) MeV	[11]
f_D	(203.7 ± 4.8) MeV	[11]

Table G.2 Individual errors for the bag parameters of the $\Delta B = 2$ matrix elements

	Λ	Intrinsic SR	Condensates	μ_ρ	$1/m_b$	μ_m	a_i
$\overline{B}_{Q_1}$	$+0.001$ -0.002	± 0.018	± 0.004	$+0.011$ -0.022	± 0.010	$+0.045$ -0.039	$+0.007$ -0.007
$\overline{B}_{Q_2}$	$+0.014$ -0.017	± 0.020	± 0.004	$+0.012$ -0.019	± 0.010	$+0.071$ -0.062	$+0.015$ -0.015
$\overline{B}_{Q_3}$	$+0.060$ -0.074	± 0.107	± 0.023	$+0.016$ -0.008	± 0.010	$+0.086$ -0.069	$+0.053$ -0.052
$\overline{B}_{Q_4}$	$+0.007$ -0.006	± 0.021	± 0.011	$+0.003$ -0.003	± 0.010	$+0.088$ -0.079	$+0.005$ -0.006
$\overline{B}_{Q_5}$	$+0.019$ -0.015	± 0.018	± 0.009	$+0.004$ -0.006	± 0.010	$+0.077$ -0.068	$+0.012$ -0.012

Table G.3 Individual errors for the bag parameters of the $\Delta B = 0$ matrix elements

	$\overline{\Lambda}$	Intrinsic SR	Condensates	μ_ρ	$1/m_b$	μ_m	a_i
$\overline{B}_1$	$+0.003$ -0.002	±0.019	±0.002	$+0.002$ -0.002	±0.010	$+0.060$ -0.052	$+0.002$ -0.003
$\overline{B}_2$	$+0.001$ -0.001	±0.020	±0.002	$+0.000$ -0.001	±0.010	$+0.084$ -0.076	$+0.001$ -0.002
$\overline{\epsilon}_1$	$+0.006$ -0.007	±0.022	±0.003	$+0.003$ -0.003	±0.010	$+0.010$ -0.012	$+0.006$ -0.007
$\overline{\epsilon}_2$	$+0.005$ -0.006	±0.017	±0.003	$+0.002$ -0.001	±0.010	$+0.001$ -0.002	$+0.003$ -0.004

Table G.4 Individual errors for the bag parameters of the $\Delta C = 2$ matrix elements

	$\overline{\Lambda}$	Intrinsic SR	Condensates	μ_ρ	$1/m_c$	μ_m	a_i
$\overline{B}_{Q_1}$	$+0.001$ -0.002	±0.013	±0.003	$+0.009$ -0.021	±0.030	$+0.039$ -0.021	±0.003
$\overline{B}_{Q_2}$	$+0.011$ -0.014	±0.015	±0.003	$+0.010$ -0.016	±0.030	$+0.092$ -0.050	±0.012
$\overline{B}_{Q_3}$	$+0.037$ -0.045	±0.059	±0.013	$+0.016$ -0.016	±0.030	$+0.116$ -0.059	±0.016
$\overline{B}_{Q_4}$	$+0.006$ -0.005	±0.017	±0.009	$+0.003$ -0.003	±0.030	$+0.131$ -0.071	±0.004
$\overline{B}_{Q_5}$	$+0.014$ -0.012	±0.014	±0.007	$+0.004$ -0.005	±0.030	$+0.127$ -0.069	±0.004

Table G.5 Individual errors for the bag parameters of the $\Delta C = 0$ matrix elements

	$\overline{\Lambda}$	Intrinsic SR	Condensates	μ_ρ	$1/m_c$	μ_m	a_i
$\overline{B}_1$	$+0.004$ -0.003	±0.017	±0.002	$+0.002$ -0.002	±0.030	$+0.068$ -0.037	$+0.003$ -0.005
$\overline{B}_2$	$+0.001$ -0.000	±0.015	±0.001	$+0.000$ -0.000	±0.030	$+0.120$ -0.065	$+0.000$ -0.001
$\overline{\epsilon}_1$	$+0.007$ -0.008	±0.024	±0.004	$+0.003$ -0.004	±0.030	$+0.012$ -0.022	$+0.006$ -0.008
$\overline{\epsilon}_2$	$+0.003$ -0.004	±0.011	±0.002	$+0.001$ -0.001	±0.030	$+0.000$ -0.000	$+0.001$ -0.002

Table G.6 Individual errors for the B_s mixing observables

Parameter	Error contribution		
	$\Delta M_s^{SM}/\text{ps}^{-1}$	$\Delta \Gamma_s^{PS}/\text{ps}^{-1}$	$a_{sl}^{s,PS}/10^{-5}$
$\delta(\overline{B}_{Q_1})$	±1.1	±0.005	±0.01
$\delta(\overline{B}_{Q_3})$	±0.0	±0.005	±0.01
$\delta(\overline{B}_{R_0})$	±0.0	±0.003	±0.00
$\delta(\overline{B}_{R_1})$	±0.0	±0.000	±0.00
$\delta(\overline{B}_{R_1'})$	±0.0	±0.000	±0.00
$\delta(\overline{B}_{R_2})$	±0.0	±0.016	±0.00
$\delta(\overline{B}_{R_3})$	±0.0	±0.001	±0.02
$\delta(\overline{B}_{R_3'})$	±0.0	±0.000	±0.05
$\delta(f_{B_s})$	±0.5	±0.002	±0.00
$\delta(\mu_1)$	±0.0	$+0.007$ -0.018	$+0.04$ -0.08
$\delta(\mu_2)$	±0.1	$+0.000$ -0.002	±0.01
$\delta(m_b)$	±0.0	$+0.000$ -0.001	±0.01
$\delta(m_c)$	±0.0	$+0.000$ -0.001	±0.06
$\delta(\alpha_s)$	±0.0	±0.000	±0.04
$\delta(\text{CKM})$	$+1.4$ -1.3	±0.006	$+0.21$ -0.22

Table G.7 Individual errors for the B_d mixing observables

Parameter	Error contribution		
	$\Delta M_d^{SM}/\mathrm{ps}^{-1}$	$\Delta \Gamma_d^{PS}/10^{-3}\,\mathrm{ps}^{-1}$	$a_{sl}^{d,PS}/10^{-4}$
$\delta(\overline{B}_{Q_1})$	$+0.04$ -0.03	± 0.16	± 0.02
$\delta(\overline{B}_{Q_3})$	± 0.00	$+0.17$ -0.16	± 0.03
$\delta(\overline{B}_{R_0})$	± 0.00	± 0.11	± 0.01
$\delta(\overline{B}_{R_1})$	± 0.00	± 0.01	± 0.00
$\delta(\overline{B}_{R_1'})$	± 0.00	± 0.01	± 0.00
$\delta(\overline{B}_{R_2})$	± 0.00	± 0.54	± 0.00
$\delta(\overline{B}_{R_3})$	± 0.00	± 0.00	± 0.04
$\delta(\overline{B}_{R_3'})$	± 0.00	± 0.01	± 0.09
$\delta(f_B)$	± 0.03	± 0.11	± 0.00
$\delta(\mu_1)$	± 0.00	$+0.24$ -0.62	$+0.17$ -0.07
$\delta(\mu_2)$	± 0.00	$+0.00$ -0.08	$+0.01$ -0.03
$\delta(m_b)$	± 0.00	$+0.01$ -0.03	$+0.01$ -0.03
$\delta(m_c)$	± 0.00	$+0.01$ -0.02	± 0.13
$\delta(\alpha_s)$	± 0.00	± 0.01	± 0.08
$\delta(\mathrm{CKM})$	± 0.08	$+0.38$ -0.37	$+0.47$ -0.44

Table G.8 Individual errors for the ratio $\tau(B^+)/\tau(B^0)$ in the PS mass scheme

Parameter	Error contribution
$\delta(\overline{B}_1)$	± 0.002
$\delta(\overline{B}_2)$	± 0.000
$\delta(\overline{\epsilon}_1)$	$+0.016$ -0.015
$\delta(\overline{\epsilon}_2)$	± 0.004
$\delta(\rho_3)$	± 0.001
$\delta(\rho_4)$	± 0.000
$\delta(\sigma_3)$	± 0.013
$\delta(\sigma_4)$	± 0.000
$\delta(f_B)$	$+0.004$ -0.003
$\delta(\mu_1)$	$+0.000$ -0.013
$\delta(\mu_0)$	$+0.000$ -0.006
$\delta(m_b)$	$+0.000$ -0.001
$\delta(m_c)$	± 0.000
$\delta(\alpha_s)$	± 0.002
$\delta(\mathrm{CKM})$	± 0.006

Table G.9 Individual errors for the ratio $\tau(D^+)/\tau(D^0)$ in the PS mass scheme

Parameter	Error contribution
$\delta(\overline{B}_1)$	$+0.07$ -0.05
$\delta(\overline{B}_2)$	± 0.00
$\delta(\overline{\epsilon}_1)$	$+0.52$ -0.47
$\delta(\overline{\epsilon}_2)$	± 0.017
$\delta(\rho_3)$	± 0.05
$\delta(\rho_4)$	± 0.00
$\delta(\sigma_3)$	± 0.46
$\delta(\sigma_4)$	± 0.00
$\delta(f_D)$	± 0.08
$\delta(\mu_1)$	$+0.07$ -0.40
$\delta(\mu_0)$	$+0.08$ -0.21
$\delta(m_c)$	± 0.08
$\delta(m_s)$	± 0.00
$\delta(\alpha_s)$	$+0.07$ -0.06
$\delta(\text{CKM})$	± 0.00

References

1. Gamiz E, Shigemitsu J, Trottier H (2008) Four fermion operator matching with NRQCD heavy and AsqTad light quarks. Phys Rev D 77:114505. https://doi.org/10.1103/PhysRevD. 77.114505, arXiv:0804.1557
2. Monahan C, Gámiz E, Horgan R, Shigemitsu J (2014) Matching lattice and continuum four-fermion operators with nonrelativistic QCD and highly improved staggered quarks. Phys Rev D 90:054015. https://doi.org/10.1103/PhysRevD.90.054015, arXiv:1407.4040
3. Buchalla G (1997) Renormalization of $\Delta B = 2$ transitions in the static limit beyond leading logarithms. Phys Lett B 395:364–368. https://doi.org/10.1016/S0370-2693(97)00043-9, arXiv:hep-ph/9608232
4. Beneke M, Buchalla G, Greub C, Lenz A, Nierste U (2002) The B^+-B_d^0 lifetime difference beyond leading logarithms. Nucl Phys B 639:389–407. https://doi.org/10.1016/S0550-3213(02)00561-8, arXiv:hep-ph/0202106
5. Ciuchini M, Franco E, Lubicz V, Martinelli G, Scimemi I, Silvestrini L (1998) Next-to-leading order QCD corrections to $\Delta F = 2$ effective Hamiltonians. Nucl Phys B 523:501–525. https://doi.org/10.1016/S0550-3213(98)00161-8, arXiv:hep-ph/9711402
6. Buras AJ, Misiak M, Urban J (2000) Two loop QCD anomalous dimensions of flavor changing four quark operators within and beyond the standard model. Nucl Phys B 586:397–426. https://doi.org/10.1016/S0550-3213(00)00437-5, arXiv:hep-ph/0005183
7. Neubert M, Sachrajda CT (1997) Spectator effects in inclusive decays of beauty hadrons. Nucl Phys B 483:339–370. https://doi.org/10.1016/S0550-3213(96)00559-7, arXiv:hep-ph/9603202
8. Ciuchini M, Franco E, Lubicz V, Mescia F (2002) Next-to-leading order QCD corrections to spectator effects in lifetimes of beauty hadrons. Nucl Phys B 625:211–238. https://doi.org/10.1016/S0550-3213(02)00006-8, arXiv:hep-ph/0110375

9. Beneke M, Maier A, Piclum J, Rauh T (2015) The bottom-quark mass from non-relativistic sum rules at NNNLO. Nucl Phys B 891:42–72. https://doi.org/10.1016/j.nuclphysb.2014.12. 001, arXiv:1411.3132

10. Beneke M, Maier A, Piclum J, Rauh T (2016) NNNLO determination of the bottom-quark mass from non-relativistic sum rules. In: PoS RADCOR2015, p 035. https://doi.org/10.22323/1.235. 0035, arXiv:1601.02949

11. Particle Data Group collaboration, Patrignani C et al (2016) Review of Particle Physics. Chin Phys C 40:100001. https://doi.org/10.1088/1674-1137/40/10/100001

12. Alberti A, Gambino P, Healey KJ, Nandi S (2015) Precision determination of the Cabibbo-Kobayashi-Maskawa element V_{cb}. Phys Rev Lett 114:061802. https://doi.org/10.1103/PhysRev Lett.114.061802, arXiv:1411.6560

13. Chetyrkin KG, Kuhn JH, Maier A, Maierhofer P, Marquard P, Steinhauser M et al (2009) Charm and bottom quark masses: an update. Phys Rev D 80:074010. https://doi.org/10.1103/ PhysRevD.80.074010, arXiv:0907.2110

14. Wang Z-G (2015) Analysis of the masses and decay constants of the heavy-light mesons with QCD sum rules. Eur Phys J C 75:427. https://doi.org/10.1140/epjc/s10052-015-3653-9, arXiv:1506.01993

15. Gelhausen P, Khodjamirian A, Pivovarov AA, Rosenthal D (2013) Decay constants of heavy-light vector mesons from QCD sum rules. Phys Rev D 88:014015. https://doi.org/10.1103/ PhysRevD.88.014015, https://doi.org/10.1103/PhysRevD.91.099901, https://doi.org/10.1103/ PhysRevD.89.099901, arXiv:1305.5432

16. Narison S (2013) A fresh look into $m_{c,b}$ and precise $f_{D_{(s)}, B_{(s)}}$ from heavy-light QCD spectral sum rules. Phys Lett B 718:1321–1333. https://doi.org/10.1016/j.physletb.2012.10.057, arXiv:1209.2023

17. Lucha W, Melikhov D, Simula S (2011) OPE, charm-quark mass, and decay constants of D and D_s mesons from QCD sum rules. Phys Lett B 701:82–88. https://doi.org/10.1016/j.physletb. 2011.05.031, arXiv:1101.5986

Appendix H
Additional Information from "One Constraint to Kill Them All?"

In this appendix, we break down our updated SM calculation in Chap. 7 by showing the inputs used and the resulting error breakdown. We also discuss in more detail the dependence of ΔM_s on inputs from lattice QCD and the CKM matrix elements.

H.1   Numerical Input for Theory Predictions

The inputs we use in our numerical evaluations are shown in Table H.1. The values are taken from the PDG [1, 2], from non-relativistic sum rules (NRSR) [3, 4], from the CKMfitter group [5] and the non-perturbative parameters from FLAG (July 2017 online update [6]). For α_s we use RunDec [7] with 5-loop accuracy [8–12], running from M_Z down to the bottom mass scale. At the low scale we use 2-loop accuracy to determine $\Lambda^{(5)}$.

H.2   Error Budget of the Theory Predictions

In this section we compare the error budget of our new SM prediction for ΔM_s^{SM} with the ones given in 2015 by [13], in 2011 by [14] and 2006 by [15]—the results are shown in Table H.2.

We observe a considerable improvement in accuracy and a sizeable shift compared to the 2015 prediction, mostly stemming from the new lattice results for $f_{B_s}\sqrt{B}$, which is still responsible for the largest error contribution of about 6%. The next important uncertainty is the accuracy of the CKM element V_{cb}, which contributes about 2 % to the error budget. The CKM parameters were determined assuming unitarity of the 3×3 CKM matrix—if this assumption is relaxed, then the uncertainties can increase. The uncertainties due to the remaining parameters are subleading. In

© Springer Nature Switzerland AG 2019
M. J. Kirk, *Charming New Physics in Beautiful Processes?*,
Springer Theses, https://doi.org/10.1007/978-3-030-19197-9

Table H.1 Input parameters for our update of ΔM_s

Parameter	Value	Source
M_{B_s}	$(5.366\ 89 \pm 0.000\ 19)$GeV	PDG 2017
m_t	(173.1 ± 0.6)GeV	PDG 2017
$\bar{m}_t(\bar{m}_t)$	(165.65 ± 0.57)GeV	Own evaluation
$\bar{m}_b(\bar{m}_b)$	(4.203 ± 0.025)GeV	NRSR
$\alpha_s(M_Z)$	$0.1181(11)$	PDG 2017
$\alpha_s(\bar{m}_b)$	$0.2246(21)$	Own evaluation
$\Lambda^{(5)}$	(0.2259 ± 0.0068)GeV	Own evaluation
V_{us}	$0.22508^{+0.00030}_{-0.00028}$	CKMfitter
V_{cb}	$0.04181^{+0.00028}_{-0.00060}$	CKMfitter
$|V_{ub}/V_{cb}|$	$0.0889(14)$	CKMfitter
γ_{CKM}	$1.141^{+0.017}_{-0.020}$	CKMfitter
$f_{B_s}\sqrt{\hat{B}}$	(274 ± 8)MeV	FLAG

Table H.2 Individual contributions to the theoretical error of the mass difference ΔM_s within the SM and comparison with the values obtained in [13–15]. In the last row the errors are summed in quadrature

Parameter	Error contribution			
	This work (%)	ABL 2015 [13] (%)	LN 2011 [14] (%)	LN 2006 [15] (%)
$\delta(f_{B_s}\sqrt{B})$	5.8	13.9	13.5	34.1
$\delta(|V_{cb}|)$	2.1	4.9	3.4	4.9
$\delta(m_t)$	0.7	0.7	1.1	1.8
$\delta(\alpha_s)$	0.1	0.1	0.4	2
$\delta(\gamma_{CKM})$	0.1	0.1	0.3	1
$\delta(|V_{ub}/V_{cb}|)$	<0.1	0.1	0.2	0.5
$\delta(\bar{m}_b)$	<0.1	<0.1	0.1	—
$\sum \delta$	6.2	14.8	14.0	34.6

total we are left with an overall uncertainty of about 6%, in comparison to the experimental uncertainty of about 0.1%.

H.3 Non-perturbative Inputs

As a word of caution we present in Table H.3 a wider range of non-perturbative determinations of the matrix elements of the four-quark operators alongside the corresponding predictions for the mass difference.

Table H.3 Predictions for the non-perturbative parameter $f_{B_s}\sqrt{\hat{B}}$ and the corresponding SM prediction for ΔM_s. The current FLAG average is dominated by the FERMILAB/MILC value from 2016. Note that the HQET-SR result is found by combining our result for the bag parameter (Eq. 6.4.2, which was published in [22]) with a separate sum rule calculation of the decay constant

Source	$f_{B_s}\sqrt{\hat{B}}$ (MeV)	ΔM_s^{SM} (ps^{-1})
HPQCD14 [16]	(247 ± 12)	(16.2 ± 1.7)
ETMC13 [17]	(262 ± 10)	(18.3 ± 1.5)
HPQCD09 [18] = FLAG13 [19]	(266 ± 18)	(18.9 ± 2.6)
FLAG17[20]	(274 ± 8)	(20.01 ± 1.25)
Fermilab16 [21]	(274.6 ± 8.8)	(20.1 ± 1.5)
HQET-SR [22, 23]	$\left(278^{+28}_{-24}\right)$	$\left(20.6^{+4.4}_{-3.4}\right)$
HPQCD06 [24]	(281 ± 20)	(21.0 ± 3.0)
RBC/UKQCD14 [25]	(290 ± 20)	(22.4 ± 3.4)
Fermilab11 [26]	(291 ± 18)	(22.6 ± 2.8)

HPQCD presented in 2014 preliminary results for $N_f = 2 + 1$ [16]—for our numerical estimate in Table H.3 we extracted the numbers from Fig. 7.3 in those proceedings. When finalised, this new calculation will supersede the 2006 [24] and 2009 [18] values. The ETMC $N_f = 2$ number stems from 2013 [17], it is obtained with only two active flavours in the lattice simulation. The Fermilab/MILC $N_f = 2 + 1$ number is from 2016 [21] and it supersedes the 2011 value [26]. This value currently dominates the FLAG average. The numerical effect of these new inputs on mixing observables was partially studied in Chap. 3. The previous FLAG average from 2013 [19] was considerably lower. There is also a large $N_f = 2 + 1$ value from RBC-UKQCD presented at LATTICE 2015 (update of [25]). However, this number is obtained in the static limit and currently missing $1/m_b$ corrections are expected to be very sizeable.[3] The HQET sum rules estimate for the bag parameter from Chap. 6 can also be combined with the decay constant from lattice.

It would be very desirable to see a convergence of these determinations, and in particular an independent confirmation of the Fermilab/MILC result which currently dominates the FLAG average.

H.4 CKM-Dependence

The second most important input parameter for the prediction of ΔM_s is the CKM parameter V_{cb}. There is a long-standing discrepancy between the inclusive determination and values obtained from studying exclusive B decays, see [27]. Recent studies

---

[3]Private communication with Tomomi Ishikawa.

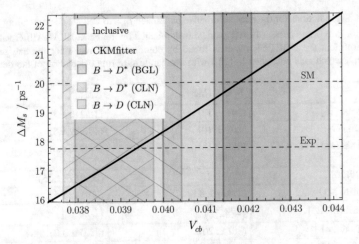

Fig. H.1 Dependence of the SM prediction of ΔM_s on the value of V_{cb}

have found that a problem with the use of a certain form factor parameterisation in the experimental analysis might be the cause of the low exclusive value. The form factor models are denoted by CLN [28] and BGL [29]. Traditionally experiments had used CLN, but it appears that this might underestimate some uncertainties. Using the BGL parameterisation instead one finds (see [30–33]) values that lie considerably closer to the inclusive one. Currently, there are various determinations of V_{cb} available:

$$V_{cb}^{\text{Inclusive}} = 0.04219 \pm 0.00078 \, [34]$$
$$V_{cb}^{B \to D} = 0.03918 \pm 0.00094 \pm 0.00031 \, [34]$$
$$V_{cb}^{B \to D^*, \, \text{CLN}} = 0.03871 \pm 0.00047 \pm 0.00059 \, [34]$$
$$V_{cb}^{B \to D^*, \, \text{BGL}} = 0.0419^{+0.0020}_{-0.0019} \, [30] \, .$$

In Fig. H.1 we plot the dependence of the SM prediction of ΔM_s on V_{cb}, and show the regions predicted by the above inclusive and exclusive determinations. We use the CKMfitter result for V_{cb} (see Table H.1) for our new SM prediction of ΔM_s (see Eq. 7.2.3 and the (upper) horizontal dashed line denoted with "SM"), and the corresponding error band is shown in orange. The predictions obtained by using the inclusive value of V_{cb} only are given by the blue region. For completeness we also show the regions obtained by using the various exclusive extractions of V_{cb}. The disfavoured CLN values result in much lower values for the mass difference (hatched areas), while the BGL value agrees well with the inclusive region, albeit with a higher uncertainty. The experimental value of ΔM_s is shown by the (lower) horizontal dashed line denoted with "Exp".

The preference for the inclusive determination agrees with the value obtained from the CKM fit (which we use in our SM estimate), as well as with the fit value that is found if the direct measurements of V_{cb} are not included in the fit [5]:

$$V_{cb}^{\text{CKMfitter (no direct)}} = 0.04235^{+0.00074}_{-0.00069} . \tag{H.4.1}$$

We also note that the CKMfitter determinations take into account loop-mediated processes, where potentially NP could be present and affect the determination. Taking only tree-level inputs, they find[4]:

$$|V_{us}| = 0.22520^{+0.00012}_{-0.00038} , \tag{H.4.2}$$

$$|V_{cb}| = 0.04175^{+0.00033}_{-0.00172} , \tag{H.4.3}$$

$$|V_{ub}/V_{cb}| = 0.092^{+0.004}_{-0.005} , \tag{H.4.4}$$

$$\gamma_{\text{CKM}} = 1.223^{+0.017}_{-0.030} , \tag{H.4.5}$$

and using these inputs we find

$$\Delta M_s^{\text{SM,2017 (tree)}} = 19.9 \pm 1.5 \text{ps}^{-1} , \tag{H.4.6}$$

which shows an overall consistency with the prediction in Eq. 7.2.3.

References

1. Particle Data Group collaboration, Patrignani C et al (2016) Review of Particle Physics. Chin Phys C 40:100001. https://doi.org/10.1088/1674-1137/40/10/100001
2. Particle Data Group collaboration. http://pdg.lbl.gov/
3. Beneke M, Maier A, Piclum J, Rauh T (2015) The bottom-quark mass from non-relativistic sum rules at NNNLO. Nucl Phys B 891:42–72. https://doi.org/10.1016/j.nuclphysb.2014.12.001, arXiv:1411.3132
4. Beneke M, Maier A, Piclum J, Rauh T (2016) NNNLO determination of the bottom-quark mass from non-relativistic sum rules. In: PoS RADCOR2015, p 035. https://doi.org/10.22323/1.235.0035. arXiv:1601.02949
5. CKMfitter collaboration, ICHEP 2016 results. http://ckmfitter.in2p3.fr/www/results/plots_ichep16/num/ckmEval_results_ichep16.html
6. FLAG collaboration. http://flag.unibe.ch/MainPage
7. Herren F, Steinhauser M (2018) Version 3 of RunDec and CRunDec. Comput Phys Commun 224:333–345. https://doi.org/10.1016/j.cpc.2017.11.014, arXiv:1703.03751
8. Baikov PA, Chetyrkin KG, Kühn JH (2017) Five-loop running of the QCD coupling constant. Phys Rev Lett 118:082002. https://doi.org/10.1103/PhysRevLett.118.082002, arXiv:1606.08659
9. Herzog F, Ruijl B, Ueda T, Vermaseren JAM, Vogt A (2017) The five-loop beta function of Yang-Mills theory with fermions. JHEP 02:090. https://doi.org/10.1007/JHEP02(2017)090, arXiv:1701.01404
10. Luthe T, Maier A, Marquard P, Schroder Y (2017) Complete renormalization of QCD at five loops. JHEP 03:020. https://doi.org/10.1007/JHEP03(2017)020, arXiv:1701.07068
11. Luthe T, Maier A, Marquard P, Schroder Y (2017) The five-loop Beta function for a general gauge group and anomalous dimensions beyond Feynman gauge. JHEP 10:166. https://doi.org/10.1007/JHEP10(2017)166, arXiv:1709.07718

---

[4]Private communication with Sébastien Descotes-Genon.

12. Chetyrkin KG, Falcioni G, Herzog F, Vermaseren JAM (2017) Five-loop renormalisation of QCD in covariant gauges. JHEP 10:179 . https://doi.org/10.1007/JHEP12(2017)006, https://doi.org/10.3204/PUBDB-2018-02123, https://doi.org/10.1007/JHEP10(2017)179, arXiv:1709.08541

13. Artuso M, Borissov G, Lenz A (2016) CP violation in the B_s^0 system. Rev Mod Phys 88:045002. https://doi.org/10.1103/RevModPhys.88.045002, arXiv:1511.09466

14. Lenz A, Nierste U (2011) Numerical updates of lifetimes and mixing parameters of B mesons. In: CKM unitarity triangle. Proceedings, 6th international workshop, CKM 2010, Warwick, UK, Sept 6–10, 2010. arXiv:1102.4274, http://inspirehep.net/record/890169/files/arXiv:1102.4274.pdf

15. Lenz A, Nierste U (2007) Theoretical update of $B_s - \bar{B}_s$ mixing. JHEP 06:072. https://doi.org/10.1088/1126-6708/2007/06/072, arXiv:hep-ph/0612167

16. Dowdall RJ, Davies CTH, Horgan RR, Lepage GP, Monahan CJ, Shigemitsu J, B-meson mixing from full lattice QCD with physical u, d, s and c quarks. arXiv:1411.6989

17. ETM collaboration, Carrasco N et al (2014) B-physics from $N_f = 2$ tmQCD: the standard model and beyond. JHEP 03:016. https://doi.org/10.1007/JHEP03(2014)016, arXiv:1308.1851

18. HPQCD collaboration, Gamiz E, Davies CTH, Lepage GP, Shigemitsu J, Wingate M (2009) Neutral B Meson Mixing in Unquenched Lattice QCD. Phys Rev D 80:014503. https://doi.org/10.1103/PhysRevD.80.014503, arXiv:0902.1815

19. Aoki S et al (2014) Review of lattice results concerning low-energy particle physics. Eur Phys J C 74:2890. https://doi.org/10.1140/epjc/s10052-014-2890-7, arXiv:1310.8555

20. Aoki S et al (2017) Review of lattice results concerning low-energy particle physics. Eur Phys J C 77:112. https://doi.org/10.1140/epjc/s10052-016-4509-7, arXiv:1607.00299

21. Fermilab Lattice, MILC collaboration, Bazavov A et al (2016) $B_{(s)}^0$-mixing matrix elements from lattice QCD for the standard model and beyond. Phys Rev D 93:113016. https://doi.org/10.1103/PhysRevD.93.113016, arXiv:1602.03560

22. Kirk M, Lenz A, Rauh T (2017) Dimension-six matrix elements for meson mixing and lifetimes from sum rules. JHEP 12:068. https://doi.org/10.1007/JHEP12(2017)068, arXiv:1711.02100

23. Gelhausen P, Khodjamirian A, Pivovarov AA, Rosenthal D (2013) Decay constants of heavy-light vector mesons from QCD sum rules. Phys Rev D 88:014015. https://doi.org/10.1103/PhysRevD.88.014015, https://doi.org/10.1103/PhysRevD.91.099901, https://doi.org/10.1103/PhysRevD.89.099901, arXiv:1305.5432

24. Dalgic E, Gray A, Gamiz E, Davies CTH, Lepage GP, Shigemitsu J et al (2007) $B_s^0 - \bar{B}_s^0$ mixing parameters from unquenched lattice QCD. Phys Rev D 76:011501. https://doi.org/10.1103/PhysRevD.76.011501, arXiv:hep-lat/0610104

25. Aoki Y, Ishikawa T, Izubuchi T, Lehner C, Soni A (2015) Neutral B meson mixings and B meson decay constants with static heavy and domain-wall light quarks. Phys Rev D 91:114505. https://doi.org/10.1103/PhysRevD.91.114505, arXiv:1406.6192

26. Bouchard CM, Freeland ED, Bernard C, El-Khadra AX, Gamiz E, Kronfeld AS et al (2011) Neutral B mixing from $2 + 1$ flavor lattice-QCD: the standard model and beyond. In: PoS LATTICE2011, p 274. https://doi.org/10.22323/1.139.0274, arXiv:1112.5642

27. Particle Data Group collaboration, Semileptonic B-hadron decays, determination of V_{cb}, V_{ub}. http://pdg.lbl.gov/2017/reviews/rpp2017-rev-vcb-vub.pdf

28. Caprini I, Lellouch L, Neubert M (1998) Dispersive bounds on the shape of anti-B → D(*) lepton anti-neutrino form-factors. Nucl Phys B 530:153–181. https://doi.org/10.1016/S0550-3213(98)00350-2, arXiv:hep-ph/9712417

29. Boyd CG, Grinstein B, Lebed RF (1995) Constraints on form-factors for exclusive semileptonic heavy to light meson decays. Phys Rev Lett 74:4603–4606. https://doi.org/10.1103/PhysRevLett.74.4603, arXiv:hep-ph/9412324

30. Grinstein B, Kobach A (2017) Model-independent extraction of $|V_{cb}|$ from $\bar{B} \to D^* \ell \bar{\nu}$. Phys Lett B 771:359–364. https://doi.org/10.1016/j.physletb.2017.05.078, arXiv:1703.08170

31. Bigi D, Gambino P, Schacht S (2017) $R(D^*)$, $|V_{cb}|$, and the Heavy Quark Symmetry relations between form factors. JHEP 11:061. https://doi.org/10.1007/JHEP11(2017)061, arXiv:1707.09509

32. Bernlochner FU, Ligeti Z, Papucci M, Robinson DJ (2017) Tensions and correlations in $|V_{cb}|$ determinations. Phys Rev D 96:091503. https://doi.org/10.1103/PhysRevD.96.091503, arXiv:1708.07134

33. Jaiswal S, Nandi S, Patra SK (2017) Extraction of $|V_{cb}|$ from $B \to D^{(*)} \ell \nu_\ell$ and the standard model predictions of $R(D^{(*)})$. JHEP 12:060. https://doi.org/10.1007/JHEP12(2017)060, arXiv:1707.09977

34. HFLAV collaboration, Amhis Y et al (2017) Averages of b-hadron, c-hadron, and τ-lepton properties as of summer 2016. Eur Phys J C 77:895. https://doi.org/10.1140/epjc/s10052-017-5058-4, arXiv:1612.07233

Printed in the United States
By Bookmasters